Numerical and Experimental Analysis of Advanced Concrete Materials

Numerical and Experimental Analysis of Advanced Concrete Materials

Editors

Nikolina Živaljić
Hrvoje Smoljanović
Ivan Balić

MDPI

Basel • Beijing • Wuhan • Barcelona • Belgrade • Novi Sad • Cluj • Manchester

Editors
Nikolina Živaljić
Faculty of Civil Engineering, Architecture and Geodesy, University of Split
Split, Croatia

Hrvoje Smoljanović
Faculty of Civil Engineering, Architecture and Geodesy, University of Split
Split, Croatia

Ivan Balić
Faculty of Civil Engineering, Architecture and Geodesy, University of Split
Split, Croatia

Editorial Office
MDPI
St. Alban-Anlage 66
4052 Basel, Switzerland

This is a reprint of articles from the Special Issue published online in the open access journal *Materials* (ISSN 1996-1944) (available at: https://www.mdpi.com/journal/materials/special_issues/simulation_concrete).

For citation purposes, cite each article independently as indicated on the article page online and as indicated below:

Lastname, A.A.; Lastname, B.B. Article Title. *Journal Name* **Year**, *Volume Number*, Page Range.

ISBN 978-3-0365-8396-9 (Hbk)
ISBN 978-3-0365-8397-6 (PDF)
doi.org/10.3390/books978-3-0365-8397-6

Contents

materials

MDPI

Article

Incorporation of Silica Particles Attached to Nylon 66 Electrospun Nanofibers with Cement

Tri N. M. Nguyen [1], Do Hyung Lee [2] and Jung J. Kim [3,*]

1 Campus in Ho Chi Minh City, University of Transport and Communications, No. 450-451 Le Van Viet Street, Tang Nhon Phu A Ward, Thu Duc City, Ho Chi Minh City 700000, Vietnam
2 Department of Civil, Railroad and Unmanned Systems Engineering, PaiChai University, 155-40 Baejaero, Seo-gu, Daejeon 35345, Korea
3 Department of Civil Engineering, Kyungnam University, 7 Kyungnamdaehakro, Changwon-si 51767, Korea
* Correspondence: jungkim@kyungnam.ac.kr; Tel.: +82-552-496-421; Fax: +82-505-999-2165

Abstract: In this study, a modified version of electrospun nylon 66 nanofibers by silica particles were blended into ordinary Portland cement to investigate the microstructure and some mechanical properties of cementitious material. The addition of silica into the nanofibers improved the tensile and compressive properties of the hardened cement pastes. The observations from the mechanical strength tests showed an increase of 41%, 33% and 65% in tensile strength, compressive strength, and toughness, respectively, when modifying the cement pastes with the proposed nanofibers. The observations from scanning electron microscopy and transmission electron microscopy showed the morphology and microstructure of the fibers as well as their behaviors inside the cement matrix. Additionally, X-ray diffraction and thermal gravimetric analysis clarified the occurrence of the extra pozzolanic reaction, as well as the calcium hydroxide consumption by the attached silica inside the cement matrix. Finally, the observations from this study showed the successful fabrication of the modified nanofibers and the feasibility of improving the tensile and compressive behaviors of cement pastes using the proposed electrospun nanofibers.

Keywords: nylon 66; silica; electrospun nanofibers; cement; mechanical strength; microstructure

Citation: Nguyen, T.N.M.; Lee, D.H.; Kim, J.J. Incorporation of Silica Particles Attached to Nylon 66 Electrospun Nanofibers with Cement. *Materials* **2022**, *15*, 7011. https://doi.org/10.3390/ma15197011

Academic Editors: Nikolina Zivaljic, Hrvoje Smoljanović and Ivan Balić

Received: 14 September 2022
Accepted: 1 October 2022
Published: 10 October 2022

1. Introduction

Today, cement is a basic material that plays an important role in the construction sector. However, there is an increased demand for higher cement strength, due to the feasibility of reducing structural size and cement consumption, which can lead to significant ecological improvement and financial efficiency. Therefore, it has posed a considerable challenge for researchers and manufacturers. Cement matrix strength can be affected by different factors, such as the component ratio, the cement hydration process, the change in the cement matrix microstructure, the curing condition, the addition of additives, and so on [1,2]. Traditionally, silica has been used as a supplement to increase the strength of cementitious materials. The small and fine silica particles provide the advantage of filling up the porous structure of the matrix, leading to the formation of a compact structure. In addition, from a chemical perspective, silica particles are a highly reactive mineral with the ability to participate in the pozzolanic reaction with calcium hydroxide (CH) generated from the cement hydration process [3–7]. Consequently, calcium silicate hydrate (CSH) is formed, which, as demonstrated, comprises approximately 60% of the solid volume of the hydration product, and is crucial in improving the long-term strength and durability of the cement matrix [1]. A study on the effect of nanosilica (NS) on the compressive strength of cement paste reported a maximum increase of 43.8% at 0.6% NS by weight of cement [8]. In addition, Flores et al. [9] reported an increase of 25% in the compressive strength of pastes containing NS or silica fume compared to that of plain paste when replacing cement with an additive proportion of 2.5% by mass. They also concluded that these supplementary factors influence the acceleration of the hydration reactions in the early stages. Mohammed et al. [10] reported an increase from 15% to 50% in the compressive strength of cement pastes containing 1% NS depending on the curing time and water-to-binder ratio.

Furthermore, recent findings showed significant performance of different nanosized fibers in improving the cement matrix. The findings by Liu et al. [11] showed the effectiveness of nanocarbon black and nickel nanofiber in improving the electrical and piezoresistive behaviors of cement. Therefore, the result shows the feasibility of utilizing that kind of cement composite in the sensor industry. The study by Haider et al. [12] reported the influence of chitin nanofiber on delaying setting time and increasing viscosity. Based on those findings, chitin nanofiber can become a promising candidate for tailoring cement for specific applications. From a mechanical strength perspective, the study [12] reported that after 28 days, the compressive strength and the flexural strength of the cement paste containing chitin nanofiber increased by 12% and 41%, respectively, compared to that of the plain paste. In addition, the role of carbon nanofiber in improving Young's modulus of cement paste by 15.5% compared to that of plain paste was reported [13]. The study [14] also reported an improvement of 18% in terms of compressive strength and 21% in terms of flexural strength when introducing cellulose nanofibers into the cement paste. Alternatively, a new approach to strengthen ordinary Portland cement (OPC) using electrospun nanofibers has been suggested in previous works [15–18]. The work [16] presented an approach to reinforce cement materials using electrospun nylon 66 nanofibers, followed by blending the as-spun nanofibers into cement. The results showed the bridging effect of the nylon 66 nanofibers inside the matrix. Furthermore, the increase in tensile strength of up to 30% confirmed the feasibility of this approach for enhancing cement characteristics. Continuing this strategy, the present study focuses on investigating the influence of electrospun nylon 66 nanofibers modified with silica particles (nanosilica) (NS-N66 NFs) on the mechanical characteristics of cement and its microstructural properties. The as-spun nanofibers were blended into the cement powder using an improved collector kit. Tensile strength and compressive strength tests were conducted to examine the changes in the mechanical characteristics of OPC. In addition, microstructural analysis methods, such as field emission scanning electron microscopy (FE–SEM), field emission transmission electron microscopy (FE–TEM), X-ray diffraction (XRD), and thermal gravimetric analysis (TGA), were employed to analyze the microstructural characteristics of the modified nanofibers and the nanofiber-blended cement pastes.

2. Materials and Methods

2.1. Materials

Type I ordinary Portland cement was purchased from Ssangyong Co., Korea [19]. The properties of the cement are listed in Table 1. Nanosilica grade 940-U (CAS No. 69012-64-2) was purchased from Elkem Microsilica (Norway). Figure 1 presents the morphology of the silica particles and their diameter distribution, which ranges from 30 nm to 510 nm, with a mean diameter of approximately 120 nm. Nylon 66 pellets and other chemicals for its dissolution (formic acid and chloroform) were prepared as previously described [15,16]. All materials were used as received.

Table 1. Chemical composition and physical properties of the cement.

CaO	Al_2O_3	SiO_2	SO_3	MgO	Fe_2O_3	Ig. Loss	Specific Surface Area (cm^2/g)	Compressive Strength, 28-Day (MPa)
61.33	6.40	21.01	2.30	3.02	3.12	1.40	2800	36

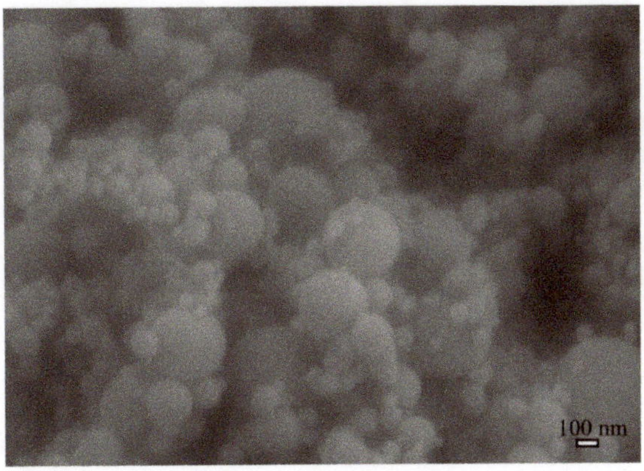

Figure 1. Morphology of silica particles.

2.2. Polymer Solution

The polymer solutions were prepared in two steps: (1) The solvent was prepared by stirring formic acid and chloroform in a volume ratio of 4:1. (2) Precursors, silica particles, and nylon 66 were added to the prepared solvent to obtain the polymer solutions at a precursor-to-solvent ratio of 1:4 by weight of the polymer solution. This proportion was determined after conducting a trial-and-error process to determine the optimal and spinnable viscosity of the polymer solution using a size 20 Taylor cone. To be more specific, 13 wt% of silica was added to two bottles containing 80 wt% of solvent. Then, the solutions were ultrasonicated for 2 h under laboratory conditions to break the van der Waals forces between the nanomaterial molecules and obtain homogeneous dissolution in the final solutions [20–22]. Finally, 7 wt% of nylon 66 was added to the solution and stirred for 3 h to obtain homogeneous polymer solutions. Due to the relaxation of the polymer chain, the polymer solutions were kept for 24 h under laboratory conditions before conducting the electrospinning process.

2.3. Electrospinning Process, Hardened Cement Paste Preparation, and Testing Methods

Electrospun nanofiber fabrication and the blending process to obtain the composite cementitious materials containing electrospun nanofibers have been reported in previous works [15,16]. After jetting out from the needle, the nanofibers are blended into cement by an improved collector as a single fiber, and the nanofiber length is broken during the blending process. As in those studies, we retained all of the input parameters of the electrospinning process and the electrospun nanofiber proportion of 5% of the cement mass here as well.

Owing to the changes in the microstructure of the matrix and the difficulty in controlling the spontaneous cracking that occurs during the curing process, it is difficult to conduct a study on the mechanical properties of hardened cement pastes. Therefore, the cement paste samples were designed with small dimensions according to the specifications in the ASTM C307 [23] and ASTM C 109/109M [24] methods for conducting tensile strength and compressive strength tests, respectively. Five briquette samples and five cubic samples for each cement composite were prepared with a constant water-to-binder proportion of 0.5 [25]. Then, the samples were cured in water at an ambient temperature of 23 ± 2 °C and a relative humidity of 50% for 28 days prior to testing.

In this study, the tensile strength and compressive strength tests were evaluated using a 5 kN mortar tensile strength test device in compliance with ASTM C307 and a 1000 kN hydraulic universal testing machine in compliance with ASTM C109/109M, respectively.

The FE–SEM analysis was conducted using a Zeiss Merlin Compact system. Input parameters, a working distance of 9 mm, and an acceleration voltage of 5 kV were used.

In addition, to increase the observation quality, a 5 Å platinum cover was coated on the sample surfaces.

The FE–TEM analysis was conducted using an FEI Tecnai F30 Twin system under an acceleration voltage of 300 kV.

XRD analysis was conducted using the D8 Advance system with a scanning speed of 0.4 s/step, a step size of 0.02° (2θ)/step from 5° to 70°, and CuKα radiation (40 kV, 40 mA).

TGA was conducted using a TA instrument SDT-Q600 under a nitrogen atmosphere with a flow rate of 100 mL/min, a heating velocity of 10 °C/min, and a heating range from 25 °C to 1000 °C.

3. Results and Discussion

3.1. Mechanical Strength

The results of the tensile and compressive strength tests indicate that the electrospun nylon 66 nanofibers modified with silica particles are effective in improving some select mechanical properties of the hardened cement paste (see Figure 2). Overall, the results showed an upward trend in the mechanical strength of the cement blended with NS-N66 NFs. Compared to that of the unmodified hardened paste, modified cement pastes exhibited a 41% increase in its tensile strength. Previously, an increase of 28% in the tensile strength of the hardened cement pastes blended with nylon 66 nanofibers (N66 NFs) was observed [16]. Notably, this was the first time the role of N66 NFs increasing the tensile strength of cementitious materials became apparent. In the present work, with the addition of silica particles, the tensile strength showed the better result compared to that reported in the literature. A summary of these works, including the polymer components, electrospun nanofiber proportions, and mechanical results, is presented in Table 2.

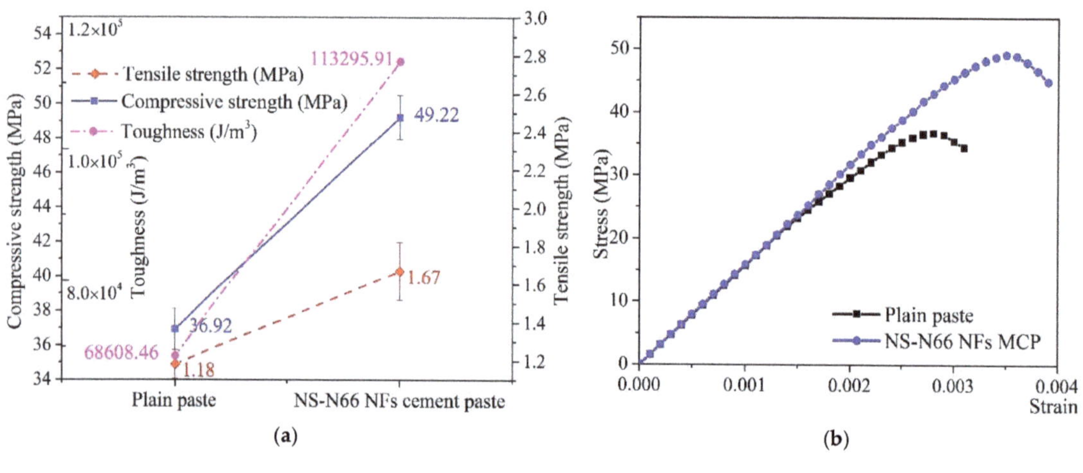

Figure 2. (**a**) The mechanical behaviors of the cement pastes at 28 days; (**b**) The constitutive curves, where "MCP" implies "modified cement paste".

In the case of compressive strength, an increase of 33% was observed with the addition of NS-N66 NFs, as compared to that of the unmodified paste (Figure 2a). In the previous study, the compressive strength only slightly increased, 8%, when similar proportions of N66 NFs were added to the pastes [16]. In addition, the toughness, calculated from the area under the constitutive curves of the hardened cement pastes based on the theory by Timoshenko and Gere [26], increased by 65% upon introducing NS-N66 NFs into the paste (Figure 2b). Based on our previous findings [16], the N66 NFs were ineffective in improving the compressive properties of cement. In contrast, the effectiveness of nylon 66 nanofibers modified by silica particles in improving the characteristics of cement paste in this study is apparent. We hypothesized that silica particles were attached alongside the nylon 66 nanofibers; thus, they filled the porous space inside the matrix and reacted with the CH created from the hydration process of the cement. As a consequence, a more compact structure was formed and a larger amount of CSH was created compared to that in the matrix of the unmodified paste; this, in turn, improved the compressive strength. The following sections discuss microstructure analyses to clarify this hypothesis.

Table 2. Comparison of the mechanical strength results with [16] (the increase in %).

	Previous Study [16]		Present Study
	N66 NFs MCP (1)	**N66 NFs MCP (2)**	**NS-N66 NFs MCP**
Solvent (Volume %)	Formic acid: dichloromethane (4:1)	Formic acid: Chloroform (4:1)	Formic acid: Chloroform (4:1)
Polymer solution (Weight %)	N66 10 wt%: solvent 90 wt%	N66 10 wt%: solvent 90 wt%	NS 13 wt%: N66 7 wt%: solvent 80 wt%
Tensile strength	32	28	41
Compressive strength	6	8	33
Toughness	42	49	65

3.2. *Morphology and Microstructure of Nanofibers*

Figure 3 presents the morphological and microstructural characteristics of the electrospun nylon 66 nanofibers modified by silica particles (NS-N66 NFs). In general, based on the SEM images in Figure 3a,b, the morphology of the fibers was observed, in which the silica particles were attached alongside the nylon 66 nanofibers. The resultant fibers contained thin nylon 66 nanofibers with a mean diameter of approximately 160–170 nm and silica particles with diameters of approximately 30–550 nm along the fiber axis. The size of the silica particles that exist along the fiber axis is almost consistent with that observed in the SEM image of NS, as presented above. However, despite the same electrospinning conditions, the diameter of the resultant nanofibers was slightly smaller than that observed in previous studies [16] due to the reduction of the nylon 66 content [27–29]. Figure 3d presents the TEM image of the modified electrospun nanofiber. From this observation, it is confirmed that there are only silica particles inside the nanofibers.

3.3. *Microstructural Characteristic of Hardened Cement Pastes*

Figure 4 presents the microstructural characteristics of the electrospun nanofiber-blended cement matrix. In general, numerous nanofibers were observed inside the matrix. Figure 4a indicates that the nanofibers grew from the matrix and bridged the hydration products of the cement. This phenomenon showed the bridging and filling effects of the electrospun nanofibers inside the matrix. Moreover, compared with the smooth surface of the raw nylon 66 nanofibers containing silica particles along its axis, as presented above, the morphology of the nanofibers inside the cement matrix was completely transformed. The surface of the nanofibers became rough with granular structures, which was evidence of the pozzolanic reaction. It is suggested that nanosilica can consume the CH produced from the hydration of the cement, which creates more CSH in the final hydration products. In addition, some studies have confirmed the efficiency of silica-based additives in reducing the pore size of the cement matrix [3–7]. As a result, an improvement in the strength and impermeability of the cement matrix was observed. Thus, the silica particles attached alongside the nylon 66 nanofibers reacted with CH, the CSH was formed, gathered, and transformed the morphology of the nanofibers (Figure 4b). The occurrence of the pozzolanic reaction showed the good incorporation of the proposed electrospun nanofibers and cement materials. Above all, the SEM images indicate that the nylon 66 nanofibers modified by silica inside the cement matrix perform well. Specifically, owing to the bridging and filling effects of the proposed nanofibers, the microstructure of the cement matrix became more compact, resulting in improved mechanical characteristics.

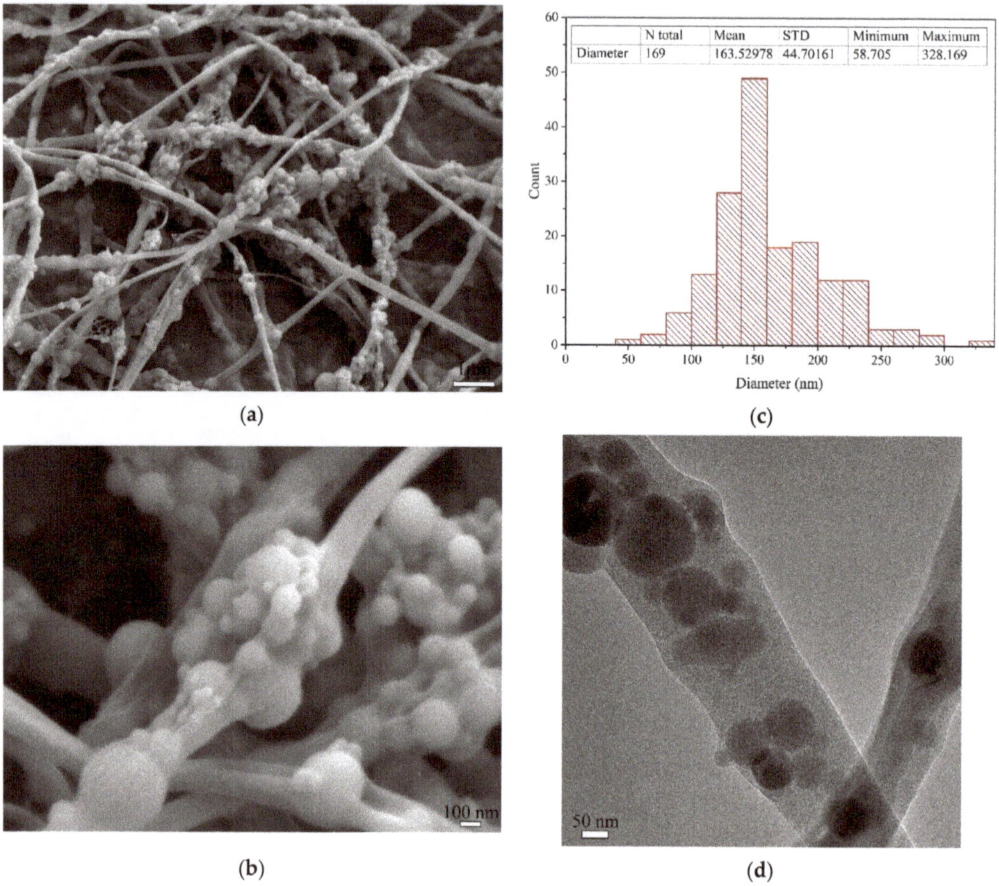

Figure 3. Morphological and microstructural characteristics of NS-N66 NFs. (**a**,**b**) SEM images of NS-N66 NFs at low and high magnification; (**c**) Diameter statistic; (**d**) TEM image of NS-N66 NFs.

Figure 4. Microstructural characteristics of the cement matrix blended with NS-N66 NFs at low and high magnification (the arrows denote the fiber locations inside the cement matrix).

3.4. XRD Analysis

The XRD patterns of the plain cement and the electrospun nanofiber-blended cement are shown in Figure 5. Rietveld refinement [30] and phase determination were conducted based on the phase structure references of the hydration products, namely, calcium silicate hydrate (CSH), calcium hydroxide (CH), gypsum ($CaSO_4.2H_2O$), hillebrandite ($Ca_2SiO_3(OH)_2$) and some dehydrated clinker grains, belite (C_2S), and alite (C_3S). The hydration products of all the samples were consistent and suitable with the results obtained from Jiang et al. [31] (Figure 5). Thus, the presence of the CH component can be correlated to the peaks between 17.94° and 18.11°, 28.62° to 28.74°, 34.02° to 34.15°, 47.03° to 47.17°, 50.73° to 50.84°, 54.29° to 54.39°, 62.51° to 62.59°, and 64.29° to 64.37° 2θ. The CSH component can be recognized by the peak between 29.40° and 29.46° 2θ. The peaks between 31.06° and 31.12°, 32.12° to 32.18°, 32.51° to 32.61°, and 41.10° to 41.20° 2θ corresponded to hillebrandite, belite, alite, and gypsum, respectively. Figure 6 shows the proportion of the hydration products calculated from the results of the Rietveld refinement using the Profex program. The CH content of the blended cement pastes decreased, while their CSH content increased. This suggests that the CH produced from the cement hydration process reacted with the NS particles attached along the nylon 66 nanofibers [31,32].

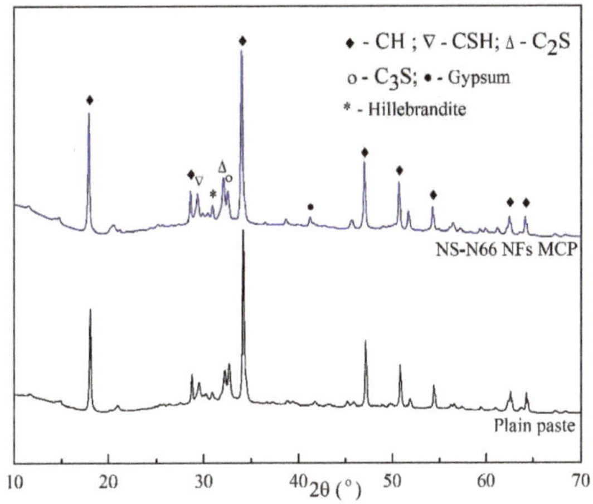

Figure 5. XRD patterns of the hardened cement pastes.

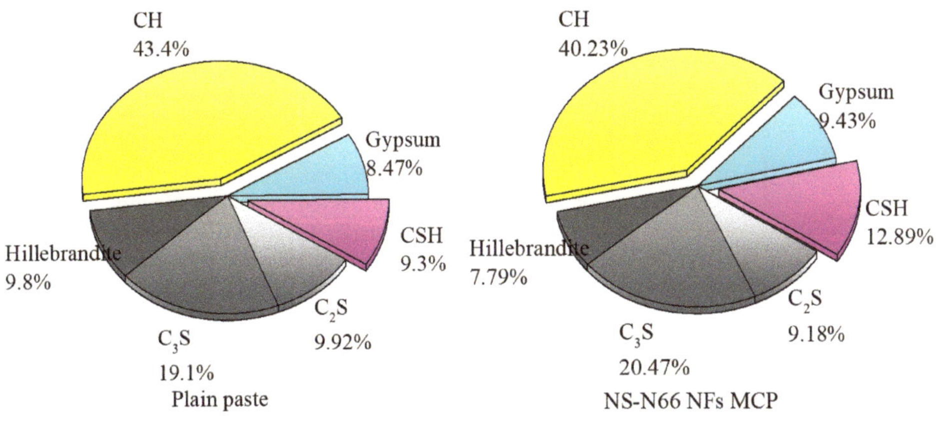

Figure 6. Component proportions of the cement matrices observed from the XRD results.

The decrease–increase tendency of the CH and CSH contents of the three samples was consistent with the order of the observations from the mechanical strength tests. The XRD analysis showed that the CH phase in the matrices was consumed. The increase in the CSH content was observed thanks to the pozzolanic reaction between the attached NS and the generated CH phase from the hydration processes.

3.5. Thermal Analysis

Figure 7 shows the TGA-derivative thermogravimetry (DTG) results of the NS-N66 NFs. The range from around 250 °C to 270 °C can be attributed to the evaporation of the adsorbed water from the nanosilica surface, and the range above 500 °C can be attributed to the condensation of the siloxyl groups. In addition, there was an incomplete decomposition process above 500 °C, that is, the silanol condensation was not complete at temperatures higher than 800 °C, which is in good agreement with the results from Das et al. [33]. Furthermore, the drop in mass from 310 °C to 480 °C can be attributed to the decomposition of the nylon 66 nanofiber, which is consistent with the results previously reported [16]. Thus, the decomposition of NS-N66 NFs occurred from 310 °C to 480 °C.

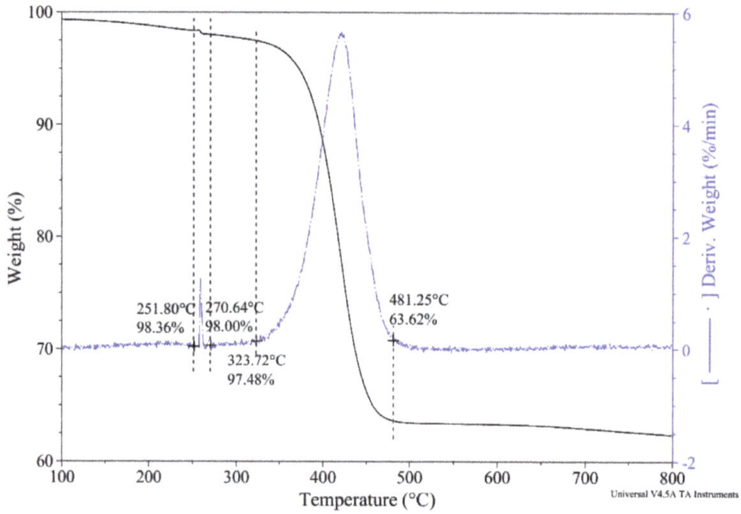

Figure 7. TGA-DTG results of NS-N66 NFs.

Figure 8 shows the TGA curves of the hardened cement pastes with and without the electrospun nanofibers after curing for 28 days. It is worth noting that the weight loss below 145 °C corresponds to the evaporation of free water inside the matrix when conducting the thermal analysis under nitrogen and free carbon dioxide conditions [34]. Therefore, the TGA results of the three samples were compared with 100% weight at 145 °C. The results also demonstrate the expected thermal behavior of the three main phases in the matrix: CH, CSH, and calcite. For instance, the CSH phase can be identified by the weight loss observed in the temperature range from 145 °C to 200 °C due to its dehydration process. The weight loss from 400 °C to 500 °C corresponds to the dehydration process of the CH phase, while that from 550 °C to 900 °C corresponds to the decarbonation of the calcite phase [34–36]. Chang et al. [37] reported that calcite was formed during the curing process when the CH phase gradually reacted with CO_2 from the ambient environment. As can be observed from the TGA results, the CSH content from the nanofiber-blended cement pastes was higher than that of the plain paste, whereas that of CH was lower. It is also worth noting that despite the decomposition range of the CH phase being similar to the decomposition range of the electrospun nanofibers from 400 °C to 500 °C, as presented above, the CH content of the hardened modified cement pastes still decreased compared to that of the hardened plain paste, as indicated by TGA analysis. Consistent with the XRD results, these observations showed a decreasing tendency of the CH and CSH contents in the presence of NS-N66 NFs in the cement. Therefore, the attached silica particles on the nylon 66 nanofibers influenced the final product of the cement paste. Above all, these observations confirmed the initial hypothesis regarding the role of the pozzolanic agent in the attached silica particles.

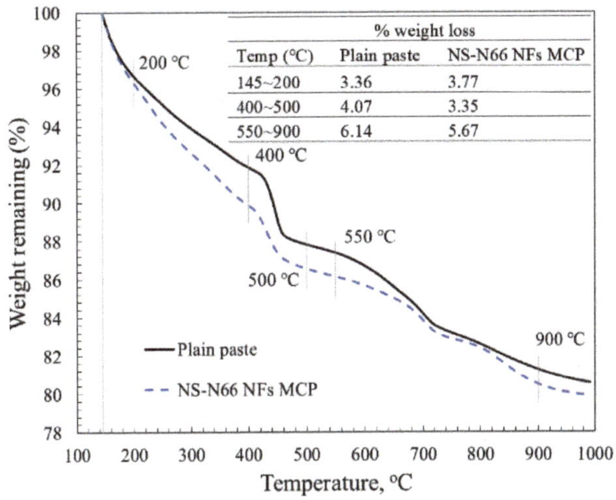

Figure 8. TGA-DTG results of the hardened cement pastes.

4. Conclusions and Perspective

This study determined the influence of electrospun nylon 66 nanofibers containing silica particles on the microstructure and selected mechanical properties of the hardened cement pastes. The following conclusions were drawn:

An increase of 41%, 33%, and 65% in tensile strength, compressive strength, and toughness, respectively, was observed when modifying the cement pastes with the proposed nanofibers containing silica particles.

The nanosilica particles attached along the axis of the nylon 66 nanofibers were effective in increasing the generated CSH content from the pozzolanic reaction. In addition, the bridging and filling effects of the nanofibers inside the matrix were observed, which increased the mechanical strength of the hardened cement pastes.

The XRD and TGA results demonstrated the decrease–increase tendency of the CH and CSH contents when blending the cement with NS-N66 NFs. Therefore, the role of the attached silica particles is clarified.

Finally, electrospun nylon 66 nanofibers containing silica particles is a promising candidate for strengthening cementitious materials. It can be seen that an improvement in the mechanical properties of cement can lead to extending the durability and service life of concrete. However, for the practical application of this product, further research needs to be conducted to improve the performance of the composite cement fabrication process, as well as investigate the effectiveness of these nanofibers on other properties of cement paste and concrete. Additionally, the analysis of the life cycle and the life cycle cost of the concrete structure utilizing the proposed cement is necessary, which can evaluate the competitiveness of the proposed cement solution and help authorities in the decision-making process [38]. In addition, further comparative studies on the mechanical properties and microstructure of cement paste modified with silica particles, silica particles and nanofibers, and the proposed nanofibers containing silica particles are also interesting topics to analyze the performance of the proposed cement in this study.

Author Contributions: Conceptualization: J.J.K.; methodology: J.J.K. and T.N.M.N.; software: T.N.M.N.; validation: D.H.L.; formal analysis: T.N.M.N.; investigation: D.H.L. and J.J.K.; data curation: J.J.K. and T.N.M.N.; writing—original draft preparation: T.N.M.N.; writing—review and editing: T.N.M.N., D.H.L., and J.J.K.; visualization: D.H.L.; supervision: D.H.L. and J.J.K.; project administration: J.J.K.; funding acquisition: J.J.K. All authors have read and agreed to the published version of the manuscript.

Funding: This research was funded by the National Research Foundation (NRF) of Korea: 2020R1A2B5B01001821.

Institutional Review Board Statement: Not applicable.

Informed Consent Statement: Not applicable.

Data Availability Statement: The data presented in this study are available upon request from the corresponding author.

Conflicts of Interest: The authors declare no conflict of interest.

References

1. Mehta, P.K.; Monteiro, P.J.M. *Concrete: Microstructure, Properties, and Materials*, 4th ed.; McGraw-Hill Education: New York, NY, USA, 2006.
2. Neville, A.M. *Properties of Concrete*, 5th ed.; (September 2, 2012); Prentice Hall: Hoboken, NY, USA, 2005.
3. Muller, A.C.A.; Scrivener, K.L.; Skibsted, J.; Gajewicz, A.M.; McDonald, P.J. Influence of silica fume on the microstructure of cement pastes: New insights from ^{1}H NMR relaxometry. *Cem. Concr. Res.* **2015**, *74*, 116–125. [CrossRef]
4. Hou, P.; Qian, J.; Cheng, X.; Shah, S.P. Effects of the pozzolanic reactivity of nanoSiO_2 on cement-based materials. *Cem. Concr. Compos.* **2015**, *55*, 250–258. [CrossRef]
5. Stefanidou, M.; Papayianni, I. Influence of nano-SiO_2 on the Portland cement pastes. *Compos. Part B Eng.* **2012**, *43*, 2706–2710. [CrossRef]
6. Land, G.; Stephan, D. The influence of nano-silica on the hydration of ordinary Portland cement. *J. Mater. Sci.* **2012**, *47*, 1011–1017. [CrossRef]
7. Kim, J.J.; Rahman, M.K.; Al-Majed, A.A.; Al-Zahrani, M.M.; Taha, M.M.R. Nanosilica effects on composition and silicate polymerization in hardened cement paste cured under high temperature and pressure. *Cem. Concr. Compos.* **2013**, *43*, 78–85. [CrossRef]
8. Shih, J.-Y.; Chang, T.-P.; Hsiao, T.-C. Effect of nanosilica on characterization of Portland cement composite. *Mater. Sci. Eng. A* **2006**, *424*, 266–274. [CrossRef]
9. Flores, Y.C.; Cordeiro, G.C.; Filho, R.D.T.; Tavares, L.M. Performance of Portland cement pastes containing nano-silica and different types of silica. *Constr. Build. Mater.* **2017**, *146*, 524–530. [CrossRef]
10. Mohammed, A.; Rafiq, S.; Mahmood, W.; Noaman, R.; AL-Darkazali, H.; Ghafor, K.; Qadir, W. Microstructure characterizations, thermal properties, yield stress, plastic viscosity and compression strength of cement paste modified with nanosilica. *J. Mater. Res. Technol.* **2020**, *9*, 10941–10956. [CrossRef]
11. Liu, L.; Xu, J.; Yin, T.; Wang, Y.; Chu, H. Improving electrical and piezoresistive properties of cement-based composites by combined addition of nano carbon black and nickel nanofiber. *J. Build. Eng.* **2022**, *51*, 104312. [CrossRef]
12. Haider, M.M.; Jian, G.; Zhong, T.; Li, H.; Fernandez, C.A.; Fifield, L.S.; Wolcott, M.; Nassiri, S. Insights into setting time, rheological and mechanical properties of chitin nanocrystals- and chitin nanofibers-cement paste. *Cem. Concr. Compos.* **2022**, *132*, 104623. [CrossRef]
13. Shi, T.; Liu, Y.; Zhao, X.; Wang, J.; Zhao, Z.; Corr, D.J.; Shah, S.P. Study on mechanical properties of the interfacial transition zone in carbon nanofiber-reinforced cement mortar based on the PeakForce tapping mode of atomic force microscope. *J. Build. Eng.* **2022**, *61*, 105248. [CrossRef]
14. Oh, J.-A.; Aakyiir, M.; Liu, Y.; Qiu, A.; Meola, T.R.; Forson, P.; Araby, S.; Zhuge, Y.; Lee, S.-H.; Ma, J. Durable cement/cellulose nanofiber composites prepared by a facile approach. *Cem. Concr. Compos.* **2022**, *125*, 104321. [CrossRef]
15. Nguyen TN, M.; Yoo, D.; Kim, J.J. Cementitious Material Reinforced by Carbon Nanotube-Nylon 66 Hybrid Nanofibers: Mechanical strength and microstructure analysis. *Mater. Today Commun.* **2020**, *23*, 100845. [CrossRef]
16. Nguyen, T.N.M.; Moon, J.; Kim, J.J. Microstructure and mechanical properties of hardened cement paste including Nylon 66 nanofibers. *Constr. Build. Mater.* **2020**, *232*, 117–132. [CrossRef]
17. Nguyen, T.N.M.; Lee, D.H.; Kim, J.J. Effect of Electrospun Nanofiber Additive on Selected Mechanical Properties of Hardened Cement Paste. *Appl. Sci.* **2020**, *10*, 7504. [CrossRef]
18. Kim, J.J.; Nguyen TN, M.; Nguyen, X.T.; Ta, D.H. Reinforcing Cementitious Material Using Singlewalled Carbon Nanotube-Nylon 66 Nanofibers. *Transp. Commun. Sci. J.* **2020**, *71*, 46–55.
19. *ASTM C150/C150M-20*; Standard Specification for Portland Cement. ASTM International: West Conshohocken, PA, USA, 2020.

20. Cai, Y.; Hou, P.; Cheng, X.; Du, P.; Ye, Z. The effects of nanoSiO_2 on the properties of fresh and hardened cement-based materials through its dispersion with silica fume. *Constr. Build. Mater.* **2017**, *148*, 770–780. [CrossRef]
21. Zou, B.; Chen, S.J.; Korayem, A.H.; Collins, F.; Wang, C.M.; Duan, W.H. Effect of ultrasonication energy on engineering properties of carbon nanotube reinforced cement pastes. *Carbon* **2015**, *85*, 212–220. [CrossRef]
22. Konsta-Gdoutos, M.S.; Metaxa, Z.S.; Shah, S.P. Highly dispersed carbon nanotube reinforced cement based materials. *Cem. Concr. Res.* **2010**, *40*, 1052–1059. [CrossRef]
23. *ASTM C307-18*; Standard Test Method for Tensile Strength of Chemical-Resistant Mortar, Grouts, and Monolithic Surfacings. ASTM International: West Conshohocken, PA, USA, 2018.
24. *ASTM C109/C109M-20b*; Standard Test Method for Compressive Strength of Hydraulic Cement Mortars (Using 2-in. or [50 mm] Cube Specimens). ASTM International: West Conshohocken, PA, USA, 2020.
25. *ASTM C305-20*; Standard Practice for Mechanical Mixing of Hydraulic Cement Pastes and Mortars of Plastic Consistency. ASTM International: West Conshohocken, PA, USA, 2020.
26. Timoshenko, S.P.; Gere, J.M. *Mechanics of Materials*; PWS-Kent Publishing: Boston, MA, USA, 1984.
27. Zhu, G.; Zhao, L.Y.; Zhu, L.T.; Deng, X.Y.; Chen, W.L. Effect of Experimental Parameters on Nanofiber Diameter from Electrospinning with Wire Electrodes. *Mater. Sci. Eng.* **2017**, *230*, 1–12. [CrossRef]
28. Chowdhury, M.; Stylios, G. Effect of Experimental Parameters on the Morphology of Electrospun Nylon 6 fibres. *Int. J. Basic Appl. Sci.* **2010**, *10*, 70–78.
29. Theron, S.A.; Zussman, E.; Yarin, A.L. Experimental investigation of the governing parameters in the electrospinning of polymer solutions. *Polymer* **2004**, *45*, 2017–2030. [CrossRef]
30. Rietveld, H.M. A profile refinement method for nuclear and magnetic structures. *J. Appl. Crystallogr.* **1969**, *2*, 65–71. [CrossRef]
31. Jiang, W.; Li, X.; Lv, Y.; Jiang, D.; Liu, Z.; He, C. Mechanical and hydration properties of low clinker cement containing high volume superfine blast furnace slag and nano silica. *Constr. Build. Mater.* **2020**, *238*, 117683. [CrossRef]
32. Maagi, M.T.; Jun, G. Effect of the particle size of nanosilica on early age compressive strength in oil-well cement paste. *Constr. Build. Mater.* **2020**, *259*, 120393. [CrossRef]
33. Das, S.K.; Khan MM, R.; Guha, A.K.; Naskar, N. Bio-inspired Fabrication of Silver Nanoparticles on Nanostructure Silica, Characterization and Application as Highly Efficient Hydrogenation Catalyst. *Green Chem.* **2013**, *15*, 2548–2557. [CrossRef]
34. Taylor, H.F.W. *Cement Chemistry*, 2nd ed.; Thomas Telford Publishing: New York, NY, USA, 1997.
35. Kim, J.J.; Foley, E.M.; Taha, M.M.R. Nano-mechanical characterization of synthetic calcium–silicate–hydrate (C–S–H) with varying CaO/SiO2 mixture ratios. *Cem. Concr. Compos.* **2013**, *36*, 65–70. [CrossRef]
36. Foley, E.M.; Kim, J.J.; Taha, M.M.R. Synthesis and nano-mechanical characterization of calcium-silicate-hydrate (C-S-H) made with 1.5 CaO/SiO2 mixture. *Cem. Concr. Res.* **2012**, *42*, 1225–1232. [CrossRef]
37. Chang, C.F.; Chen, J.W. The experimental investigation of concrete carbonation depth. *Cem. Concr. Res.* **2006**, *36*, 1760–1767. [CrossRef]
38. Siverio Lima, M.S.; Hajibabaei, M.; Hesarkazzazi, S.; Sitzenfrei, R.; Buttgereit, A.; Queiroz, C.; Haritonovs, V.; Gschösser, F. Determining the Environmental Potentials of Urban Pavements by Applying the Cradle-to-Cradle LCA Approach for a Road Network of a Midscale German City. *Sustainability* **2021**, *13*, 12487. [CrossRef]

materials

MDPI

Article

Strength, Frost Resistance, and Resistance to Acid Attacks on Fiber-Reinforced Concrete for Industrial Floors and Road Pavements with Steel and Polypropylene Fibers

Željko Kos [1,*], Sergii Kroviakov [2], Vitalii Kryzhanovskyi [2] and Daria Hedulian [2]

[1] Department of Civil Engineering, University North, University Centre of Varaždin, 104, Brigade 3, 42000 Varazdin, Croatia
[2] Department of Highways and Airfields, Odessa State Academy of Civil Engineering and Architecture, Didrihsona Street 4, 65029 Odessa, Ukraine
* Correspondence: zeljko.kos@unin.hr; Tel.: +385-98757989

Abstract: A comparison of the effect of steel and polypropylene fibers on the strength, frost resistance, abrasion, and corrosion resistance in an acidic environment of fiber-reinforced concrete for industrial floors and road pavements was carried out. Steel fibers with a length of 50 mm and a diameter of 1 mm and polypropylene fibers with a length of 36 mm and a diameter of 0.68 mm were used. The amount of steel fiber varied from 15 to 25 kg/m^3, and the amount of polypropylene fiber varied from 2 to 3 kg/m^3. It has been established that steel fiber more significantly increases the concrete compressive strength, and both types of dispersed reinforcement increase the flexural strength equally by 27–34%. Also, dispersed reinforcement reduces the concrete abrasion resistance by 15–35% and increases its frost resistance by 50 cycles, which helps to improve the durability of industrial floors and road pavements. The use of steel fiber in an amount of 20 kg/m^3 and polypropylene fiber in an amount of 2.5 kg/m^3 also increases the concrete corrosion resistance in an acidic environment. In general, dispersed reinforcement with both fiber types has approximately the same technological effect concerning the mentioned applications. However, the use of polypropylene fibers is economically more profitable since an increase in the cost of 1 m^3 of concrete with steel fiber reinforcement is from \$22.5 to \$37.5, and an increase in cost with polypropylene fiber is from \$10 to \$15.

Keywords: industrial floors; road pavements; fiber-reinforced concrete; polypropylene fiber; steel fiber; frost resistance; corrosion resistance

Citation: Kos, Ž.; Kroviakov, S.; Kryzhanovskyi, V.; Hedulian, D. Strength, Frost Resistance, and Resistance to Acid Attacks on Fiber-Reinforced Concrete for Industrial Floors and Road Pavements with Steel and Polypropylene Fibers. *Materials* **2022**, *15*, 8339. https://doi.org/10.3390/ma15238339

Academic Editors: Nikolina Zivaljic, Hrvoje Smoljanović and Ivan Balić

Received: 20 October 2022
Accepted: 17 November 2022
Published: 23 November 2022

1. Introduction and Background

Concrete industrial floors must provide a quality surface for production processes and comfortable, safe human activities. At the same time, they are operated under conditions of multidirectional loads and dynamic influences [1,2]. Rigid pavements are operated under similar conditions while additionally being affected by freezing and thawing [3–5].

According to the current standard in Ukraine [6], concrete grades from C12/15 to C32/40 can be used for industrial floors, depending on the load intensity. The floor material should be used in accordance with the project requirements and taking into account the reliability and durability of the structure, as well as the rational use of resources. In the regulatory documents of European countries [7] and the USA [8], similar requirements are imposed for concrete industrial floors. For rigid road pavements in accordance with Ukrainian [4] and European [5] standards, concretes of grades from C20/25 to C32/40 are used. At the same time, the main characteristic of road concrete is flexural strength, which should be from 3.0 to 4.1 MPa.

The use of fiber reinforcement is a well-known and well-established method for improving important mechanical properties for industrial floor and road pavement concretes.

For dispersed reinforcement, different types of fibers are used, most often steel, polypropylene, and basalt [1,3,9–11]. Also, a promising option for increasing the resistance of concrete to the formation and development of micro cracks is carbon nanotube reinforcement [12,13]. However, for the quality work of carbon nanotube reinforcement, it is important to ensure their uniform dispersion in the concrete mixture. The efficiency of using each fiber type is determined not only by the value of the change in the concrete properties due to reinforcement but also by the change in the cost of concrete with dispersed reinforcement [2]. The corrosion resistance of the fiber is also important because it is impossible to protect the fibers with a sufficient layer of concrete from contact with an aggressive environment [14,15]. Alternatively, textile concretes made of carbon or AR glass fibers are increasingly used to eliminate steel reinforcement in the concrete and thus increase durability and save material [16,17].

According to many studies, reinforcement with steel fiber makes it possible to most significantly increase the concrete flexural strength and, at the same time, compressive strength. It was shown in [18] that steel fibers are superior to glass and polypropylene fibers in increasing the compressive and flexural strength even with the same volume content of dispersed reinforcement (1% of the volume of concrete of the coating). In [19], the addition of steel fibers made it possible to obtain high-strength concrete with a compressive strength of 70 MPa and improved flexural strength after post-cracking. According to [20], concrete reinforced with steel fibers has a higher flexural strength than fiber-reinforced concrete with polypropylene fibers but has lower compressive strength and workability. In [21], due to the use of 100 kg/m^3 of steel fiber, the compressive strength of concrete pavement was increased by 10–12 MPa, and the flexural strength increased from 7–9 MPa to 15–17 MPa. However, the economic efficiency of the use of steel fiber in pavements, as well as the impact of the fiber on the environment, depends on its amount in the concrete composition [22].

The use of polypropylene fibers is also an effective method for improving the concrete properties of industrial floors and rigid pavements. For example, it was shown in [23] that dispersed reinforcement reduces the initial shrinkage of pavements and, as a result, prevents crack formation. In [24], polypropylene fiber reinforcement increased the flexural strength of concrete to 191%, while frost resistance improved by more than 25% and compressive strength increased by 28%. The positive effect of polypropylene fiber on frost resistance is also confirmed in [25]. In [26], using fiber-reinforced concrete with polypropylene fibers ensured the efficiency and safety of the pavement, including inside tunnels. According to [27], as the age of the concrete pavement increases, the positive effect of reinforcement with polypropylene fibers grows, and the maximum allowable content of fibers is 1.5% of the concrete volume. In [1,9], a positive effect of dispersed reinforcement using polypropylene fiber on the physical and mechanical characteristics of concrete industrial floors is noted. Separately, it is emphasized that the use of polypropylene fiber is more cost-effective compared to steel fiber.

In addition to the above important parameters of the concrete industrial floor's durability, such as shrinkage, frost resistance, and abrasion, another important characteristic is the acid resistance of concrete. The authors [28] point out the importance of selecting concrete components for satisfactory operation in aggressive environments, in particular, the type of cement and microfiller. In turn, the researchers [29,30] emphasize the positive effect of steel fiber in the amount of 2% of the concrete volume on the corrosion resistance of fiber-reinforced concrete on ordinary Portland cement under the action of acid rain corrosion. Researchers [31] have established the optimal amount of polypropylene fiber (0.3% of the concrete volume) to improve the acid resistance of fiber-reinforced concrete. It should be noted that in recent years there have been very few studies on the effect of fiber material type on the resistance of concrete in an acidic environment.

Thus, comparing the effectiveness of steel and polypropylene fibers in concrete industrial floors and road pavements is a crucial task. It is important to consider the corrosion resistance of concrete reinforced with different fiber types since different substances, including acid solutions, can act on pavements. At the same time, it is important to compare the effect of different fiber types on the strength, durability, and cost of concrete [1,9,26].

2. Materials and Research Methods

For the fiber-reinforced concrete mixing, the following materials were used:

- Portland cement CEM II/A-S 42.5 R, manufactured by CRH Ukraine in accordance with [32,33];
- Granite breakstone with a maximum aggregate size of 20 mm, in accordance with [34–37];
- Quartz sand with a fineness modulus of 2.75 in accordance with [35–37];
- Polycarboxylate superplasticizer MC-PowerFlow 3200 produced by MC-Bauchemie, Bottrop, Germany, in accordance with [38];
- Steel anchor fiber with 50 mm long and 1 mm in diameter, and ultimate tensile strength 1150 MPa, produced by Stalkanat-Silur, Odessa, Ukraine, in accordance [39], Figure 1a.
- Polypropylene fiber "Baumesh" with 36 mm long and 0.68 mm in diameter and ultimate tensile strength 530 MPa, produced by Bautech-Ukraine (Odessa) in accordance [40] Figure 1b.

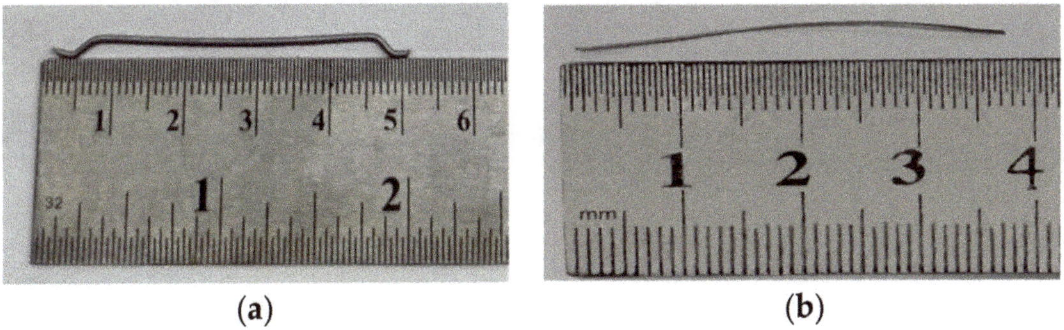

Figure 1. Steel anchor fiber (**a**), polypropylene fiber (**b**).

Concrete without dispersed reinforcement was studied as a control composition. Three fiber-reinforced concrete compositions with steel and polypropylene fibers were produced. The amount of steel fiber varied from 15 to 25 kg/m^3, and the amount of polypropylene fiber varied from 2 to 3 kg/m^3. Such a quantity of each type of dispersed reinforcement was adopted based on the recommendations of manufacturers and production practices.

The compositions of the investigated fiber-reinforced concrete with different fiber types and control concrete are shown in Table 1.

The workability of all the studied concretes and fiber-reinforced concrete mixtures was the same S4 with cone slump of 17–18 cm and determined according to [41–43]. To achieve equal workability, when fiber was introduced into the concrete mixture, the amount of superplasticizer was changed from 3.40 kg/m^3 (0.94% of the cement content) to 4.08 kg/m^3 (1.13% of the cement content). At the same time, the amount of water in the mixture remained constant for the correct comparison of the fiber-reinforced concrete properties, and the W/C ratio was 0.5.

For the preparation of fiber-reinforced concrete, a rotary mixer was used. The production and curing of the samples were carried out in accordance with the standards [44–46]. For each batch, 30 cubes 10 cm × 10 cm × 10 cm (6 cubes for compressive strength test at 7 and 28 days of age, 12 cubes for frost resistance test, and 12 cubes for acid attack resistance test), and 3 prisms 10 cm × 10 cm × 40 cm were manufactured. After the flexural strength test, cubes with dimensions of 7 cm × 7 cm × 7 cm were cut out of the prisms (6 specimens for each concrete mixture) for the abrasion resistance test.

Table 1. Compositions of the studied concrete and fiber concrete.

<table>
<tr><th rowspan="2">No.</th><th rowspan="2">Marking</th><th colspan="7">Compositions, kg/m³</th></tr>
<tr><th>Cement</th><th>Crushed Stone</th><th>Sand</th><th>Steel Fiber</th><th>Polypropylene Fiber</th><th>MC-Power Flow 3200</th><th>Water</th></tr>
<tr><td>1</td><td>Control concrete</td><td rowspan="7">360</td><td rowspan="4">1110</td><td>780</td><td>-</td><td>-</td><td>3.40</td><td rowspan="7">180</td></tr>
<tr><td>2</td><td>Fiber concrete with steel fiber 15 kg/m³</td><td>779</td><td>15</td><td>-</td><td rowspan="3">3.64</td></tr>
<tr><td>3</td><td>Fiber concrete with steel fiber 20 kg/m³</td><td>778</td><td>20</td><td>-</td></tr>
<tr><td>4</td><td>Fiber concrete with steel fiber 25 kg/m³</td><td>777</td><td>25</td><td>-</td></tr>
<tr><td>5</td><td>Fiber concrete with polypropylene fiber 2.0 kg/m³</td><td>1105</td><td>770</td><td>-</td><td>2.0</td><td rowspan="3">4.08</td></tr>
<tr><td>6</td><td>Fiber concrete with polypropylene fiber 2.5 kg/m³</td><td>1103</td><td>767</td><td>-</td><td>2.5</td></tr>
<tr><td>7</td><td>Fiber concrete with polypropylene fiber 3.0 kg/m³</td><td>1102</td><td>763</td><td>-</td><td>3.0</td></tr>
</table>

For all the studied fiber-reinforced concrete, compressive strength at the age of 7 and 28 days [46,47], flexural strength at the age of 28 days [46,48,49], abrasion resistance [50,51], frost resistance [52,53] and corrosion resistance in an acidic environment [54], water absorption [55] were determined. Flexural strength was determined by a four-point load scheme (Figure 2), and load transfer was carried out at a speed of 1.77 kN/min. The determination of the frost resistance of the studied fiber-reinforced concrete was carried out according to an accelerated method [52,53]. Samples were saturated with a 5% sodium chloride solution. Control samples after extraction from the solution are tested for compressive strength. The main samples were frozen to a temperature of −50 °C for 4 h, then the temperature was raised to a temperature of −10 °C for 1 h, then the samples were subjected to complete thawing (one cycle). The main samples, after the required number of freeze-thaw cycles, are tested for compressive strength, and weight loss is also determined. The obtained results are compared with control samples, the loss of compressive strength should not exceed 5%, and the weight loss should not exceed 3%. According to the method for determining the resistance to acid attack [54], for each concrete composition, cube samples 10 cm × 10 cm × 10 cm in size were divided into two series, the first series was placed in water, the second in an acidic environment with pH = 3, obtained by adding sulfuric acid into distilled water. The samples were kept this way for 6 months, after which their properties were determined.

Figure 2. Flexural strength test using a 4-point load scheme.

3. Research Results and Analysis

3.1. Compressive Strength, Flexural Strength, and Water Adsorption

It was experimentally established that the average concrete density of the base composition (no. 1) and fiber-reinforced concrete with polypropylene fiber (no. 2–4) was approximately the same and was in the range of 2415–2425 kg/m^3. Fiber concretes with steel fibers (no. 5–7) were expected to have a slightly higher average density of 2430 to 2440 kg/m^3.

Table 2 presents the results of testing the fiber-concrete compressive and flexural strength at 7 and 28 days.

Table 2. Values of compressive and flexural strength ('±' indicates standard deviation, 'CoV' indicates the coefficient of variation).

No. of Mixture	Compressive Strength 7 Day, MPa		-	Compressive Strength 28 Day, MPa		-	Flexural Strength 28 Day, MPa		-
	Samples	Average	CoV	Samples	Average	CoV	Samples	Average	CoV
1	28.88 28.60 29.26	28.91 ± 0.33	1.15	35.63 34.93 36.50	35.69 ± 0.79	2.0	3.54 3.50 3.55	3.53 ± 0.03	0.75
2	30.88 31.89 30.21	30.99 ± 0.86	2.73	38.96 39.33 39.90	39.40 ± 0.47	1.0	4.49 4.50 4.41	4.47 ± 0.05	1.10
3	29.55 28.97 30.21	32.15 ± 0.67	2.10	42.04 42.61 41.42	42.02 ± 0.60	1.42	4.59 4.61 4.62	4.59 ± 0.04	0.33
4	33.35 32.53 34.20	33.36 ± 0.84	2.50	43.70 42.66 43.13	43.16 ± 0.52	1.21	4.73 4.78 4.75	4.75 ± 0.04	0.53
5	29.55 28.97 30.21	29.58 ± 0.62	2.10	39.43 38.19 38.95	38.86 ± 0.63	1.61	4.48 4.51 4.47	4.49 ± 0.02	0.46
6	31.85 30.88 31.69	31.47 ± 0.52	1.65	39.91 39.14 40.49	39.85 ± 0.68	1.70	4.58 4.58 4.59	4.58 ± 0.01	0.13
7	32.34 32.87 31.51	32.24 ± 0.69	2.13	41.76 42.75 41.01	41.84 ± 0.87	2.09	4.78 4.76 4.63	4.72 ± 0.08	1.72

An important indicator of the difference in the concrete structure of the control and fiber-reinforced concrete composition is that the water adsorption of concrete without fiber (no. 1) was 4.2%. For all fiber-reinforced concrete, the water absorption was lower and was in the range of 3.6–3.8%. Since water absorption is an indicator of the open porosity of concrete, it can be concluded that fiber-reinforced concretes (no. 2–7) have a lower open porosity than concrete without fibers. This can be explained by an increase in the amount of superplasticizer MC-PowerFlow 3200 when using fiber (Table 1), which affected the capillary-pore structure, as well as by the direct effect of fibers on the pores distribution and capillaries in the composite.

The effect of the steel and polypropylene fibers amount on the compressive strength is shown in Figure 3, on the flexural strength in Figure 4. The diagrams in Figures 3 and 4 are built using a combined x-axis, which simultaneously shows the scale of the amount of steel fiber up to 25 kg/m^3 and polypropylene fiber up to 3.0 kg/m^3.

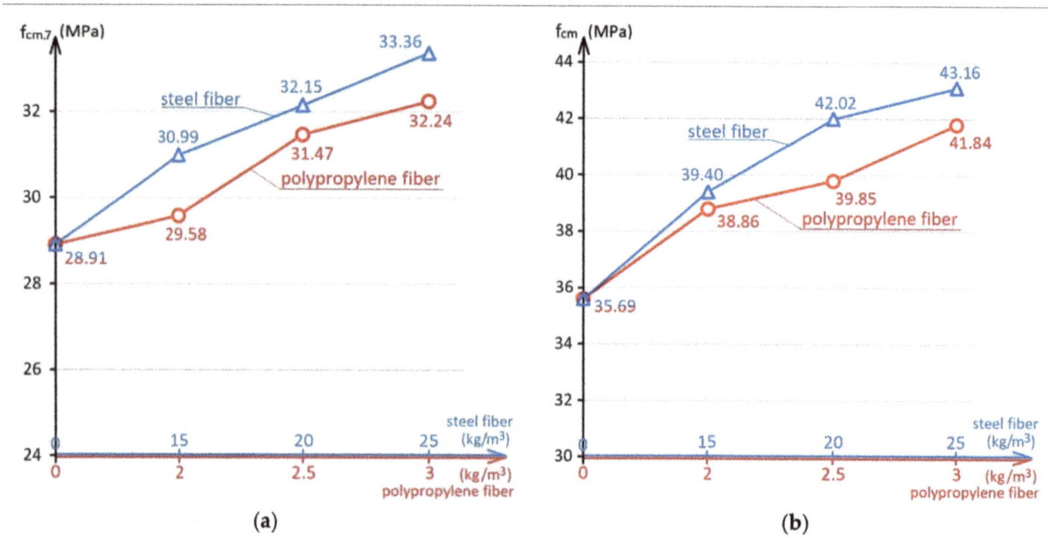

Figure 3. The influence of the steel and polypropylene fibers amount on the concrete compressive strength at 7 days (**a**) and compressive strength at 28 days (**b**).

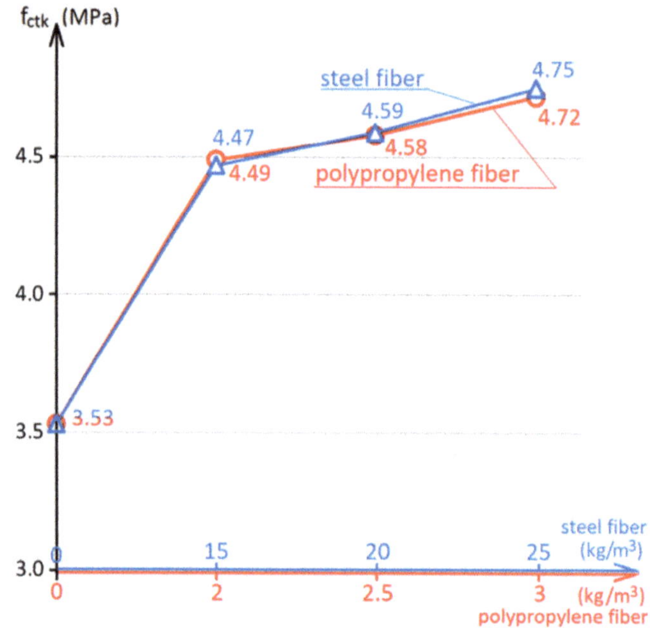

Figure 4. The influence of the steel and polypropylene fibers amount on the concrete flexural strength at the age of 28 days.

As can be seen from the diagram in Figure 3b, dispersed reinforcement increases the concrete compressive strength by 10–21% with steel fibers and by 9–17% with polypropylene fibers. Also, in the studies, it was determined that on the 7th day (Figure 3a), the concrete and fiber-reinforced concrete compressive strength was 76 to 81% of their strength at the design age. That is, dispersed reinforcement does not affect the rate of concrete hardening, which is expected.

For concrete industrial floors and pavements, the more important mechanical property is flexural strength since this property determines the bearing capacity of the pavement structure [4,5,18,27]. As shown in the diagram in Figure 4, the use of fiber reinforcement increases the concrete flexural strength by 27–34%. Improving the strength properties of concrete with fiber reinforcement can be explained by the fact that normal concrete is a brittle material with bad resistance to tensile stresses, and fiber introduction can take up internal tensile stresses by itself. The equal effect on the concrete flexural strength with less polypropylene fiber compared with a higher quantity of steel fiber interpreted that per unit of volume, there is a higher quantity of polypropylene fibers. Therefore, the distribution of tensile stresses occurs more evenly over the cross-section.

The use of polypropylene fibers increases strength by the same amount as the use of steel fibers. However, an important economic effect is that when using steel fiber its amount in fiber-reinforced concrete is from 15 to 25 kg/m^3, and with using polypropylene from 2 to 3 kg/m^3. The cost of steel fiber is about \$1.5 per 1 kg, and the cost of polypropylene fiber is about \$5 per 1 kg. Accordingly, the increase in the cost of 1 m^3 of concrete due to reinforcement with steel fibers is from \$22.5 to \$37.5, and the increase in price due to reinforcement with polypropylene fibers is from \$10 to \$15.

3.2. Abrasion Resistance

Industrial floors and road pavements are subjected to abrasive action during operation. Therefore, the concrete durability of such structures is largely determined by the abrasion resistance of concrete. Figure 5 shows the effect of the amount of steel and polypropylene fibers on the concrete abrasion resistance. The lower abrasion of the material, the higher its wear resistance. Figure 6 shows samples of composition No. 5 before the abrasion resistance test.

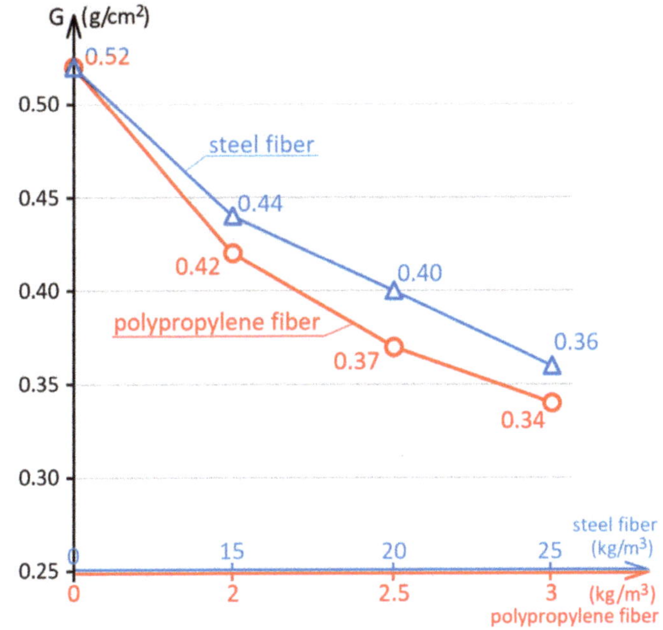

Figure 5. The influence of the steel and polypropylene fibers amount on the concrete abrasion resistance.

Figure 6. Abrasion resistance test. Composition samples No. 5 on a wheel tester.

Analysis of the diagram in Figure 5 shows that the concrete abrasion resistance of the control composition without fibers is 0.52 g/cm^2. This level of concrete abrasion resistance does not meet the requirements of the Ukrainian national standard [56], according to which the abrasion of concrete pavement should be no more than 0.50 g/cm^2. Due to the introduction of steel fiber, the abrasion of concrete is reduced by 15–31%, and with the introduction of polypropylene fiber by 19–35%.

Using 15–20 kg/m^3 steel fibers or 2 kg/m^3 polypropylene fibers makes it possible to achieve the G3 abrasion grade (0.41–0.5 g/cm^2), which already allows the use of concrete for rigid road pavements [56]. By increasing the amount of steel fibers to 25 kg/m^3, as well as by using 2.5–3 kg/m^3 polypropylene fibers, the concrete abrasion grade improves to G2 (0.31–0.4 g/cm^2), which increases the durability of road pavements and industrial floors. It is also possible to note the best wear resistance of concretes with polypropylene fibers, despite the significantly lower amount of dispersed reinforcement in concrete by weight. This can be explained by the better adhesion of polypropylene fibers to the concrete matrix compared to steel fibers. Under abrasive action, steel fibers lose their adhesion to the concrete matrix more easily, and polypropylene fibers, due to better cooperation with the matrix and retention of individual blocks of the composite, give the concrete greater wear resistance [57,58].

3.3. *Frost Resistance*

The durability of rigid road pavements in the climatic conditions of European and many other countries is significantly affected by the frost resistance of concrete [20,21,25,59]. For concrete industrial floors, frost resistance is of lesser importance, except for the floors of open warehouses. However, frost resistance is an indirect indicator of resistance to temperature fluctuations, which can occur in several technological processes.

The value of frost resistance of the studied concretes was determined in accordance with [52,53] using an accelerated method for concretes of road and airfield pavements, Figure 7. Other methods require a lot of time for the experiment and are more laborious. For example, using basic techniques according to [52,53], to achieve concrete frost resistance F100 cycles, it is necessary to carry out 100 freeze-thaw cycles; in turn, one cycle freezes at a temperature of -18 ± 2 °C and then completely thaws the samples. It is important to note that the accuracy of the accelerated technique is rather limited since it discretely distinguishes only grades F100, F150, F200, F300, etc. In general, this discreteness does not prevent us from evaluating the effect of dispersed reinforcement on frost resistance, but it does not allow distinguishing the F index in the range between F200 and F300.

Figure 7. Frost resistance test.

It has been experimentally established that the frost resistance of control concrete composition (No. 1) is grade F150. For all other studied fiber-reinforced concretes (No. 2–9), the frost resistance grade was F200. The experimental data are presented in Table 3. Figure 8 shows specimens after frost resistance testing, (**a**): specimen with dispersed reinforcement with polypropylene fiber; (**b**): specimen with dispersed reinforcement with steel fiber.

(**a**) (**b**)

Figure 8. The studied specimens after freeze-thaw cycles.(**a**) fiber concrete with polypropylene fiber; (**b**) fiber concrete with steel fiber.

Table 3. Data of fiber-reinforced concrete after frost resistance test ('±' indicates standard deviation, 'CoV' indicates coefficient of variation).

Marking of Mixture	Compressive Strength of Control Specimens, MPa			Compressive Strength of Main Specimens after 200 Cycles of Freezing and Thawing, MPa			Average Strength Reduction, %	Frost Resistance, Cycles
	Samples	Average	CoV	Samples	Average	CoV		
Control concrete	39.5 38.3 39.6 40.5 36.3 38.5	38.80 ± 1.46	3.76	36.1 37.7 36.2 37.5 37.0 36.5	36.80 ± 0.67	1.83	5.3	F150
Fiber concrete with steel fiber 15 kg/m^3	40.8 40.5 40.0 39.1 39.8 40.9	40.20 ± 0.69	1.71	38.0 38.2 39.1 38.9 38.2 38.1	38.40 ± 0.46	1.20	4.7	F200
Fiber concrete with steel fiber 20 kg/m^3	44.2 44.4 43.7 42.5 42.8 42.6	43.40 ± 0.84	1.94	40.9 41.8 41.5 41.2 42.0 40.8	41.40 ± 0.48	1.17	4.8	F200
Fiber concrete with steel fiber 25 kg/m^3	43.9 44.8 44.8 44.5 44.8 45.2	44.70 ± 0.44	0.98	43.1 42.7 42.4 42.8 41.8 43.0	42.60 ± 0.48	1.12	4.9	F200
Fiber concrete with polypropylene fiber 2.0 kg/m^3	40.5 40.3 39.5 38.3 39.6 37.3	39.30 ± 1.23	3.13	38.3 37.9 36.2 37.8 37.4 37.3	37.50 ± 0.73	1.93	4.8	F200
Fiber concrete with polypropylene fiber 2.5 kg/m^3	41.6 41.2 42.0 40.8 41.7 41.1	41.40 ± 0.44	1.07	39.9 39.5 38.8 38.7 40.2 40.0	39.50 ± 0.64	1.61	4.8	F200
Fiber concrete with polypropylene fiber 3.0 kg/m^3	43.3 44.2 44.1 43.1 42.9 42.1	43.30 ± 0.79	1.82	40.8 41.6 40.7 41.9 41.5 41.7	41.40 ± 0.50	1.20	4.6	F200

That is, dispersed reinforcement with both steel and polypropylene fibers made it possible to increase the frost resistance of concrete to approximately the same extent. The positive effect of using fibers is explained by the ability of fibers as a stabilizing factor to prevent the separation of concrete as a composite material into separate structural blocks under the influence of freezing and thawing [20,59]. As shown above, the fiber also affects the capillary-pore structure of concrete, thereby increasing the concrete frost resistance. The influence occurs both under the action of the fibers themselves as well as changes in the dosage of the superplasticizer due to the experimental conditions. At the same time, as noted earlier, the use of polypropylene fiber due to its lower consumption per 1 m^3 of material is more economically beneficial.

3.4. Resistance to Acid Attack

For industrial floors in a number of chemical industries and other enterprises using acids in the production process, as well as warehouses where substances with acidic properties are stored, an important indicator of the quality of concrete is its resistance to an acidic environment. For road pavements, acid resistance can provide concrete durability when de-icers are used.

As noted above, for all the studied fiber-reinforced concrete, the resistance to the acid attack in an acidic environment was determined. For each concrete composition, one part of the samples was kept in water for 6 months, and the second in an acidic environment with pH = 3, Figure 9. The strength of the studied concretes and fiber-reinforced concretes after 6 months of storage in an acidic environment and the magnitude of the decrease in the strength of concrete in an acidic environment are shown in Table 4. Figure 10 shows samples during testing with (a): fiber-reinforced concrete saturated with water; (b): fibrous concrete in an acidic environment, pH = 3. Figure 11 shows samples of fiber-reinforced concrete after 6 months of acid exposure with (a): steel fiber-reinforced concrete; (b): polypropylene fiber-reinforced concrete, pH = 3.

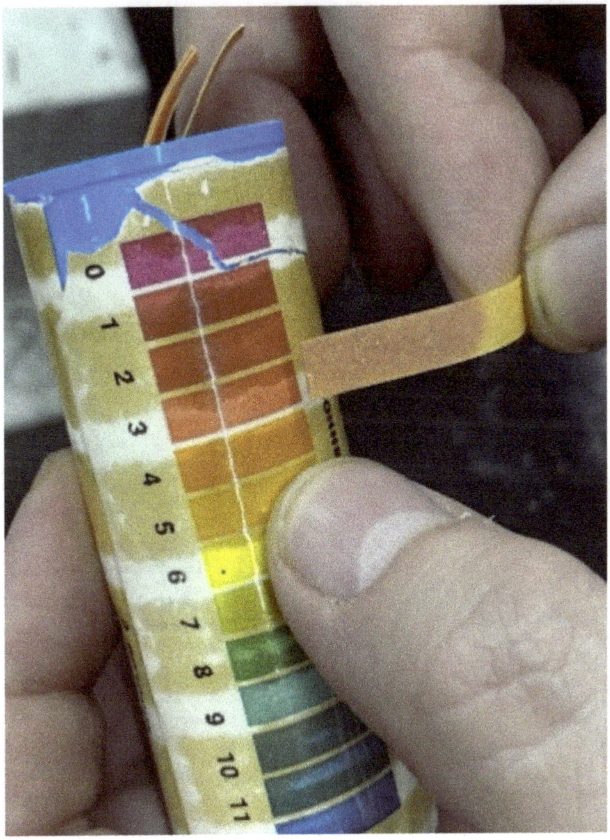

Figure 9. Acid environment test, pH = 3.

Table 4. Strength of concrete and fiber-reinforced concrete after 6 months exposure to water and in an acidic environment ('±' indicates standard deviation, 'CoV' indicates coefficient of variation).

Marking of Mixture	Compressive Strength after Soaking in Water, MPa			Compressive Strength after Aging in an Acidic Environment, MPa			Strength Reduction, %
	Samples	Average	CoV	Samples	Average	CoV	
Control concrete	45.8 45.6 45.0 45.0 45.5 45.2	45.4 ± 0.29	0.73	34.6 34.4 34.7 34.2 34.5 34.1	34.4 ± 0.23	0.67	24
Fiber concrete with steel fiber 15 kg/m^3	50.0 49.8 49.1 49.3 49.5 49.2	49.5 ± 0.35	0.72	36.3 36.5 36.6 36.9 36.1 36.1	36.4 ± 0.31	0.86	26
Fiber concrete with steel fiber 20 kg/m^3	53.1 53.0 52.1 52.5 52.4 52.3	52.6 ± 0.4	0.76	40.5 40.9 40.2 40.1 40.7 40.0	40.4 ± 0.36	0.89	23
Fiber concrete with steel fiber 25 kg/m^3	54.3 54.0 53.6 53.4 53.7 53.8	53.8 ± 0.32	0.59	41.3 41.6 42.3 42.0 41.1 41.4	41.6 ± 0.45	1.09	23
Fiber concrete with polypropylene fiber 2.0 kg/m^3	47.5 48.2 48.4 48.6 48.8 47.5	48.2 ± 0.55	1.15	35.9 36.4 36.2 35.7 36.5 36.6	36.2 ± 0.35	0.98	25
Fiber concrete with polypropylene fiber 2.5 kg/m^3	49.7 51.1 49.9 49.1 49.5 51.4	50.1 ± 0.92	1.84	38.8 39.6 39.8 39.6 38.9 39.5	39.4 ± 0.41	1.05	21
Fiber concrete with polypropylene fiber 3.0 kg/m^3	52.0 52.3 53.5 53.2 51.9 51.7	52.4 ± 0.74	1.42	41.7 41.8 41.4 42.6 42.7 41.6	42.0 ± 0.55	1.30	20

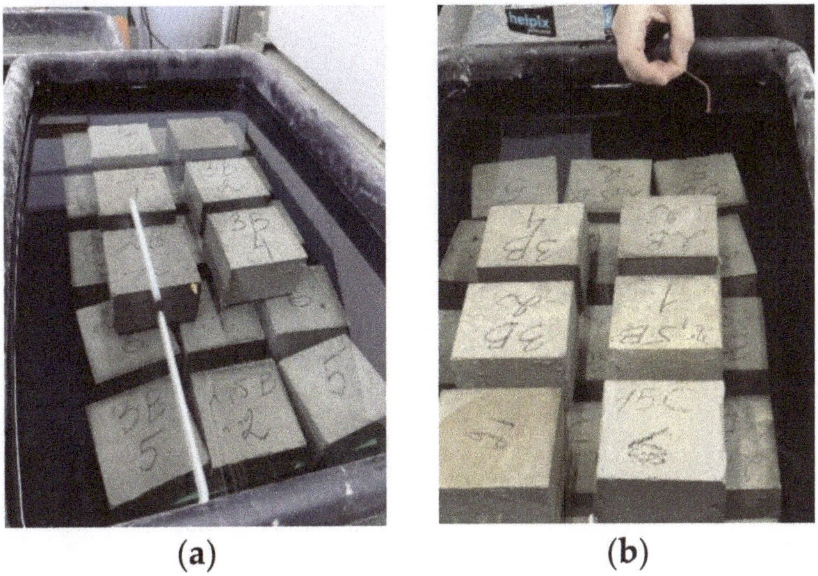

Figure 10. Acid environment test, pH = 3. Total amount of specimens 84 cubes (12 cubes for each batch). (**a**) control specimens; (**b**) specimens in acid environment.

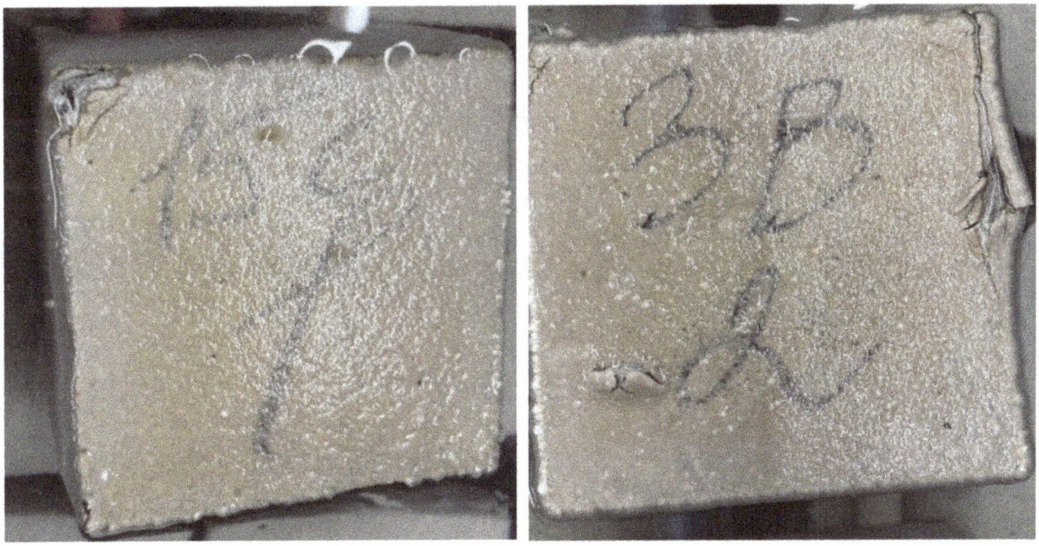

Figure 11. Specimens after 6 months exposure in an acidic environment, pH = 3. Total amount of specimens 84 cubes (12 cubes for each batch).

Figure 12 shows the effect of the amount of two types of fiber on the concrete strength after curing in water and in an acidic environment.

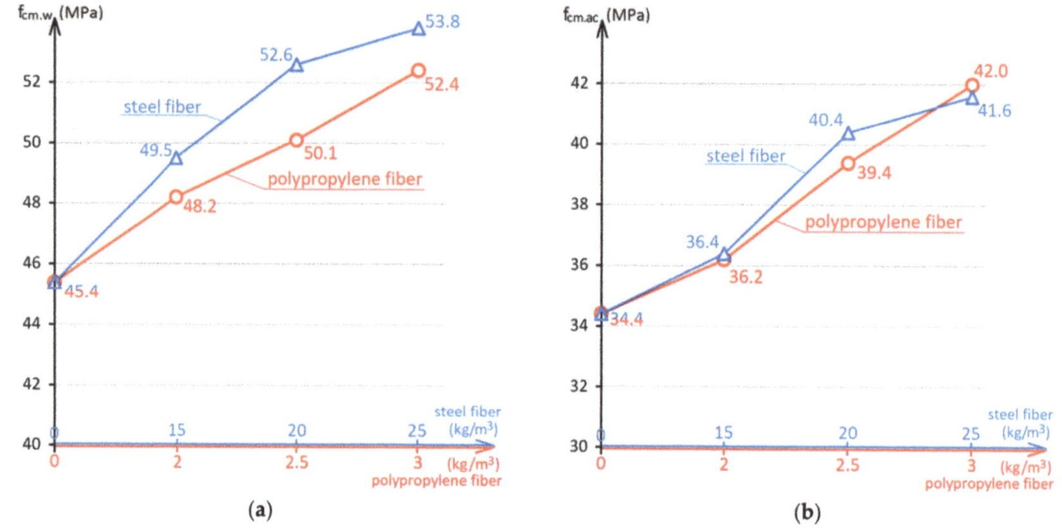

Figure 12. The influence of the steel and polypropylene fibers amount on the concrete strength after 6 months exposure to water (**a**) and in an acidic environment with pH = 3 (**b**).

Analysis of the diagram in Figure 12a concludes that after 6 months of exposure in a humid environment, the effect of dispersed reinforcement with steel and polypropylene fibers on the concrete compressive strength is similar to the effect of fibers on strength at the standard age.

Under the influence of an acidic environment, the nature of the fiber effect on the concrete strength changes (Figure 12b). Dispersed reinforcement also has a positive effect on concrete strength and increases its value by 5–22%. However, if steel fiber showed a slightly higher efficiency in a non-aggressive water environment, then after operation in an acidic environment, the degree of different types of fiber influence on the concrete strength is equalized. It is also characteristic that in an acidic environment at low fiber dosages (15 kg/m^3 for steel and 2 kg/m^3 for polypropylene fibers), dispersed reinforcement has little effect on the strength and does not increase the resistance to the acid attack on concrete. The decrease in the concrete strength of the control composition was 24%, concrete with a small amount of fiber was 26% and 25%, respectively. With a higher amount of fiber, dispersion-reinforced concretes have a strength degree reduction of 23% with using steel fibers and 20–21% with using polypropylene fibers.

Such an effect of fiber on resistance to acid attack in an acidic environment is explained by the fact that as a result of the reaction of concrete components with acids, calcium carbonate is formed. It is insoluble in water and accumulates in pores and microcracks, which leads to concrete cracking [60,61]. In the experiment, the mass of samples 10 cm × 10 cm × 10 cm in size after exposure to an acidic environment was, on average, 15 g more than the mass of similar samples exposed to water. Dispersed reinforcement prevents the development of cracks caused by internal stresses in the composite [62]. Reducing the water absorption of concrete with fiber (open porosity) also contributes to an increase in the resistance to the acid attack of the material. But with a small amount of fiber, reinforcement is not effective enough, and internal stresses are more capable of causing destructive effects on the material. It can also be noted that the efficiency of polypropylene fibers in an acidic environment turned out to be at least not lower than steel ones, which can be explained by better adhesion to the concrete matrix and chemical inertness to acids.

In general, dispersed reinforcement with both steel and polypropylene fibers improves the concrete resistance to acid attack for industrial floors and road pavements.

4. Conclusions

The conducted research allowed us to carry out a comprehensive comparison of the effectiveness of concrete dispersed reinforcement for industrial floors and road pavements with steel and polypropylene fibers.

It has been established that both types of dispersed reinforcement increase the strength, frost resistance, abrasion resistance, and resistance to acid attack on fiber-reinforced concrete. Steel fibers increase the concrete compressive strength slightly more effectively. In other analyzed indicators of the concrete quality (flexural strength, abrasion resistance, frost resistance, and resistance to acid attack), the use of polypropylene fibers can improve at the same level as the use of steel fibers. To ensure increased resistance to an acid attack on concrete in an acidic environment, the amount of steel fiber should not be lower than 20 kg/m^3, and the amount of polypropylene should not be lower than 2.5 kg/m^3.

Dispersed reinforcement with both types of fibers increases the strength characteristics and durability of pavements under operating conditions typical of most European countries. However, due to a significant difference in the effective dosage of different fiber types, the increase in the cost of 1 m^3 of concrete mix using dispersed reinforcement of polypropylene fibers is $12.5–$22.5 less than the increase in cost with dispersed reinforcement with steel fibers. Other advantages of polypropylene fibers compared to steel fibers are less wear and tear of technological equipment during the preparation and placing of the fiber-reinforced concrete mixture, as well as lower energy costs in manufacturing fibers.

Our data complement [29] notes that the introduction of steel and basalt fibers increased the acid resistance of concrete. Results [62] show that dispersed steel fiber reinforcement bridge cracks and inhibits the development of concrete chemical erosion. Experiment [63] reflects the positive effect of using multiscale polypropylene fiber with different geometry properties on concrete acid resistance.

It is important to emphasize that in our experiment, a more aggressive acid attack environment was used, which can be useful for expanding the range of data on the effect of dispersed reinforcement on the concrete acid attack resistance.

Author Contributions: Conceptualization, Ž.K. and S.K.; methodology, S.K., V.K. and D.H.; validation, Ž.K., V.K. and S.K.; formal analysis, Ž.K. and S.K.; resources, V.K. and D.H.; data curation, Ž.K. and S.K.; writing—original draft preparation, Ž.K. and V.K.; writing—review and editing, S.K. and D.H.; visualization, S.K. and V.K.; supervision, Ž.K. and V.K.; project administration, Ž.K. and S.K.; funding acquisition, Ž.K. All authors have read and agreed to the published version of the manuscript.

Funding: This research received no external funding.

Institutional Review Board Statement: Not applicable.

Informed Consent Statement: Not applicable.

Data Availability Statement: Not applicable.

Conflicts of Interest: The authors declare no conflict of interest.

References

1. Chajec, A.; Sadowski, L. The Effect of Steel and Polypropylene Fibers on the Properties of Horizontally Formed Concrete. *Materials* **2020**, *13*, 5827. [CrossRef] [PubMed]
2. Sadrolodabaee, P.; Claramunt, J.; Ardanuy, M.; de la Fuente, A.A. Textile Waste Fiber-Reinforced Cement Composite: Comparison between Short Random Fiber and Textile Reinforcement. *Materials* **2021**, *14*, 3742. [CrossRef] [PubMed]
3. Affan, M.; Ali, M. Experimental investigation on mechanical properties of jute fiber reinforced concrete under freeze-thaw conditions for pavement applications. *Constr. Build. Mater.* **2022**, *323*, 126599. [CrossRef]
4. Highways. Part I. Design. Part II. Construction. Kyiv, Ukraine, DBN V.2.3-4:2015. 2015. Available online: http://online.budstandart.com/ru/catalog/doc-page.html?id_doc=62131 (accessed on 18 September 2022).
5. *BS EN 13877-2:2013*; Concrete Pavements. Functional Requirements for Concrete Pavements. British Standard Institution: London, UK, 2013.

6. Floors. CITP, Gosstroy (USSR), SNiP 2.03.13-88. 1988. Available online: http://online.budstandart.com/ru/catalog/doc-page.html?id_doc=4122 (accessed on 18 September 2022).
7. TR 34. Concrete Industrial Ground Floors–Fourth Edition, London, UK. 2016. Available online: https://www.concretebookshop.com (accessed on 18 September 2022).
8. ACI 360R-06. Design of Concrete Slabs on Ground, Michigan, USA. 2006. Available online: https://www.concrete.org/publications/internationalconcreteabstractsportal/m/details/id/18324 (accessed on 18 September 2022).
9. Abbasi Dezfouli, A.; Orak, M. Effect of Using Different Fibers on Slab on Grades. *J. Civ. Eng. Mater. Appl.* **2019**, *3*, 91–99.
10. Pawelska-Mazur, M.; Kaszynska, M. Mechanical Performance and Environmental Assessment of Sustainable Concrete Reinforced with Recycled End-of-Life Tyre Fibres. *Materials* **2021**, *14*, 256. [CrossRef]
11. Mishutin, A.; Kroviakov, S.; Kryzhanovskyi, V.; Chintea, L. Fiber-reinforced concrete for rigid road pavements modified with polycarboxylate admixture and metakaolin. *Electron. J. Fac. Civ. Eng. Osijek-E-GFOS* **2021**, *12*, 1–10. [CrossRef]
12. Ramezani, M.; Kim, Y.H.; Sun, Z. Mechanical properties of carbon-nanotube-reinforced cementitious materials: Database and statistical analysis. *Mag. Concr. Res.* **2020**, *72*, 1047–1071. [CrossRef]
13. Ramezani, M.; Kim, Y.H.; Sun, Z. Elastic modulus formulation of cementitious materials incorporating carbon nanotubes: Probabilistic approach. *Constr. Build. Mater.* **2020**, *274*, 122092. [CrossRef]
14. Ahmed, S.F.U.; Mihashi, H. A review on durability properties of strain hardening fibre reinforced cementitious composites (SHFRCC). *Cem. Concr. Compos.* **2007**, *29*, 365–376. [CrossRef]
15. Frazão, C.M.V.; Barros, J.A.O.; Camões, A.; Alves, A.M.V.C.P.; Rocha, L. Corrosion effects on pullout behavior of hooked steel fibers in self-compacting concrete. *Cem. Concr. Res.* **2016**, *79*, 112–122. [CrossRef]
16. Beßling, M.; Groh, M.; Koch, V.; Auras, M.; Orlowsky, J.; Middendorf, B. Repair and Protection of Existing Steel-Reinforced Concrete Structures with High-Strength, Textile-Reinforced Mortars. *Buildings* **2022**, *12*, 1615. [CrossRef]
17. Beßling, M.; Orlowsky, J. Quantification of the Influence of Concrete Width per Fiber Strand on the Splitting Crack Failure of Textile Reinforced Concrete (TRC). *Polymers* **2022**, *14*, 489. [CrossRef] [PubMed]
18. Hussain, I.; Ali, B.; Akhtar, T.; Jameel, M.S.; Raza, S.S. Comparison of mechanical properties of concrete and design thickness of pavement with different types of fiber-reinforcements (steel, glass, and polypropylene). *Case Stud. Constr. Mater.* **2020**, *13*, e00429. [CrossRef]
19. Jang, S.J.; Yun, Y.J.; Yun, H.D. Influence of Fiber Volume Fraction and Aggregate Size on Flexural Behavior of High Strength Steel Fiber-Reinforced Concrete (SFRC). *Appl. Mech. Mater.* **2013**, *372*, 223–226. [CrossRef]
20. Hasani, M.; Nejad, F.M.; Sobhani, J.; Chini, M. Mechanical and durability properties of fiber reinforced concrete overlay: Experimental results and numerical simulation. *Constr. Build. Mater.* **2020**, *268*, 121083. [CrossRef]
21. Kos, Ž.; Kroviakov, S.; Kryzhanovskyi, V.; Grynyova, I. Research of Strength, Frost Resistance, Abrasion Resistance and Shrinkage of Steel Fiber Concrete for Rigid Highways and Airfields Pavement Repair. *Appl. Sci.* **2022**, *12*, 1174. [CrossRef]
22. Achilleos, C.; Hadjimitsis, D.; Neocleous, K.; Pilakoutas, K.; Neophytou, P.O.; Kallis, S. Proportioning of Steel Fibre Reinforced Concrete Mixes for Pavement Construction and Their Impact on Environment and Cost. *Sustainability* **2011**, *3*, 965–983. [CrossRef]
23. Mazzoli, A.; Monosi, S.; Plescia, E.S. Evaluation of the early-age-shrinkage of Fiber Reinforced Concrete (FRC) using image analysis methods. *Constr. Build. Mater.* **2015**, *101*, 596–601. [CrossRef]
24. Al-Lebban, M.F.; Khazaly, A.I.; Shabbar, R.; Jabal, Q.A.; Al Asadi, L.A.R. Effect of Polypropylene Fibers on some Mechanical Properties of Concrete and Durability against Freezing and Thawing Cycles. *Key Eng. Mater.* **2021**, *895*, 130–138. [CrossRef]
25. Salemi, N.; Behfarnia, K. Effect of nano-particles on durability of fiber-reinforced concrete pavement. *Constr. Build. Mater.* **2013**, *48*, 934–941. [CrossRef]
26. Nobili, A.; Lanzoni, L.; Tarantino, A. Experimental investigation and monitoring of a polypropylene-based fiber reinforced concrete road pavement. *Constr. Build. Mater.* **2013**, *47*, 888–895. [CrossRef]
27. Lu, C. Mechanical Properties of Polypropylene Fiber Reinforced Concrete Pavement. *Adv. Mater. Res.* **2013**, *739*, 264–267. [CrossRef]
28. Barbhuiya, S.; Kumala, D. Behaviour of a Sustainable Concrete in Acidic Environment. *Sustainability* **2017**, *9*, 1556. [CrossRef]
29. Jianqiao, Y.; Hongxia, Q.; Geifei, Z.; Xinke, W. Research on damage and deterioration of fiber concrete under acid rain environment based on GM(1,1)-Markov. *Materials* **2021**, *14*, 6326.
30. Sanytsky, M.; Kropyvnytska, T.; Fic, S.; Ivashchyshyn, H. Sustainable low-carbon binders and concretes. *E3S Web Conf.* **2020**, *166*, 06007. [CrossRef]
31. Maruthachalam, D.; Vishnuram, B.G.; Gurunathan, M.; Padmanaban, I. Durability properties of fibrillated polypropylene fibre-reinforced high performance concrete. *J. Struct. Eng.* **2011**, *38*, 1–9.
32. *BS EN 196-1:2016*; Methods of Testing Cement. Determination of Strength. British Standard Institution: London, UK, 2016.
33. *BS EN 196-2:2013*; Methods of Testing Cement. Chemical Analysis of Cement. British Standard Institution: London, UK, 2013.
34. Crushed Stone and Gravel Are Dense Natural for Building Materials, Products, Structures and Works, DSTU B V.2.7-75-98. 1999. Available online: http://online.budstandart.com/ru/catalog/doc-page?id_doc=4674 (accessed on 1 October 2022).
35. *BS EN 12620:2013*; Aggregates for Concrete. British Standard Institution: London, UK, 2013.
36. *ASTM C 33/C33M-18*; Standard Specifications for Concrete Aggregates. ASTM International: West Conshohocken, PA, USA, 2018.
37. Building Materials. Sand Dense Natural for Construction Materials, Products, Designs and Works. Specifications, DSTU B V.2.7-32-95. 1996. Available online: http://online.budstandart.com/ru/catalog/doc-page?id_doc=4053 (accessed on 1 October 2022).

38. *BS EN 934-2:2009*; Admixtures for Concrete, Mortar and Grout Concrete admixtures. Definitions, Requirements, Conformity, Marking and Labelling. British Standard Institution: London, UK, 2009.
39. *BS EN 14889-1:2006*; Fibres for Concrete—Part 1: Steel Fibres–Definitions, Specifications and Conformity. British Standard Institution: London, UK, 2006.
40. *BS EN 14889-1:2006*; Fibres for Concrete—Part 2: Polymer Fibres–Definitions, Specifications and Conformity. British Standard Institution: London, UK, 2006.
41. *BS EN 12350-2:2019*; Testing Fresh Concrete. Slump Test. British Standard Institution: London, UK, 2019.
42. *ASTM C143/C143M-12*; Standard Test Method for Slump of Hydraulic-Cement Concrete. ASTM International: West Conshohocken, PA, USA, 2012.
43. Concrete Mixtures. Test Methods, DSTU B V.2.7-114:2002. 2002. Available online: http://online.budstandart.com/ua/catalog/doc-page?id_doc=4913 (accessed on 1 October 2022).
44. *BS EN 12390-2:2019*; Testing Hardened Concrete. Making and Curing Specimens for Strength Tests. British Standard Institution: London, UK, 2019.
45. *ASTM C192/C192M-19*; Standard for Making and Curing Concrete Test Specimens in the Laboratory. ASTM International: West Conshohocken, PA, USA, 2019.
46. Building Materials. Concrete. Methods of Determining the Strength of Control Samples, DSTU B V.2.7-2014:2009. Available online: http://online.budstandart.com/ru/catalog/doc-page.html?id_doc=25943 (accessed on 1 October 2022).
47. *BS EN 12390-3:2009*; Testing Hardened Concrete. Compressive Strength of Test Specimens. British Standard Institution: London, UK, 2009.
48. *BS EN 12390-5:2009*; Testing Hardened Concrete. Flexural Strength of Test Specimens. British Standard Institution: London, UK, 2009.
49. *ASTM C78/C78M-16*; Standard Test Method for Flexural Strength of Concrete (Using Simple Beam with Third-Point Loading). ASTM International: West Conshohocken, PA, USA, 2016.
50. *ASTM C944/C944M-19*; Standard Test Method for Abrasion Resistance of Concrete or Mortar Surfaces by the Rotating Cutter Method. ASTM International: West Conshohocken, PA, USA, 2019.
51. Building Materials. Concrete. Methods for Determining Abrasion Resistance, DSTU B V.2.7-212:2009. 2010. Available online: http://online.budstandart.com/ru/catalog/doc-page?id_doc=25953 (accessed on 1 October 2022).
52. Concrete. Accelerated Methods for Determining Frost Resistance during Repeated Freezing and Thawing. General Requirements, DSTU B V.2.7-47-96. 1996. Available online: http://online.budstandart.com/ua/catalog/doc-page?id_doc=4061 (accessed on 1 October 2022).
53. Concrete. Accelerated Methods for Determining Frost Resistance during Repeated Freezing and Thawing, DSTU B V.2.7-49-96. 1996. Available online: http://online.budstandart.com/ru/catalog/doc-page?id_doc=4950 (accessed on 1 October 2022).
54. Corrosion Protection in Construction. Concretes. General Requirements for Testing, DSTU B GOST 27677:2011. 2011. Available online: http://online.budstandart.com/ua/catalog/doc-page?id_doc=28066 (accessed on 1 October 2022).
55. Building Materials. Concretes. Methods of Determining Average Density, Humidity, Water Absorption, Porosity and Water-Proofness, DSTU B V.2.7-170:2008. 2009. Available online: http://online.budstandart.com/ua/catalog/doc-page?id_doc=24882 (accessed on 1 October 2022).
56. Cement-Concrete Road Mixes and Cement-Concrete Road Mixes. Specifications, DSTU 8858:2019. 2019. Available online: http://online.budstandart.com/ua/catalog/doc-page.html?id_doc=82987 (accessed on 1 October 2022).
57. Abbass, W.; Khan, M.I.; Mourad, S. Evaluation of mechanical properties of steel fiber reinforced concrete with different strengths of concrete. *Constr. Build. Mater.* **2018**, *168*, 556–569. [CrossRef]
58. Soulioti, D.V.; Barkoula, N.M.; Paipetis, A.; Matikas, T.E. Effects of Fibre Geometry and Volume Fraction on the Flexural Behaviour of Steel-Fibre Reinforced Concrete. *Strain–Int. J. Exp. Mech.* **2009**, *47*, e535–e541. [CrossRef]
59. Öztürk, O. Comparison of frost resistance for the fiber reinforced geopolymer and cementitious composites. *Mater. Today: Proc.* **2022**, *65*, 1504–1510. [CrossRef]
60. Huber, B.; Hilbig, H.; Drewes, J.E.; Müller, E. Evaluation of concrete corrosion after short- and long-term exposure to chemically and microbially generated sulfuric acid. *Cem. Concr. Res.* **2017**, *94*, 36–48. [CrossRef]
61. Taheri, S.; Delgado, G.P.; Agbaje, O.B.A.; Giri, P.; Clark, S.M. Corrosion Inhibitory Effects of Mullite in Concrete Exposed to Sulfuric Acid Attack. *Corros. Mater. Degrad.* **2020**, *1*, 282–295. [CrossRef]
62. Marcos-Meson, V.; Fischer, G.; Edvardsen, C.; Skovhus, T.L.; Michel, A. Durability of Steel Fibre Reinforced Concrete (SFRC) exposed to acid attack–A literature review. *Constr. Build. Mater.* **2018**, *200*, 490–501. [CrossRef]
63. Liang, N.; Mao, J.; Yan, R.; Liu, X.; Zhou, X. Corrosion resistance of multiscale polypropylene fiber-reinforced concrete under sulfate attack. *Case Stud. Constr. Mater.* **2022**, *16*, e01065. [CrossRef]

materials

MDPI

Article

Simulation and Experimental Substantiation of the Thermal Properties of Non-Autoclaved Aerated Concrete with Recycled Concrete Powder

Xiaosong Ma [1,2,3,4], Hao Li [1], Dezhi Wang [1,*], Chunbao Li [2] and Yongqi Wei [3,4,*]

1 School of Civil and Water Conservancy Engineering, Ningxia University, Yinchuan 750021, China
2 Department of Civil Engineering, China University of Petroleum (East China), Qingdao 266580, China
3 Key Laboratory of Advanced Civil Engineering Materials of Ministry of Education, Tongji University, Shanghai 201804, China
4 School of Materials Science and Engineering, Tongji University, Shanghai 201804, China
* Correspondence: wangdzh@nxu.edu.cn (D.W.); yq.wei@tongji.edu.cn (Y.W.)

Abstract: Non-autoclaved aerated concrete (NAAC) is a two-phase material with a concrete matrix and air, exhibits good thermal insulation performance and shows good potential in the insulating construction industry. In this study, recycled concrete fine powder was used as an auxiliary cementing material, and the NAAC with different porosity and distribution was fabricated by the non-autoclaved method at different curing temperatures. The effect of porosity on the thermal conductivity and mechanical strength of NAAC is analyzed by experimental tests. A prediction method of thermal conductivity combining pore structure reconstruction and numerical simulation was proposed, which is established by two steps. Firstly, the pore size distributions of NAAC with different porosities were characterized by stereology image analyses. Secondly, the thermal conductivity prediction model based on the pore structure information was established by a COMSOL steady-state heat transfer module. The thermal conductivity results of COMSOL simulations were compared with the experiments and other theoretical models to verify the reliability of the model. The model was used to evaluate the effect of porosity, pore size distribution and the concrete matrix's thermal conductivity on the thermal conductivity of NAAC; these are hard to measure when only using laboratory experiments. The results show that with the increase in curing temperature, the porosity of NAAC increases, and the number and volume proportion of macropores increase. The numerical results suggest that the error between the COMSOL simulations and the experiments was less than 10% under different porosities, which is smaller than other models and has strong reliability. The prediction accuracy of this model increases with the increase in NAAC porosity. The steady thermal conductivity of NAAC is less sensitive to the distribution and dispersion of pore size in a given porosity. With the increase in porosity, the thermal conductivity of NAAC is linearly negatively correlated with that of the concrete matrix, and the correlation is close to 1.

Keywords: non-autoclaved aerated concrete; thermal conductivity; COMSOL simulation; pore size distribution; image-based analysis

Citation: Ma, X.; Li, H.; Wang, D.; Li, C.; Wei, Y. Simulation and Experimental Substantiation of the Thermal Properties of Non-Autoclaved Aerated Concrete with Recycled Concrete Powder. *Materials* **2022**, *15*, 8341. https://doi.org/10.3390/ma15238341

Academic Editors: Hrvoje Smoljanović, Ivan Balić and Nikolina Zivaljic

Received: 4 November 2022
Accepted: 21 November 2022
Published: 23 November 2022

1. Introduction

With the burgeoning call towards environmental-friendliness and sustainable development strategies, it is necessary to explore effective solutions for energy efficiency and carbon emission reduction in the construction. Heating and cooling costs of a building are the main parts of operating costs during the life cycle of the building and also one of the main sources of carbon emissions [1]. Hence, excellent insulation building materials are considered important in the building's design and construction.

Aerated concrete (AC) is a lightweight porous material made by introducing expansion agents into a slurry mixed with binders, supplementary cementitious material, fine aggregate, admixture and water [2–4]. Due to its small bulk density, good thermal insulation and fire resistance, AC attracted extensive interest in the field of energy-efficient and thermal insulation [5–7]. According to the production method, the AC can be divided into autoclaved aerated concrete (AAC) and non-autoclaved aerated concrete (NAAC). The production of NAAC does not require an environment of saturated steam under pressure, which is expensive and requires high energy consumption [8–10]. Simultaneously, a large amount of recycled material can be reused as raw materials for AC production, revealing the great potential in increasing the utilization rate of industrial waste [11–14]. As one of the widely used building materials in modern construction, the production of concrete has generated a large amount of waste concrete due to remodeling and demolition processes [15,16]. Waste concrete is reused in low-value operations or end up in landfills, which can cause a serious waste of resources. Furthermore, the current production of cement clinkers is accompanied by a large amount of energy consumption and carbon dioxide emissions. Many researchers proved that recycled concrete fines can partially replace traditional cementitious materials in concrete production [17–20]. Consequently, incorporating recycled concrete aggregates in NAAC production is considered to be a cost-effective means for solving the construction waste disposal challenges and achieving environmental sustainability.

Many researchers studied the characteristics of concrete produced with recycled concrete aggregates, including mechanical strength, crack resistance, porosity and water absorption [21,22]. Nevertheless, the research on the thermal conductivity of NAAC mixed with recycled concrete fine powder is very limited. In the past few decades, most studies have focused on examining the influence of the mix ratio on the thermal conductivity of porous concrete in order to obtain improved thermal insulation properties [23–27]. For example, Qu et al. [23] showed that the thermal conductivity of AAC is mainly influenced by its dry density. Narayanan [24] and Jerman et al. [25] reported that the moisture content is a key factor affecting the thermal conductivity of AC, and the higher value of moisture contents led to higher thermal conductivities. Jeong et al. [26] investigated the effect of water/cement ratios on the thermal conductivity of porous concrete with coal bottom ash. As a result of the research study, the authors found that thermal conductivities were significantly reduced by increases in the water/cement ratio.

The NAAC is a two-phase material with a concrete matrix and air, and one of its most important features is its internal porous structure. The macropores are formed from the chemical reaction between expansion agent and cementitious mixtures. The gas overflow after the addition of the expansion agent forms millions of uniformly sized and evenly distributed air voids in the concrete matrix, which are much bigger than the ordinary concrete [2,28,29]. In recent years, a large number of studies proved that the thermal insulating ability of porous concrete is significantly affected by its pore structure [30,31]. To intensively study the heat transfer performance of porous materials, many studies attempted to link thermal conductivity with its pore structure parameters [32,33]. The constitutive models of two-phase materials between the pore structure and thermal behaviors either have been established by many researchers in an analytical or empirical way to estimate the effective thermal conductivity [34–36]. For example, the parallel–series model and Campbell-Allen model are commonly used thermal conductivity models. Li et al. [35] established a predicting model for a simple two-phase inclusion-matrix system based on an effective medium and mean-field theories and focused on the effect of pore shape on thermal conductivity. Othuman et al. [36] proposed an analytical model of thermal conductivity with respect to foam concrete based on the assumed internal structure (porous) and constituents (cement, water and air). These models can quantitatively establish the relationship between porosity and thermal conductivity by using the thermal conductivity of two phases (concrete matrix and air) in porous concrete and the corresponding volume fraction. However, due to the fact that the microstructure of porous concrete is complex and the hypotheses of two-phase homogeneous material is oversimplified, the real properties of porous concrete are not been expressed exactly. Generally, all of the above models simply focused on changes in thermal conductivities caused by pore volume differences but neglected the effect of parameters such as pore size and distribution.

In fact, it is usually difficult to take the geometric details of the pores into consideration in experimental studies. Based on the reasonable calculation model and enough accurate data of the NAAC's internal features, numerical simulations can quantitatively establish the relationship between pore structure and thermal conductivity, and these simulations have become one of the important methods for the prediction of effective thermal conductivity. For example, Miled et al. [37] used five homogenization models to predict foam concrete's thermal conductivity, and the experimental data and FEM simulations are used to identify the best model. Qin et al. [38] used the steady-state method to numerically simulate the thermal conductivity of the foam glass. Ding et al. [39] used COMSOL simulations to study the relationship between thermal conductivity and the porosity of concrete materials. For most existing numerical models, the pore structure in NAAC is reproduced by random methods; thus, the influence of production process and material properties differences on the real pore structure of NAAC is usually ignored. However, this issue is of great importance for creating appropriate pore structure for NAAC to predict thermal conductivities; thus, the prediction results of the numerical model based on the random pore generation method usually lack stability. A large number of pores in NAAC are macro-pores, and common pore structure characterization methods such as scanning electron microscope (SEM), mercury injection method, and gas permeation method are usually difficult to apply. With the development of computer and vision technology, image-based analyses have become one of the most promising techniques for identifying the concrete's internal pore structure [40,41]. The digital image-processing techniques can use projected images or cross-sectional images to estimate gradation and analyze the data by using professional automatic image analyzers.

Herein, this study focuses on confirming the feasibility of accurately simulating the thermal conductivity of NAAC mixed with recycled concrete fine powders by COMSOL. Based on the real pore structure's parameters (porosity, pore size and distribution) obtained by the stereo-image method, the model is established to avoid the effects of changing production processes and material properties on the pore structure and prediction of thermal conductivity. Firstly, the NAAC was prepared with recycled concrete fine powder as the auxiliary cementing material, and NAAC with different porosities and pore size distributions was produced by changing curing temperatures. The effect of porosity on the thermal conductivity and mechanical strength of NAAC is analyzed by experimental tests. Furthermore, the pore size distribution of NAAC with different porosity were characterized by stereology-image analyses, and the thermal conductivity prediction model was established by COMSOL steady-state analyses. The thermal conductivity results of COMSOL simulations, experiments and other classical theoretical models are compared; it is proved that the finite element model based on real pore structures obtained by the image method has good accuracy. Finally, the effects of porosity, pore size distribution and concrete matrix thermal conductivities on thermal conductivity of NAAC were studied, and the results offer a relationship between the pore structure and thermal conductivity of NAAC. This work provides a possible method for predicting the thermal conductivity of NAAC by examining pore structure reconstruction and using numerical simulations.

2. Materials and Methods

2.1. Materials

The raw materials of NAAC used in this study mainly include P.O 42.5 cement, lime, gypsum, fly ash, aluminum powder-based chemical expansion agent, foam stabilizer and recycled concrete fines. The cement was obtained from Ningxia Saima Cement Co. Ltd. (Ningxia, China), and its basic properties are shown in Table 1.

Table 1. Basic properties of cement.

Cement Grade	Specific Surface Area (m^2/kg)	Setting Time (min)		Water Consumption of Normal Consistency (%)	LOI (%)
		Initial	Final		
P.O 42.5	350	182	234	27.2	4.7

The waste concrete was screened and crushed into concrete aggregates, and then the recycled aggregates were ground by a ball mill for 50 min to obtain recycled fines. The phase composition and microstructures of the recycled concrete powder were characterized by X-ray diffraction (XRD; D/max 2550VB3, Rigaku, Japan) and scanning electron microscopes (SEM; TM4000PlusII, Hitachi, Japan), as shown in Figure 1. It can be seen that the recycled concrete fines are irregular polygons with uneven size. Grade I fly ash was used as supplementary cementitious material, and the chemical components of cement, fly ash and recycled concrete fines used in this study are presented in Table 2. According to the chemical composition, CaO and SiO_2 are the main components of the powder material with a small number of impurities. The results show that the chemical composition of recycled concrete fines is similar to that of cement, and it can be used as a siliceous material to prepare NAAC.

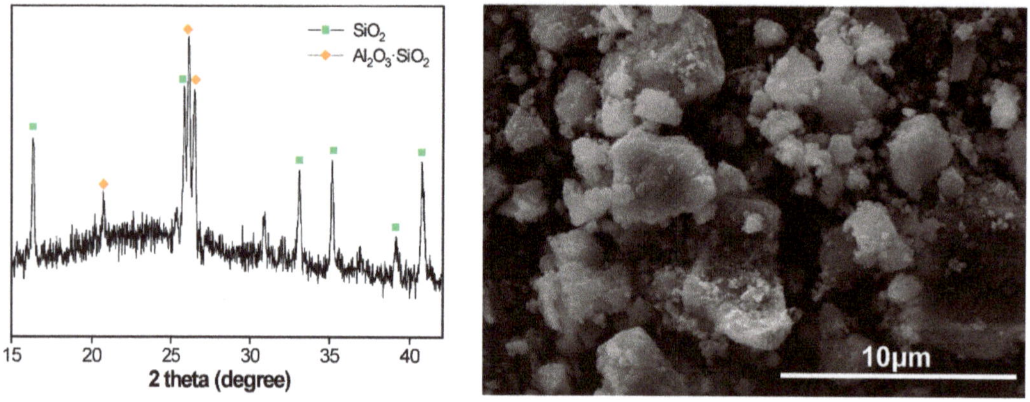

(a) (b)

Figure 1. (**a**) XRD pattern and (**b**) SEM image of recycled concrete fines.

Table 2. Chemical components of raw materials.

Constituents (wt.%)	CaO	SiO_2	Al_2O_3	Fe_2O_3	SO_3	Na_2O	LOI
Cement	43.5	14.1	5.19	2.17	4.91	21.3	8.83
Fly ash	5.7	51.5	24.8	7.0	1.5	1.7	7.8
Recycled concrete fines	34.1	23.9	3.86	2.32	1.32	27.7	6.8

The expansion agent was a GLS-65 aluminum powder-based expansion agent provided by Huaian Xinshengfa New Material Co. Ltd. (Huaian, China), and its technical index is shown in Table 3. Sodium stearate was used as a foam stabilizer, and its purity was of analytical grade. Lime and gypsum were provided by Ningxia Jinxin Environmental Technology Co. Ltd. (Ningxia, China), which complies with the Chinese standard of JC/T 621—2009.

Table 3. Characteristics of aluminum powder-based expansion agent.

Label	Type	Solid Content (%)	Active Aluminum in Solid (%)	Rate of Gas Generation (%)		
				4 min	16 min	30 min
GLS-65	Water-based	65	85	40–60	≥90	≥99

2.2. Specimen Preparation and Mix Design

The NAAC was prepared in the following sequence (as shown in Figure 2): Cement, lime, gypsum, fly ash and recycled concrete fines were added into the mixer and mixed at low speeds (48 ± 3 rpm) for 1 min. Then, 50% of the water required for mixing was introduced into the mixture of dry ingredients and mixed for 1.5–2 min to obtain a well-mixed slurry. After that, aluminum powder and sodium stearate were mixed in the remaining mixing water, and then the prepared aqueous solution was added into the resulting mixture slurry; mixing continued for 0 s. After resting for 1 min, the slurry was continually mixed at low speeds for 2 min until a homogeneous mixture without agglomerates was obtained.

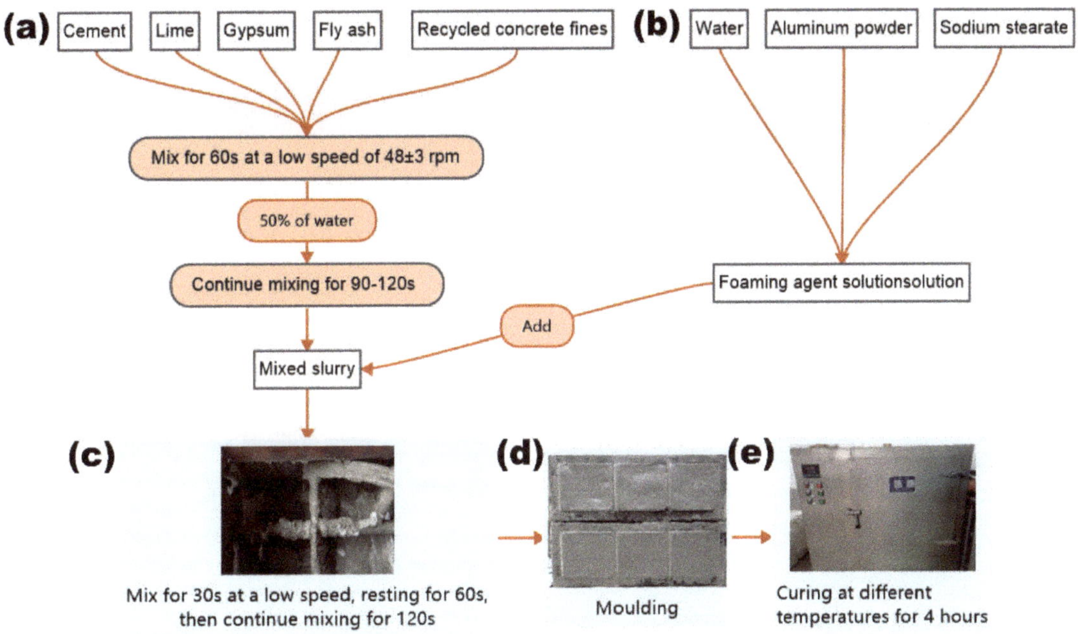

Figure 2. The preparation process of NAAC.

For the measurement of thermal conductivity, the slurry in a fresh state was poured into pre-prepared 100 mm × 100 mm × 100 mm molds brushed with lubricants and compacted with a metal rod. The molds with concrete were then placed into an oven with a relative humidity of 40 ± 5% for 4 h, and the curing temperatures were controlled at T1 (30 °C), T2 (35 °C), T3 (40 °C), T4 (45 °C) and T5 (50 °C), respectively. After the NAAC formed and expanded, the increased volume was removed by using scraper. Next, the specimens were covered with plastic sheeting and kept indoors at a temperature of 22 ± 2 °C for 24 h. Finally, the specimens were demolded and naturally cured to the specified age for performance testing. To increase the reliability of test results, three samples of each mixture were made.

The reference concrete was used for testing the thermal conductivity of the concrete matrix, and the mix proportions of the reference concrete and NAAC are shown in Table 4.

Table 4. Mix proportions of each sample.

Mix Types	W/C	Cement (%)	RCFP (%)	Gypsum (%)	FA (%)	Lime (%)	Expansion Agent (%)	Foam Stabilizer (%)
Reference Concrete	0.65	20	14	18	46	2	-	-
NAAC	0.65	20	14	18	46	2	0.06	0.12

Note: W/C represents the water-cement ratio; RCFP represents the recycled concrete fine powder; FA represents the fly ash.

2.3. Pore Structure Characterization

2.3.1. Pretreatment of Specimen

In this study, the porosity and pore size distribution of NAAC were characterized by the image-based analysis method. In order to improve the measurement accuracy, the image-based analysis of NAAC's pore structure requires preliminary processing, including contrast enhancements. First, the specimens that have been molded and fully cured were cut, and the exposed surfaces after cutting should be parallel and flat. Then, each surface of the test block was polished with 400-, 2000- and 5000-mesh sandpaper to make the exposed surfaces smooth and flat, and the pore structure was clearly visible. Finally, the cutting surfaces were evenly smeared with ink and dried naturally. The gypsum powder was used to fill the pores on the cutting surface, and the excess gypsum powder was removed with a scraper to obtain a high contrast specimen surface.

2.3.2. Stereology Principle

Stereology analysis is a mathematical method for studying the relationship between s-dimensional measurements obtained from the section of an organization and its n-dimensional (where $s < n$) parameters, which allows inferring the three-dimensional structural characteristics of NAAC by analyzing its two-dimensional structural features [42,43]. The fundamental viewpoint of stereology is that the structure of a material can be considered as an image consisting of points, lines, planes and bodies. There is a quantitative relationship between the statistics of a phase at point P, line length L, plane area A and volume V. The basic stereological Equation (1) applied here is as follows.

$$V_V = A_A = L_L = P_P, \tag{1}$$

The component area fraction A_A in a NAAC section is an unbiased estimate of the component volume fraction V_V in a three-dimensional structure. Consequently, the statistical results of the two-dimensional distribution can be used to characterize the three-dimensional pore distribution.

2.3.3. Image Processing

The image-processing software and pore structure characterization software used in this study was Image-Pro Plus [44]. First, the cross sections of NAAC specimens after pretreatment were photographed and cropped using a digital microscope. The maximum resolution of the obtained RGB images was 640 × 480 using 50× magnification. Then, the Image-Pro Plus software was used for image binarization and pore segmentation to obtain the image shown in Figure 3. The red part represents pore spaces, and the black part represents pore walls.

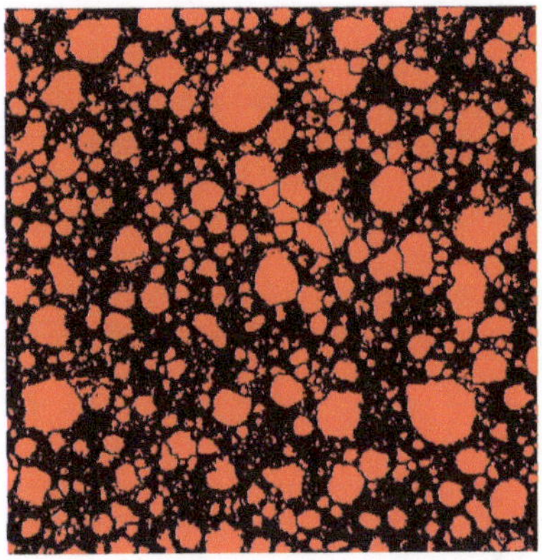

Figure 3. Cross section of NAAC after Image-Pro Plus software processing.

2.4. Thermal Conductivity Measurement

The thermal conductivity was measured by the transient hot-wire method, and the instrument used was a TC3000E heat conduction coefficient measurement device (measuring range is 0.001–50.0 W/mK, measuring accuracy is 0.0005 W/mK), as shown in Figure 4. Before formal measurements, the standard sample was used to calibrate the instrument. The specimens that have reached the specified curing age were cut into two equal volumes with flat surfaces, and the residual powder was removed. Then, the sensor was placed in the middle of the cut sample, and a 500 g weight was placed above the sample to produce the two parts after cutting them closely together so as to reduce test errors. The test voltage was 1.5 v, and each specimen was tested five times with an interval of 10 s.

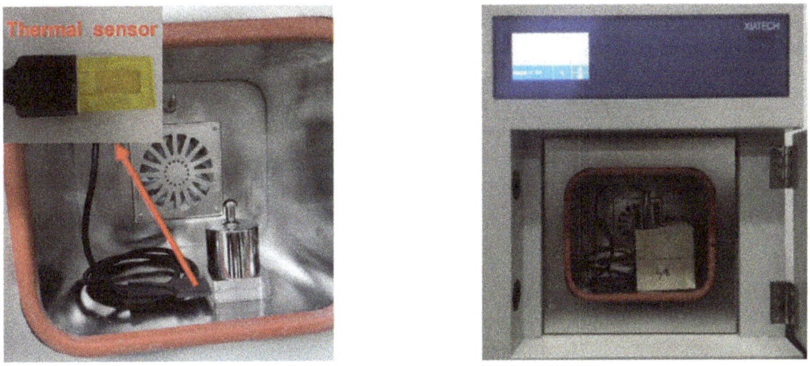

(**a**) (**b**)

Figure 4. (**a**) Automatic test instrument for measuring thermal conductivity with hot wire method; (**b**) Thermal conductivity test of NAAC.

2.5. Numerical Modeling

The COMSOL finite element software can effectively simulate the pore structure and predict the thermal conductivity of composite materials. A steady-state heat conduction-based numerical model was developed by using COMSOL to simulate the thermal conductivity of NAAC with different porosity and pore size distribution.

2.5.1. Principles of Heat Conduction and Material Parameters

Heat conduction, convection and radiation are the basic methods for heat transfers. The heat transfer process relative to porous materials from the high temperature section to the low temperature section is a result of a contribution of these three different mechanisms. The NAAC is a two-phase material with a concrete matrix and air; the internal heat transfer process comprises heat conduction in a cementitious matrix, heat conduction, heat convection and heat radiation in pores. Due to the small pore size inside the NAAC, it is difficult for the air to circulate; thus, the heat transfer due to convection can be ignored. Furthermore, radiation usually accounts for a small proportion in the heat transfer of porous concrete and can be negligible. Consequently, the heat transfer simulation of NAAC in this study only considered the heat conduction. The thermal conductivity of the concrete matrix was measured by experiments, and the parameters of the concrete matrix and air are shown in Table 5.

Table 5. Phase parameters of NAAC.

Phase	Thermal Conductivity (W/mK)	Specific Heat Capacity (J/(kg·K))	Density (kg/m^3)
Concrete matrix	0.3414	1050	2470
Air	0.023	1000	1.29

2.5.2. Calculating Model of Thermal Conductivity

As mentioned in the introduction, the existing analysis models are usually based on the random generation method for reproducing the microstructure and the pores forming randomly based on some main structural parameters. In such models, the realistic influences of production processes and material properties on pore size and distribution are usually neglected. In this paper, the three-dimensional model of the NAAC was established by COMSOL based on the porosity and pore size distribution obtained from the pore structure's characterization. Considering the computational volume and time, the specimen size used for simulation was 10 mm × 10 mm × 10 mm, and the pores are described as spheres in the solid matrix [45]. The conditions for the formation of pore structure were that the pores were within the range of the concrete specimen, and the pores did not overlap or stick together. Due to the fineness requirement of meshing, the lower limit of the pore size was set as 0.1 mm. The three-dimensional models of the reference concrete and NAAC are shown in Figure 5.

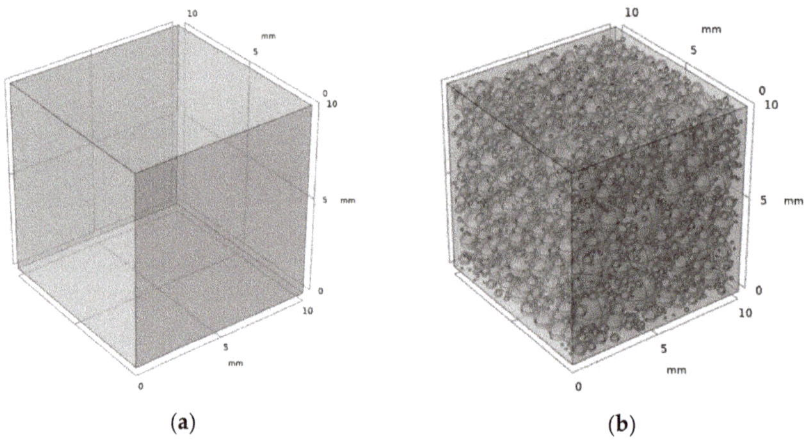

Figure 5. The three-dimensional models of the (**a**) reference concrete and (**b**) NAAC (curing temperature of 35 °C as an example).

The model was established by mimicking the actual thermal conductivity test environment, and the heat flux was set to be transferred from the upper side to the lower side of the sample without internal heat generation. The parameter settings are as follows: the room ambient temperature was set at 20 °C. The upper boundary perpendicular to the z-axis was set at 35 °C (hot boundary), and the lower boundary was set at 15 °C (cold boundary). Additionally, the other four boundaries parallel to the z-axis were set as open boundaries.

The meshing method with physical field controls in COMSOL was used, and the grid's independence test was carried out. It was determined that the simulation results of the thermal conductivity of NAAC are less sensitive to grid fineness. Based on the consideration of the simulation accuracy of small size pores and calculation time, the meshing parameters of reference concrete were set as follows: the maximum unit size of reference concrete was 1 mm, the minimum unit size was 0.18 mm, the maximum unit growth rate was 1.5, the curvature factor was 0.6, and the narrow area resolution was 0.5. The meshing parameters of NAAC were set as follows: The maximum unit size of reference concrete was 0.8 mm, the minimum unit size was 0.1 mm, the maximum unit growth rate was 1.45, the curvature factor was 0.5, and the narrow area resolution was 0.6. The meshing results of the reference concrete and NAAC are shown in Figure 6.

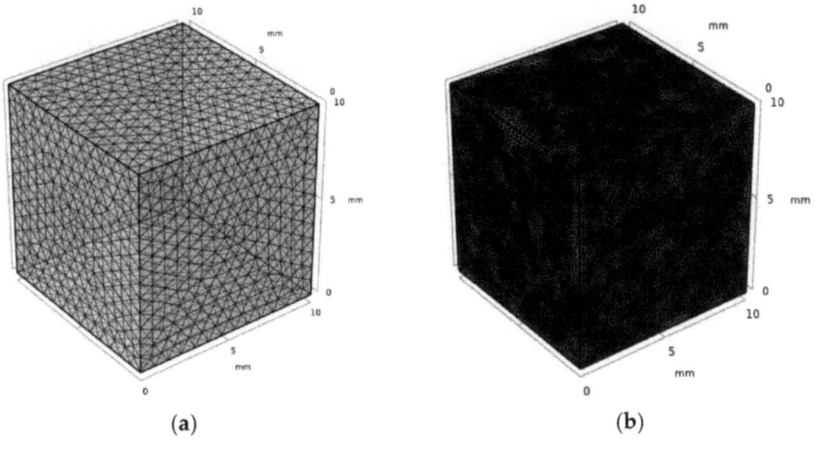

Figure 6. The meshing results of (**a**) reference concrete and (**b**) NAAC (curing temperature of 35 °C as an example).

3. Results and Discussion

3.1. Effect of Porosity on Thermal Conductivity and Compressive Strength

The porosity, compressive strength and thermal conductivity of the NAAC at different curing temperatures were tested, and the results are shown in Figure 7. The results show that the curing temperature of NAAC has a significant effect on its porosity. With the increase in curing temperature, the porosity of NAAC increases. As observed from Figure 7a, the compressive strength of NAAC firstly increases and then decreases with the increase in porosity, showing a downward trend as a whole. With the increase in the volume of pores, the compressive strength of NAAC decreases, which is in line with the basic law of porous materials. The compressive strength increases as the porosity increases from 47.5% (35 °C) to 48.8% (40 °C). This is mainly because the compressive strength of NAAC is not only affected by the porosity but also the pore size and pore distribution. The NAAC with a uniform distribution of pore structure, concentrated pore size distribution and a smaller number of connected pores and large pores show better compressive properties. Furthermore, the experimental thermal conductivity of the reference concrete (with approximately 0 porosity) is 0.3414 W/mK. It can be seen from Figure 7b that the thermal conductivity of NAAC is much lower than that of the reference concrete, and it decreases with the increase in porosity. This is mainly because the NAAC is a two-phase composite material of air and concrete. The thermal conductivity of air under standard conditions is only 0.023 W/mK, which is much lower than that of the concrete matrix (an order of magnitude difference). Therefore, the increase in the proportion of air in the unit volume of NAAC can reduce the overall thermal conductivity.

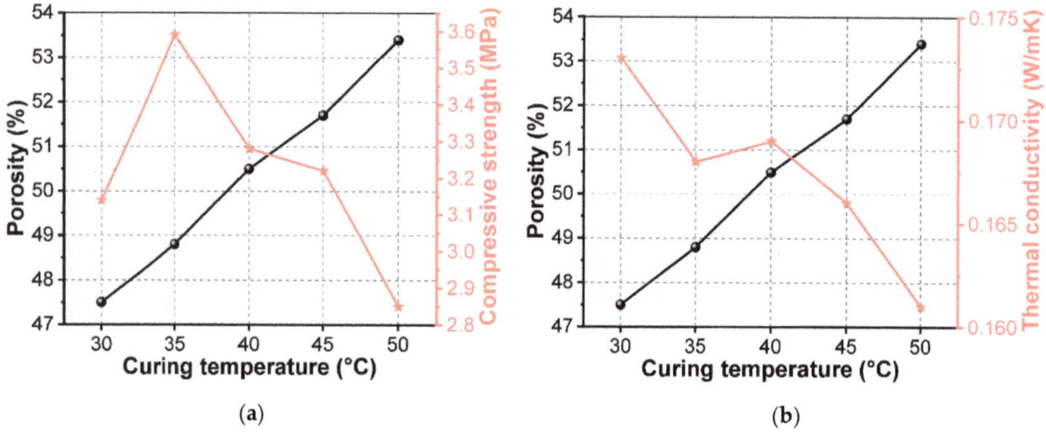

Figure 7. (**a**) Porosity and thermal conductivity of NAAC at different curing temperatures; (**b**) porosity and compressive strength of NAAC at different curing temperatures.

The thermal conductivity of NAAC are mainly attributed to the porosity and pore size distribution. The pore size distribution of NAAC mixed with recycled fines at five different curing temperatures was statistically analyzed by Image Pro Plus software, and the cumulative frequency distribution of the number of pores at each level is shown in Figure 8. It can be seen from Figure 8a–e that the number of small pores less than 0.2 mm is significantly higher than that of other pore sizes, reaching 85.19%, 78.59%, 69.88%, 76.53% and 71.93%, respectively. From Figure 8f, it can be obtained that the D90 of NAAC at five different temperatures is 0.4 mm (92.7%), 0.6 mm (92.9%), 0.6 mm (91.0%), 0.6 mm (91.3%) and 0.8 mm (91.1%) (D90 represents the pore diameter when the cumulative pore distribution is 90%), which proves that the diameter of most internal pores is less than 0.8 mm. A distinct phenomenon worth noticing is that the pore size of NAAC tends to increase significantly with the increase in curing temperatures. Furthermore, the D100 of NAAC at five temperatures is 2.2 mm, 2.6 mm, 2.6 mm, 2.6 mm and 3.0 mm, respectively. It can be seen that with the increase in curing temperatures, the maximum pore diameter of NAAC gradually increases. The reason behind this phenomenon is that with the porosity of NAAC increases with the increase in curing temperature. As a result, the probability of contact and fusion between pores increases, and there is a higher chance to generate large pore sizes. Further statistical analyses of the cumulative frequency distribution of the pore volumes at all levels were performed, and the results are shown in Figure 9. It can be seen that the D10 of NAAC at five temperatures is 0.6 mm (9.2%), 6 mm (7.2%), 0.6 mm (12.3%), 0.6 mm (11.2%) and 0.8 mm (8.2%). Moreover, the D90 is 1.6 mm (90.8%), 2.4 mm (89.2%), 2.4 mm (85.6%), 2.0 mm (90.4%) and 2.8 mm (93.1%), respectively. This comparison indicates that with the increase in pouring temperatures, the volume proportion of large pore sizes also increases. Based on the above analysis, with the increase in curing temperatures, the porosity increases and the pore size tends to be larger. Therefore, the thermal conductivity decreases, which is consistent with the above thermal conductivity test results.

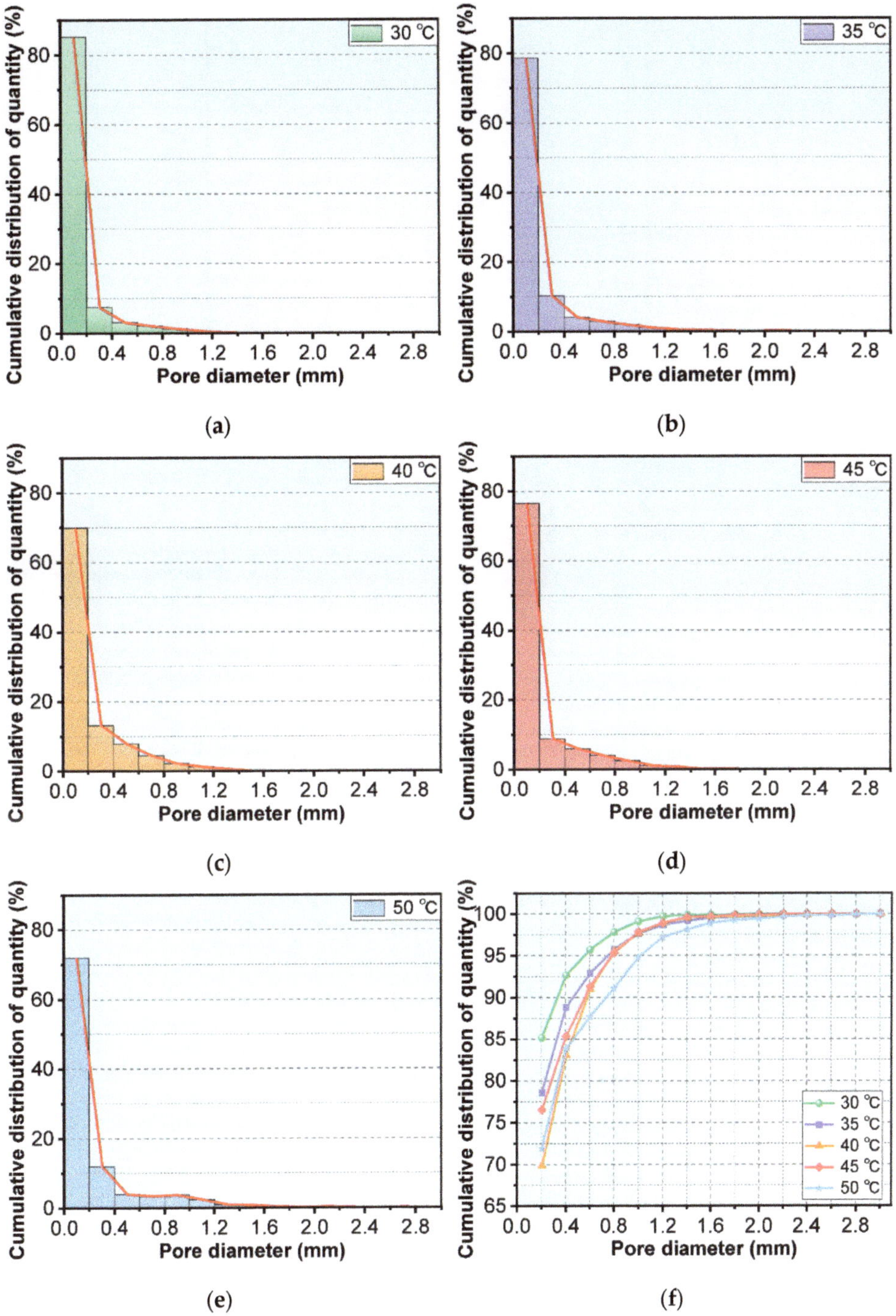

Figure 8. Pore number distribution at (**a**) 30 °C, (**b**) 35°C, (**c**) 40 °C, (**d**) 45 °C and (**e**) 50 °C; (**f**) cumulative frequency distribution of the pore number at different curing temperatures.

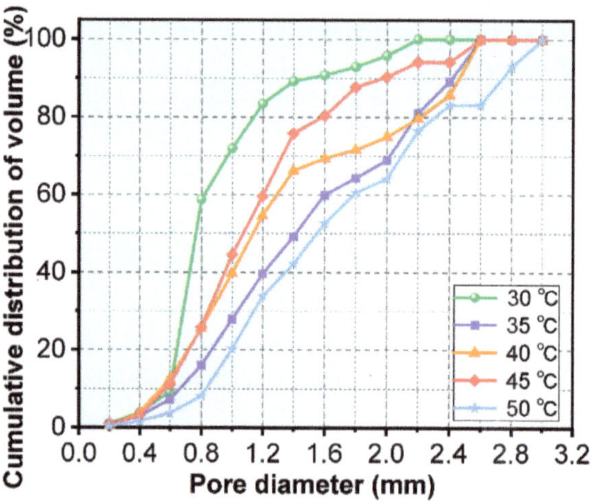

Figure 9. Cumulative frequency distribution of the pore volume at different curing temperatures.

3.2. *Simulation of Thermal Conductivity*

3.2.1. Steady-State Thermal Analyses

Based on the porosity and pore size distribution obtained from the pore structure's analysis, the model of the NAAC's pore structure was reconstructed by COMSOL. Moreover, the three-dimensional temperature distribution of NAAC specimens at different curing temperatures was obtained using COMSOL steady-state simulations, as shown in Figure 10. It can be seen that the temperature distribution of NAAC specimens obtained under different curing temperatures is similar to that of the reference concrete, and the NAAC with different porosities all reach the temperature equilibrium in the middle of the test blocks. Figure 11 shows the three-dimensional isotherm of the NAAC test blocks. It can be seen from Figure 11a that the isotherm of the reference concrete specimen (porosity is approximately 0) is straight and uniformly distributed, which indicates that heat is evenly diffused in the concrete test block. It is worth noting that with the increase in curing temperatures (porosity increases), the isotherms between the continuous medium and the dispersed medium are significantly distorted (as shown in Figure 11b–f). This is mainly due to the increase in porosity and the appearance of large pore sizes as the curing temperature increases. The change in the pore's structure will lead to an inhomogeneous interface and changes in the heat transfer path. Additionally, the thermal conductivity of air is much lower than that of concrete, and there will be a thermal bridge effect at the edges of the pore's structure.

3.2.2. Calculation of Thermal Conductivity

According to the results of the steady-state thermal analysis, the temperature gradient and heat flux of NAAC with different porosities can be obtained by the steady-state heat conduction equation, which is expressed as Equation (2). According to Fourier's law, the thermal conductivity of NAAC can be further calculated, and the calculation method is shown in Equation (3):

$$\rho C_p u \cdot \nabla T + \nabla(-k\nabla T) = Q, \quad (2)$$

$$q = -k\nabla T, \quad (3)$$

where u is the velocity field (m/s), ρ is the material density (kg/m^3), C_p is the specific heat capacity of material (Jkg^{-1}K^{-1}), ∇ is the gradient operator, T is the temperature (K), Q is the quantity of heat (J), k is the thermal conductivity (W/mK), and q is the heat flux (W/m^2).

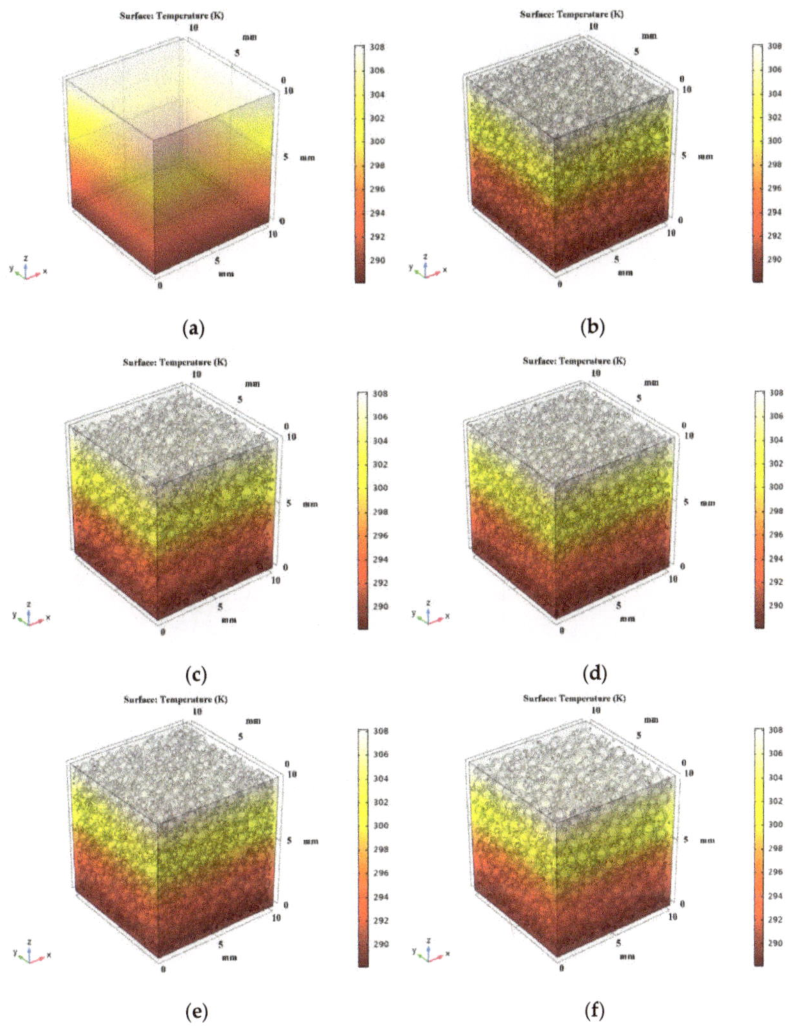

Figure 10. (**a**) Temperature distribution of reference concrete; temperature distribution of NAAC specimens at (**b**) 30 °C, (**c**) 35°C, (**d**) 40 °C, (**e**) 45 °C and (**f**) 50 °C.

The series model, the parallel model and the Campbell-Allen model derived from Ohm's law based on the series and parallel model are the classic models for thermal conductivity calculations. The thermal conductivity of NAAC with different porosities calculated by COMSOL was compared with the predicted results of the classical models, and the results are shown in Figure 12. Table 6 further shows the relative errors between the predicted thermal conductivity of the four models and the experimental test results. It can be seen that the thermal conductivity predicted by COMSOL is in the best agreement with the experimental test results, and the errors under the five types of porosity are 9.09%, 8.18%, 5.17%, 4.53% and 2.84%. A distinct phenomenon worth noticing is that the prediction accuracy of COMSOL increases with the increase in NAAC's porosity. This result indicates that the COMSOL finite element software has a better adaptability to the dense and complex pore structure of NAAC. The high accuracy of COMSOL in predicting the thermal conductivity of NAAC can be attributed to the fact that the model was established based on the information on porosity and pore size distribution obtained by the image method, and the basic parameters such as material density and specific heat capacity were fully considered [46,47]. Simultaneously, the COMSOL finite element software can well describe the distribution of the internal pore structure of NAAC and effectively mesh so as to obtain the quantitative results of the thermal conductivity. In addition to the COMSOL simulation, the parallel model also shows good predictions. The reason is that the porosity of NAAC is large and the pore size distribution is relatively concentrated, so its internal pore structure is closer to the geometric model of the parallel model than that of the series model and the Campbell-Allen model. The above results show that COMSOL simulations can be used as effective methods for predicting the thermal conductivity of NAAC, and its calculation accuracy is higher than that of simple two-phase models, such as parallel models, series models and the Campbell-Allen model. Moreover, the COMSOL simulation shows better prediction results on the thermal conductivity of NAAC with high porosity.

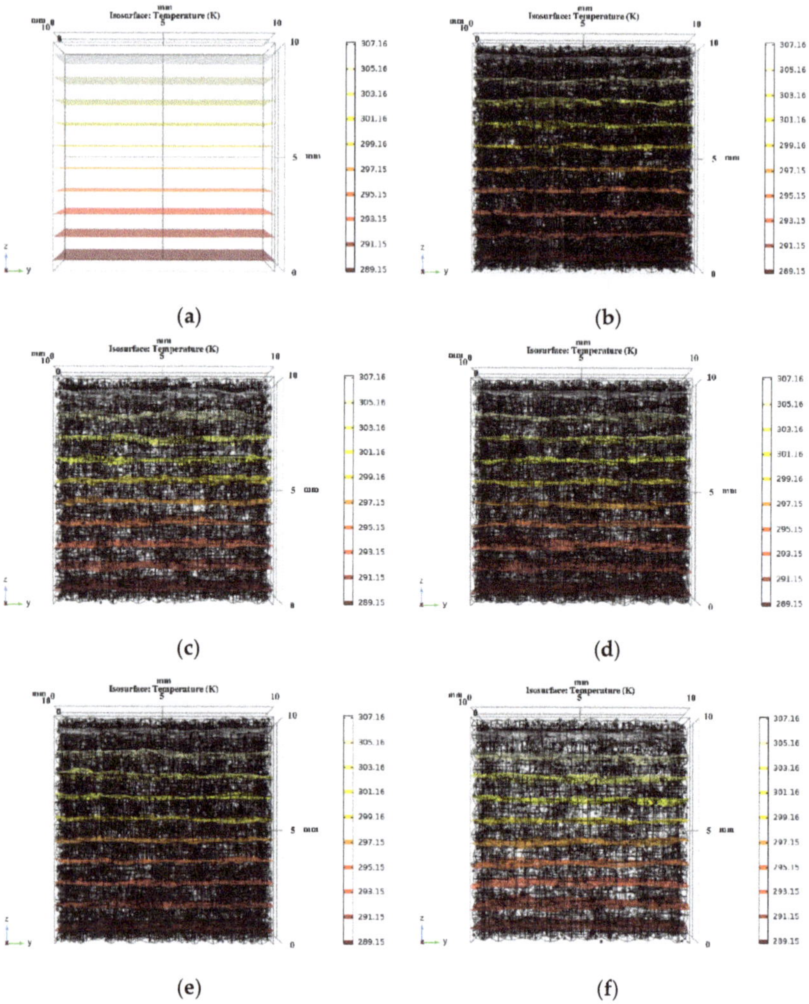

Figure 11. (**a**) Isotherms of reference concrete; isotherms of NAAC specimens at (**b**) 30 °C, (**c**) 35 °C, (**d**) 40 °C, (**e**) 45 °C and (**f**) 50 °C.

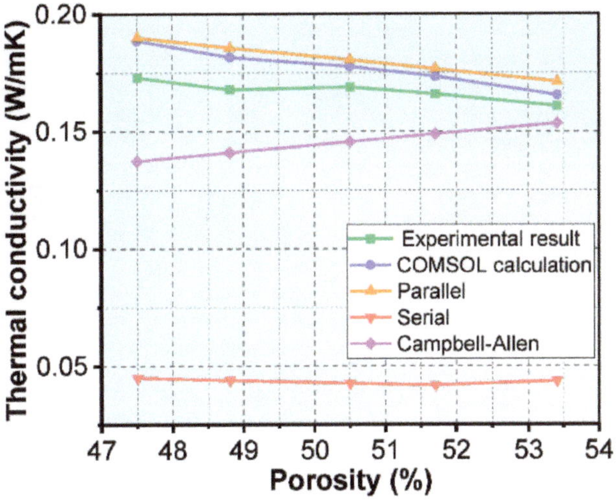

Figure 12. Comparison between COMSOL analysis, classical thermal conductivity models and test results.

Table 6. Relative errors between the thermal conductivity of the COMSOL analysis, classical thermal conductivity models and test results.

Curing Temperature (°C)	Porosity (%)	Experimental (W/mK)	COMSOL (W/mK)	Parallel (W/mK)	Serial (W/mK)	Campbell-Allen (W/mK)
30 °C	47.5	0.173	0.18873 (9.09%)	0.1901 (9.88%)	0.0451 (−73.93%)	0.1374 (−20.58%)
35 °C	48.8	0.168	0.18175 (8.18%)	0.1857 (10.54%)	0.0440 (−73.81%)	0.1410 (−16.07%)
40 °C	50.5	0.169	0.17773 (5.17%)	0.1806 (6.86%)	0.0427 (−74.73%)	0.1456 (−13.85%)
45 °C	51.7	0.166	0.17352 (4.53%)	0.1767 (6.45%)	0.0419 (−74.76%)	0.1489 (−10.30%)
50 °C	53.4	0.161	0.16558 (2.84%)	0.1713 (6.40%)	0.0437 (−72.86%)	0.1536 (−4.60%)

3.3. Study on the Factors Affecting the Thermal Conductivity

3.3.1. Effect of Pore Size Distribution on Thermal Conductivity

The pore size distribution is also an important parameter that affects the internal pore structure of NAAC. Thus, the influence of pore size distribution on the thermal conductivity of NAAC was investigated by controlling the change in pore size in an interval. First, the pore size was controlled in a small interval of 0.2 mm, and the effect of the pore size distribution interval on the thermal conductivity was investigated by changing the pore size distribution's center. The results of the COMSOL simulation are shown in Table 7, and it can be seen that the thermal conductivity of NAAC varies very slightly with the change in distribution center. Furthermore, the 1.5 mm was used as the center of the pore size distribution interval, and the size of the interval gradually changed to control the dispersion coefficient of the pore size. The prediction results of the thermal conductivity are shown in Table 8, which proves that the dispersion degree of the pore size in the pore size's distribution area centered at 1.5 mm has little effects on the thermal conductivity of NAAC. According to the above analysis, when the porosity of NAAC is determined, the thermal conductivity is less sensitive to the pore size's distribution. This conclusion is consistent with the results of relevant studies; that is, the influence of pore diameter on thermal conductivity can be ignored.

3.3.2. Effect of Concrete Matrix Thermal Conductivity on NAAC Thermal Conductivity

The thermal conductivity of the concrete matrix is usually measured by laboratory tests, and the results are usually subject to large errors. Therefore, it is necessary to consider the influence of the concrete matrix's thermal conductivity on the prediction accuracy of NAAC's thermal conductivity. The thermal conductivity of the concrete matrix was changed so that it can deviate from the experimental measurement results (with an error range from −15 to 15%), and the sensitivity of the thermal conductivity of NAAC to its variation was explored. Simulations were performed with parameters obtained for NAAC at T2 temperatures, and the simulation results are shown in Table 9.

Table 7. Effect of pore diameter distribution on thermal conductivity.

Distribution interval (mm)	0.4–0.6	0.6–0.8	0.8–1.0	1.0–1.2	1.2–1.4	1.4–1.6
Interval center (mm)	0.5	0.7	0.9	1.1	1.3	1.5
Thermal conductivity (W/mK)	0.26241	0.26281	0.26277	0.26263	0.26252	0.26270
Distribution interval (mm)	1.6–1.8	1.8–2.0	2.0–2.2	2.2–2.4	2.4–2.6	2.6–2.8
Interval center (mm)	1.7	1.9	2.1	2.3	2.5	2.7
Thermal conductivity (W/mK)	0.26257	0.26253	0.26204	0.26221	0.26236	0.2624

Table 8. Effect of pore diameter dispersion coefficient on thermal conductivity.

Distribution interval (mm)	0.2–2.8	0.4–2.6	0.6–2.4	0.8–2.2	1.0–2.0	1.2–1.8	1.4–1.6
Coefficient of dispersion	0.61710	0.45339	0.33806	0.27670	0.20859	0.13818	0.09101
Thermal conductivity (W/mK)	0.26283	0.26281	0.26277	0.26273	0.26278	0.26256	0.26266

Table 9. Effect of concrete matrix thermal conductivity on the thermal conductivity of NAAC.

Error	−15%	−10%	−5%	0%	5%	10%	15%
Thermal conductivity of concrete matrix (k_1)	0.29011	0.30717	0.32424	0.3413	0.35837	0.37543	0.39250
Thermal conductivity of NAAC (k_2)	0.15622	0.16473	0.17325	0.18175	0.19027	0.19878	0.20729

Figure 13a shows the influence of the concrete matrix's thermal conductivity on the thermal conductivity of NAAC, and it can be seen that the two are basically linear. The fitting result is $k_2 = 0.4988k_1 + 0.01152$, and it has an error coefficient, SSE, of 5.718×10^{-11} and R^2 of 1, which indicates that there is a strong linear correlation between the two. The slope of k_1 to k_2 was used as the response coefficient of the NAAC thermal conductivity to the concrete matrix's thermal conductivity, as shown in Figure 13b. It can be seen that the porosity of NAAC is linearly and negatively correlated with the degree of responses, and the rate of change is close to 1. Consequently, with the increase in porosity, the sensitivity of NAAC thermal conductivity to the concrete matrix's thermal conductivity decreases.

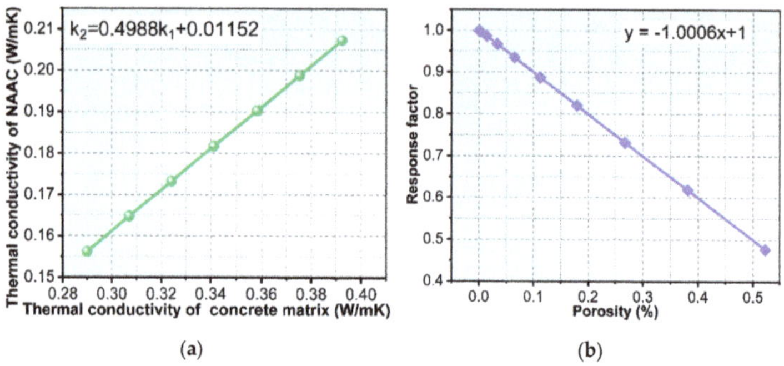

Figure 13. (**a**) Effect of concrete matrix thermal conductivity on the thermal conductivity of NAAC; (**b**) response coefficient of the NAAC thermal conductivity with different porosities to the concrete matrix's thermal conductivity.

4. Conclusions

In this study, recycled concrete fine powder was used to prepare NAAC. The NAAC with different porosities and distributions was fabricated by the non-autoclaved method at different curing temperatures, and its pore size distribution was reproduced by stereology-image analysis. Based on these pore structural parameters, the thermal conductivity prediction model was established by the COMSOL steady-state heat transfer module. The established model was used to study the effects of porosity pore size distributions and concrete matrix thermal conductivity on the thermal conductivity of NAAC. The following conclusions can be drawn:

(1) Recycled concrete fine powder can be used as a good auxiliary cementitious material to produce NAAC, and the curing temperature of the non-autoclaved method has a great effect on the pore structure of NAAC. With the increase in curing temperature, the porosity of NAAC increases and the number and volume proportion of macropores increase. The increase in porosity will lead to a decrease in thermal conductivity and the compressive strength of NAAC.

(2) The thermal conductivity results of COMSOL simulations, experiments and other classical theoretical models were compared, and it was found that the error between the COMSOL simulations and the experiments was less than 10% under different porosities, which is smaller than other models and has strong reliability. The prediction accuracy of COMSOL increases with the increase in NAAC porosity. This shows that COMSOL simulations can provide a fine description of the distribution of the complex pore structure of NAAC. Compared with the model that only considers porosity and adopting random pore distributions, the COMSOL model based on the real porosity and pore size distribution obtained by the image method can fully consider the impact of production processes and material properties on the pore structure of NAAC, and the accuracy of thermal conductivity estimates is improved.

(3) The steady thermal conductivity of NAAC is less sensitive to the distribution and dispersion of pore size in a given porosity. With the increase in porosity, the thermal conductivity of NAAC is linearly and negatively correlated with that of the concrete matrix, and the correlation is close to 1.

(4) This study provides a possible method for predicting the thermal conductivity of NAAC from the point of view of pore structure reconstruction and numerical simulations, and the established model has good accuracy. Further research can explore the use of machine learning and 3D reconstruction technology to reconstruct the three-dimensional model of a pore structure so as to establish a faster and more accurate calculation method for thermal conductivities.

Author Contributions: Conceptualization, X.M., H.L., D.W. and Y.W.; methodology, X.M., H.L. and D.W.; software, H.L.; investigation, D.W., C.L. and Y.W.; resources, D.W., C.L. and Y.W.; data curation, X.M.; writing—original draft preparation, X.M.; writing—review and editing, X.M. and D.W. and Y.W.; funding acquisition, D.W. All authors have read and agreed to the published version of the manuscript.

Funding: This research was funded by the Key Research and Development Program of Ningxia (2022BFE), the National Natural Science Foundation of China (51968060), the Ningxia Youth Talent Support Program and First Class Discipline Project of Ningxia Higher Education Institution (NXYLXK2021A03).

Institutional Review Board Statement: Not applicable.

Informed Consent Statement: Not applicable.

Data Availability Statement: The data used to support the findings of this study are available from the corresponding author upon request.

Conflicts of Interest: The authors declare that they have no known competing financial interests or personal relationships that could have appeared to influence the work reported in this paper.

References

1. Tasdemir, C.; Sengul, O.; Tasdemir, M.A. A comparative study on the thermal conductivities and mechanical properties of lightweight concretes. *Energy Build.* **2017**, *151*, 469–475. [CrossRef]
2. Ulykbanov, A.; Sharafutdinov, E.; Chung, C.; Zhang, D.; Shon, C. Performance-based model to predict thermal conductivity of non-autoclaved aerated concrete through linearization approach. *Constr. Build. Mater.* **2019**, *196*, 555–563. [CrossRef]
3. Jiang, J.; Lu, X.; Niu, T.; Hu, Y.; Wu, J.; Cui, W.; Zhao, D.; Ye, Z. Performance optimization and hydration characteristics of BOF slag-based autoclaved aerated concrete (AAC). *Cem. Concr. Compos.* **2022**, *134*, 104734. [CrossRef]
4. Lei, M.; Deng, S.; Huang, K.; Liu, Z.; Wang, F.; Hu, S. Preparation and characterization of a CO_2 activated aerated concrete with magnesium slag as carbonatable binder. *Constr. Build. Mater.* **2022**, *353*, 129112. [CrossRef]
5. Jiang, J.; Cai, Q.; Ma, B.; Hu, Y.; Qian, B.; Ma, F.; Shao, Z.; Xu, Z.; Wang, L. Effect of ZSM-5 waste dosage on the properties of autoclaved aerated concrete. *Constr. Build. Mater.* **2021**, *278*, 122114. [CrossRef]
6. Ghazi Wakili, K.; Hugi, E.; Karvonen, L.; Schnewlin, P.; Winnefeld, F. Thermal behaviour of autoclaved aerated concrete exposed to fire. *Cem. Concr. Compos.* **2015**, *62*, 52–58. [CrossRef]
7. Kalpana, M.; Mohith, S. Study on autoclaved aerated concrete: Review. *Mater. Today Proc.* **2020**, *22*, 894–896. [CrossRef]
8. Stel Makh, S.A.; Shcherban, E.M.; Beskopylny, A.N.; Mailyan, L.R.; Meskhi, B.; Beskopylny, N.; Dotsenko, N.Y.; Kotenko, M. Influence of Recipe Factors on the Structure and Properties of Non-Autoclaved Aerated Concrete of Increased Strength. *Appl. Sci.* **2022**, *12*, 6984. [CrossRef]
9. Sharafutdinov, E.; Shon, C.; Zhang, D.; Chung, C.; Kim, J.; Bagitova, S. Frost Resistance Number to Assess Freeze and Thaw Resistance of Non-Autoclaved Aerated Concretes Containing Ground Granulated Blast-Furnace Slag and Micro-Silica. *Materials* **2019**, *12*, 4151. [CrossRef]
10. Shon, C.; Mukangali, I.; Zhang, D.; Ulykbanov, A.; Kim, J. Evaluation of Non-Autoclaved Aerated Concrete for Energy Behaviors of a Residential House in Nur-Sultan, Kazakhstan. *Buildings* **2021**, *11*, 610. [CrossRef]
11. Wei, S.; Yiqiang, C.; Yunsheng, Z.; Jones, M.R. Characterization and simulation of microstructure and thermal properties of foamed concrete. *Constr. Build. Mater.* **2013**, *47*, 1278–1291. [CrossRef]
12. Shams, T.; Schober, G.; Heinz, D.; Seifert, S. Rice husk ash as a silica source for the production of autoclaved aerated concrete—A chance to save energy and primary resources. *J. Build. Eng.* **2022**, *57*, 104810. [CrossRef]
13. Shan, C.; Yang, Z.; Su, Z.; Rajan, R.; Zhou, X.; Wang, L. Preparation and characterization of waterproof autoclaved aerated concrete using molybdenum tailings as the raw materials. *J. Build. Eng.* **2022**, *49*, 104036. [CrossRef]
14. Xu, R.; He, T.; Da, Y.; Liu, Y.; Li, J.; Chen, C. Utilizing wood fiber produced with wood waste to reinforce autoclaved aerated concrete. *Constr. Build. Mater.* **2019**, *208*, 242–249. [CrossRef]
15. Akeed, M.H.; Qaidi, S.; Ahmed, H.U.; Faraj, R.H.; Mohammed, A.S.; Emad, W.; Tayeh, B.A.; Azevedo, A.R.G. Ultra-high-performance fiber-reinforced concrete. Part I: Developments, principles, raw materials. *Case Stud. Constr. Mater.* **2022**, *17*, e1290. [CrossRef]
16. Amin, M.; Tayeh, B.A.; Agwa, I.S. Effect of using mineral admixtures and ceramic wastes as coarse aggregates on properties of ultrahigh-performance concrete. *J. Clean. Prod.* **2020**, *273*, 123073. [CrossRef]
17. Song, S.; Li, X.; Wang, Z.; Wang, W. Orthogonal experimental and theoretical study on mechanical properties of fiber-reinforced recycled powder concrete. *Case Stud. Constr. Mater.* **2022**, *17*, e1546. [CrossRef]
18. Cantero, B.; Bravo, M.; de Brito, J.; Sáez Del Bosque, I.F.; Medina, C. Thermal Performance of Concrete with Recycled Concrete Powder as Partial Cement Replacement and Recycled CDW Aggregate. *Appl. Sci.* **2020**, *10*, 4540. [CrossRef]
19. Horsakulthai, V. Effect of recycled concrete powder on strength, electrical resistivity, and water absorption of self-compacting mortars. *Case Stud. Constr. Mater.* **2021**, *15*, e725. [CrossRef]
20. Ying, J.; Zhou, B.; Xiao, J. Pore structure and chloride diffusivity of recycled aggregate concrete with nano-SiO_2 and nano-TiO_2. *Constr. Build. Mater.* **2017**, *150*, 49–55. [CrossRef]
21. Sasanipour, H.; Aslani, F. Durability properties evaluation of self-compacting concrete prepared with waste fine and coarse recycled concrete aggregates. *Constr. Build. Mater.* **2020**, *236*, 117540. [CrossRef]
22. Aslani, F.; Ma, G.; Yim Wan, D.L.; Muselin, G. Development of high-performance self-compacting concrete using waste recycled concrete aggregates and rubber granules. *J. Clean. Prod.* **2018**, *182*, 553–566. [CrossRef]
23. Qu, X.; Zhao, X. Previous and present investigations on the components, microstructure and main properties of autoclaved aerated concrete—A review. *Constr. Build. Mater.* **2017**, *135*, 505–516. [CrossRef]
24. Narayanan, N.; Ramamurthy, K. Structure and properties of aerated concrete: A review. *Cem. Concr. Compos.* **2000**, *22*, 321–329. [CrossRef]
25. Jerman, M.; Keppert, M.; Výborný, J.; Černý, R. Hygric, thermal and durability properties of autoclaved aerated concrete. *Constr. Build. Mater.* **2013**, *41*, 352–359. [CrossRef]
26. Jeong, S.; Bui, Q.; Yang, I. A Comparative Study of the Thermal Conductivities of CBA Porous Concretes. *Materials* **2022**, *15*, 5204. [CrossRef]
27. Wang, W. Compressive strength and thermal conductivity of concrete with nanoclay under Various High-Temperatures. *Constr. Build. Mater.* **2017**, *147*, 305–311. [CrossRef]
28. Chen, G.; Li, F.; Jing, P.; Geng, J.; Si, Z. Effect of Pore Structure on Thermal Conductivity and Mechanical Properties of Autoclaved Aerated Concrete. *Materials* **2021**, *14*, 339. [CrossRef]

29. Hlaváček, P.; Amilauer, V.; Akvára, F.; Kopecký, L.; Aulc, R. Inorganic foams made from alkali-activated fly ash: Mechanical, chemical and physical properties. *J. Eur. Ceram. Soc.* **2015**, *35*, 703–709. [CrossRef]
30. Fomina, E.V.; Chulenyov, A.S.; Kozhukhova, N.I. Properties control in autoclave aerated concrete by choosing of pore forming Al-agent. In *IOP Conference Series: Materials Science and Engineering*; IOP Publishing Ltd: Bristol, UK, 2018; Volume 365, p. 32044. [CrossRef]
31. Chen, G.; Li, F.; Geng, J.; Jing, P.; Si, Z. Identification, generation of autoclaved aerated concrete pore structure and simulation of its influence on thermal conductivity. *Constr. Build. Mater.* **2021**, *294*, 123572. [CrossRef]
32. Ordonez-Miranda, J.; Alvarado-Gil, J.J. Effect of the pore shape on the thermal conductivity of porous media. *J. Mater. Sci.* **2012**, *47*, 6733–6740. [CrossRef]
33. Bary, B. Estimation of poromechanical and thermal conductivity properties of unsaturated isotropically microcracked cement pastes. *Int. J. Numer. Anal. Methods Geomech.* **2011**, *35*, 1560–1586. [CrossRef]
34. Zhang, W.; Min, H.; Gu, X.; Xi, Y.; Xing, Y. Mesoscale model for thermal conductivity of concrete. *Constr. Build. Mater.* **2015**, *98*, 8–16. [CrossRef]
35. Li, H.; Zeng, Q.; Xu, S. Effect of pore shape on the thermal conductivity of partially saturated cement-based porous composites. *Cem. Concr. Compos.* **2017**, *81*, 87–96. [CrossRef]
36. Othuman, M.A.; Wang, Y.C. Elevated-temperature thermal properties of lightweight foamed concrete. *Constr. Build. Mater.* **2011**, *25*, 705–716. [CrossRef]
37. Miled, K.; Limam, O. Effective thermal conductivity of foam concretes: Homogenization schemes vs experimental data and FEM simulations. *Mech. Res. Commun.* **2016**, *76*, 96–100. [CrossRef]
38. Qin, Z.; Li, G.; Tian, Y.; Ma, Y.; Shen, P. Numerical Simulation of Thermal Conductivity of Foam Glass Based on the Steady-State Method. *Materials* **2019**, *12*, 54. [CrossRef]
39. Ding, Y.; Deng, M.; Zhou, S.; Wang, Z.; Dong, J.; Wei, Y. The evolution relation between materials' porosity and thermal conductivity based on the simulation by COMSOL software. *Mater. Rep.* **2019**, *33*, 211–213.
40. Guo, Q.; Bian, Y.; Li, L.; Jiao, Y.; Tao, J.; Xiang, C. Stereological estimation of aggregate gradation using digital image of asphalt mixture. *Constr. Build. Mater.* **2015**, *94*, 458–466. [CrossRef]
41. Debnath, B.; Pratim Sarkar, P. Quantification of random pore features of porous concrete mixes prepared with brick aggregate: An application of stereology and mathematical morphology. *Constr. Build. Mater.* **2021**, *294*, 123594. [CrossRef]
42. Zhou, S.; Sheng, W.; Wang, Z.; Yao, W.; Huang, H.; Wei, Y.; Li, R. Quick image analysis of concrete pore structure based on deep learning. *Constr. Build. Mater.* **2019**, *208*, 144–157. [CrossRef]
43. Song, Y.; Shen, C.; Damiani, R.M.; Lange, D.A. Image-based restoration of the concrete void system using 2D-to-3D unfolding technique. *Constr. Build. Mater.* **2021**, *270*, 121476. [CrossRef]
44. Wang, X.F.; Yang, Z.J.; Yates, J.R.; Jivkov, A.P.; Zhang, C. Monte Carlo simulations of mesoscale fracture modelling of concrete with random aggregates and pores. *Constr. Build. Mater.* **2015**, *75*, 35–45. [CrossRef]
45. She, W.; Zhao, G.; Cai, D.; Jiang, J.; Cao, X. Numerical study on the effect of pore shapes on the thermal behaviors of cellular concrete. *Constr. Build. Mater.* **2018**, *163*, 113–121. [CrossRef]
46. Sadeghi, B.; Sadeghian, B.; Taherizadeh, A.; Laska, A.; Cavaliere, P.; Gopinathan, A. Effect of Porosity on the Thermo-Mechanical Behavior of Friction-Stir-Welded Spark-Plasma-Sintered Aluminum Matrix Composites with Bimodal Micro- and Nano-Sized Reinforcing Al_2O_3 Particles. *Metals* **2022**, *12*, 1660. [CrossRef]
47. Dehkordi, E.R.; Moodi, F.; Givkashi, M.R.; Ramezanianpour, A.A.; Khani, M. Software simulation of chloride ions penetration into composite of pre-fabricated geopolymer permanent formworks (PGPFs) and substrate concrete. *J. Build. Eng.* **2022**, *51*, 104344. [CrossRef]

materials

MDPI

Article

Experimental Study on Early Strength and Hydration Heat of Spodumene Tailings Cemented Backfill Materials

Shunchun Deng [1], Lang Liu [1,2,*], Pan Yang [1], Caixin Zhang [1], Yin Lv [1] and Lei Xie [1]

[1] School of Energy Engineering, Xi'an University of Science and Technology, Xi'an 710054, China
[2] Research Center for Mine Functional Filling Technology, Xi'an 710054, China
* Correspondence: liulang@xust.edu.cn

Abstract: Spodumene tailing is the associated solid waste of extracting lithium from spodumene. With the increase in the global demand for lithium resources, its emissions increase yearly, which will become a key factor restricting the economic development of the mining area. Mechanical and hydration reactions, as well as the microstructure of early CSTB, are studied under different tailings–cement ratios (TCR) and solid mass concentration (SC) conditions. The results show that the uniaxial compressive strength of early CSTB has a negative exponential correlation with the decrease in TCR and a positive correlation with the increase in SC: when the age of CSTB increases to 7 days, the strength increases with the rise in SC in an exponential function, and the sensitivity of strength to TCR is higher than that of SC. Compared to other tailings cemented backfill materials, the addition of spodumene tailings reduces the sulfate ion concentration and leads to a new exothermic peak (i.e., the third exothermic peak) for the hydration exotherm of CSTB. Additionally, with the increase in TCR or decrease in SC, the height of the third exothermic peak decreases and the occurrence time is advanced. At the same time, the duration of induction phase was prolonged, the period of acceleration phase was shortened, and the total amount of heat released was significantly increased. The decrease in TCR or the increase in SC led to the rise in the number of hydration products which can effectively fill the internal pores of CSTB, enhance its structural compactness, and increase its compressive strength. The above study reveals the influence of TCR and SC on the early strength, hydration characteristics, and microstructure of CSTB and provides an essential reference for the mix design of underground backfill spodumene tailings.

Keywords: spodumene tailings; cemented tailing backfill; hydration heat; X-ray diffraction; scanning electron microscope

Citation: Deng, S.; Liu, L.; Yang, P.; Zhang, C.; Lv, Y.; Xie, L. Experimental Study on Early Strength and Hydration Heat of Spodumene Tailings Cemented Backfill Materials. *Materials* **2022**, *15*, 8846. https://doi.org/10.3390/ma15248846

Academic Editor: Christian M. Julien

Received: 6 November 2022
Accepted: 6 December 2022
Published: 11 December 2022

1. Introduction

Lithium is known as "white oil" for its excellent physical and chemical properties. Lithium plays an irreplaceable role in battery energy, aerospace industries, medicine, and health [1,2]. Global lithium demand is growing steadily at an annual rate of 8–11%, and the total global market is expected to reach 1 million tons in 2025 [3,4]. One of the most critical extraction sources of lithium is spodumene. However, the content of lithium in natural spodumene is relatively low, and a large amount of spodumene tailings are generated during the extraction process [5,6]. China is a major lithium producer, accounting for about 10% of the global lithium production capacity, with nearly 2 million tons of spodumene tailings produced in 2020. Currently, the treatment of spodumene tailings is mainly handled in situ, and the continuous discharge of a large number of tailings causes environmental safety hazards to the air, soil, and water in the surrounding areas and affects human health [7–9]. With the rise of the new energy industry, the demand for lithium will continue to increase, and more spodumene tailings will be produced in the future. Effective disposal of spodumene tailings has become an urgent problem to be solved.

Scholars have undertaken a great deal of research to promote the resource utilization of spodumene tailings. This research mainly focuses on ceramic materials, geopolymers, cement mortar aggregate, etc. Yang et al. [10] found that the chemical composition of spodumene tailings contains albite and microcline. The chemical composition is consistent with green ceramic materials and can be utilized to prepare porous ceramics. Using spodumene flotation powder can absorb methylene blue and provide good flexural and compressive strength, apparent porosity, and water absorption. Patrick et al. [11] prepared incombustible lightweight porous ceramics from spodumene tailings and glass wool waste. Such ceramics can be a satisfactory substitute for lightweight building materials for high-rise buildings. However, the preparation of ceramics requires that the tailing size be kept within a specific range to reduce the grinding cost. The uneven distribution of spodumene tailing size may limit the industrial application of large-scale ceramic preparation [12,13]. The polymer composites prepared by Patrick et al. [14], which use a large number of spodumene tailings, have good thermal stability. The form of tailings added (such as grinding, fractionation, or original) will change the material's thermal shrinkage, compactness, and strength. Yang et al. [12] prepared a green geopolymer with high bending strength using alkali-activated spodumene flotation tailings and metakaolin. Additionally, this treatment method provides a novel idea for resource utilization of spodumene tailings. However, the cost of this geological polymer is relatively high, which is not conducive to its promotion. In addition, Wu et al. [15] used spodumene flotation tailings as the aggregate of cement mortar. They found that increasing tailings' particle size can effectively improve artillery's mechanical properties, providing a reference for using spodumene as the aggregate of cement mortar.

As an effective means of solid waste management, tailings cemented backfill has attracted many scholars to study different tailings of cemented backfill materials [16,17]. Fridjonsson et al. [18] sought a more thorough and complete understanding of the function of cemented tailings backfill material as a support structure in mining operations. The changes in pore size and distribution during the hydration of backfill materials are also studied. Behera et al. [19] studied the physical and chemical properties of the mixture of lead–zinc grinding tailings and fly ash and proved the feasibility of using them for underground backfill. Qiu et al. [20] used microorganisms to induce carbonate precipitation as a new auxiliary cementation technology for iron tailings, and the backfill strength of iron tailings treated by this method was significantly improved. However, the existing research focuses on using spodumene tailings in ceramic production or geopolymer materials, such methods is too expensive to achieve the purpose of large-scale use of the waste of tailings. There are no other studies using spodumene tailings as filling materials. If tailings are used for filling, they can meet the strength requirements of conventional metal mines, not only not needing to increase the cost of tailings treatment but also achieving the purpose of protecting the environment. The hydration reaction of the backfill material, especially the early hydration reaction, directly affects the strength development and the later durability performance. The early hydration reaction mechanism and the development law of the hydration reaction products are the focus of early hydration process research [21]. At the same time, the development of thr early strength of filling materials has important practical significance for on-site production, affecting the mining production cycle. The hydration reaction and microstructure evolution of the filling material is the internal driving force for strength formation. The difference in physical and chemical properties of different material components directly affects the hydration process of backfill materials [22,23]. Accurately understanding the change law of the hydration process and the evolution law of microstructure of cemented backfill materials is conducive to analyzing the development law of early strength, thus controlling the backfill material strength formation process and providing reasonable process parameters [24–27]. However, the early strength development, exothermic hydration rule, and microstructure formation mechanism of CSTB still need to be studied.

This paper explores the influence of different tailings–cement ratios (TCR) and solid mass concentration (SC) on the early strength of CSTB. Additionally, it analyzes its hydration characteristics through the isothermal calorimetry test and studies its microstructure through X-ray diffraction and scanning electron microscope tests to further reveal the development mechanism of strength. It has the further objectives of promoting the utilization of spodumene tailings for mine backfill and providing a theoretical basis for the proportioning design of CSTB materials.

2. Materials and Methods

2.1. Experimental Material

2.1.1. Tailings

Spodumene tailing was obtained from the general sand outlet of a spodumene mine and used to prepare CSTB samples. The particle size of tailings was measured by a Malvern Mastersizer 2000 laser particle size analyzer. Figure 1a shows the particle size distribution of tailings; the tailing D_{10} is 43.34 μm, D_{60} is 167.29 μm, D_{90} is 305.94 μm, and the nonuniformity coefficient Cu = 7.06. It is generally believed that when Cu $\geq$ 5, the tailings have good compactness and grading. However, if Cu is too large, it may indicate that the middle particle size is missing, and so the curvature coefficient Cc is also used for evaluation. The curvature coefficient 1 < Cc = 1.28 < 3 indicates that the tailings are well graded and are beneficial to the bonding of tailings particles in the filling process [28]. The chemical composition of tailings (Table 1) shows that the main chemical components of tailings are SiO_2, Al_2O_3, and Na_2O, that the content is more than 90%, and that higher oxide content may have a good gelling effect [29]. Figure 1b shows the X-ray diffraction (XRD) pattern. The tailings contain albite (Na_2O Al_2O_3 $6SiO_2$, PDF#10-0393), auartz (SiO_2, PDF#46-1045) and muscovite ($KAl_2(AlSi_3O_{10})(OH)_2$, PDF#07-0025). They are mainly composed of aluminates or silicates of silicon, aluminum, and sodium and sodium oxides. Above all, spodumene tailing is a comparatively ideal cemented filling material. The analysis results are consistent with the results of chemical element analysis.

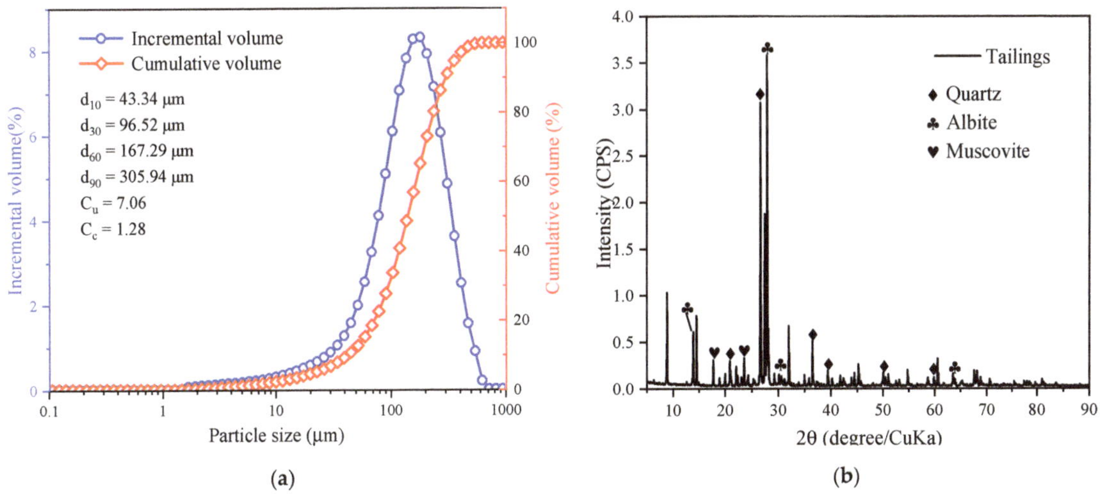

Figure 1. (**a**) Particle size distributions and (**b**) XRD patterns of tailings.

Table 1. Chemical composition of cement and tailings.

Composition	CaO	SiO_2	MgO	Fe_2O_3	Al_2O_3	SO_3	TiO_2	CuO	Na_2O	Loss
Cement	51.84	27.75	4.88	2.94	7.65	1.06	0.24	0.02	0.18	4.76
Tailings	0.66	70.62	0.12	0.50	14.52	0.24	0.02	0.01	7.38	11.65

2.1.2. Binder and Water

The binder used is ordinary Portland cement P·O 42.5 grade ordinary Portland cement, and the chemical composition analysis results are shown in Table 1. It can be seen from Table 1 that the main chemical components are CaO, SiO_2, Al_2O_3, and MgO, and that the total amount is 92.12%.

The water used for CSTB is ordinary tap water.

2.2. *Sample Preparation*

Depending on field practice and reading of the literature [30,31], the tailing cement ratio of 4:1, 6:1, 8:1, 10:1, and 12:1 were selected as the research object of this paper. Used solid mass concentrations (SC) of 68%, 70%, 72%, 74%, and 76% were selected. Table 2 shows the evaluated compositions detailing the cement, tailings, and water amount. Mix the tailings, cement, and water according to a certain proportion and pour it into the UJZ-15 planetary cement mortar mixer for stirring. Then, pour the stirred CSTB into the standard cylindrical (φ50 mm×h100 mm) abrasive tool and demold after 48 h. All test pieces, after demolding, were placed in a traditional curing environment with a curing temperature of 20 ± 2 °C and relative humidity of 95% ± 1% for curing.

Table 2. Experimental schemes.

SC/%	TCR	Material Quality/%		
		Tailing	Cement	Water
68	4:1	54.40	13.60	32
	6:1	58.29	9.71	32
	8:1	60.44	7.56	32
	10:1	61.82	6.18	32
	12:1	62.77	5.23	32
70	4:1	56.00	14.00	30
	6:1	60.00	10.00	30
	8:1	62.22	7.78	30
	10:1	63.64	6.36	30
	12:1	64.61	5.39	30
72	4:1	57.60	14.40	28
	6:1	61.71	10.29	28
	8:1	64.00	8.00	28
	10:1	65.45	6.55	28
	12:1	66.46	5.54	28
74	4:1	59.20	14.80	26
	6:1	63.43	10.57	26
	8:1	65.78	8.22	26
	10:1	67.27	6.73	26
	12:1	68.31	5.69	26
76	4:1	60.80	15.20	24
	6:1	65.14	10.86	24
	8:1	67.56	8.44	24
	10:1	69.09	6.91	24
	12:1	70.15	5.85	24

2.3. Methods

Through mechanical experiments, isothermal calorimetry, XRD, and scanning electron microscopy (SEM) experiments, the compressive strength, hydration process, and mineralogical composition of CSTB are analyzed.

According to the standard testing method of mechanical properties on ordinary concrete (GB/T 50081-2002), the CSTB specimen with a certain curing age was tested by WDW (20-1000) electronic universal testing machine. The loading rate is 1 mm/min, and the final strength of CSTB is the average of the strength of the three samples. According to ASTM C 1702 [32], put 15 g dry sample (containing a certain proportion of tailings and cement) powder into a glass ampoule, add water and stir for 2 min, and seal with a plastic plug. Then, put in the TAM air isothermal calorimeter (Thermometric AB, New Castle, DE, USA,) to measure the normalized heat flow (N) and cumulative heat (Q) of CSTB. The test temperature is set at 20 °C, and the test time is 72 h. SEM observed the microstructure and morphology of CSTB. A small part was taken from the inner center of the cured specimen to be broken and immersed in ethanol to prevent hydration. The material was then quickly dried in a drying oven at 60 °C. The dried samples were taken into small pieces, and the samples were sprayed with gold and then tested by a TESCAN MIRA4 field emission environment scanning electron microscope. XRD can determine the type of hydration products. Take a small number of dried samples and grind them in an agate mortar. After grinding until there is no particle feeling, use the Bruker AXS D8 X-ray diffractometer produced by Bruker company in Germany to determine the type of hydration products of CSTB by XRD. The tube current and voltage of XRD equipped with a Cu-Kα X-ray source were 40 mA and 40 kV, respectively. The diffraction angle (2 θ) is 5–90°, and the scanning speed is 6°/min.

3. Results and Analysis

3.1. Compressive Strength Analysis

3.1.1. Effect of TCR on Compressive Strength of CSTB

Figure 2 shows the relationship between the early strength of CSTB and TCR. It can be observed from Figure 2 that the compressive strength of CSTB is negatively correlated with TCR when the SC is constant. With the continuous decrease in TCR, the intensity of CSTB gradually increased, and this change did not change with age. Taking SC = 76% as an example, when TCR = 12, 10, 8, 6, and 4, the strength is 0.163 MPa, 0.330 MPa, 0.529 MPa, 0.749 MPa, and 1.832 MPa, respectively, at the curing age of 3 days, and the power is increased by 0.167 MPa, 0.199 MPa, 0.22 MPa, and 1.083 MPa, respectively. This is because, with the increase in TCR, the lower the cement content in CSTB is and the smaller the hydration reaction will be. Therefore, the fewer hydration products that will be generated through the hydration reaction, the weaker the connection between tailings particles is, and the more pores there are, resulting in a CSTB intensity reduction [33].

It is worth noting that the lower the TCR is, the more pronounced the strength increase will be under the same SC. The test results are fitted exponentially according to the change law of CSTB strength with TCR. Except for the linear correlation coefficient R^2 of the strength fitting curve with SC of 68% at 7 days being too low, the other correlation coefficients R^2 are all close to 1. This indicates a robust exponential correlation between TCR and CSTB compressive strength under a constant SC, which can be expressed by Equation (1). It is consistent with the conclusions drawn by other scholars [34,35]:

$$y = a + be^{-x/c} \quad (1)$$

where y is the compressive strength, MPa; a, b and c are constants depending on TCR and SC factors, and x is TCR.

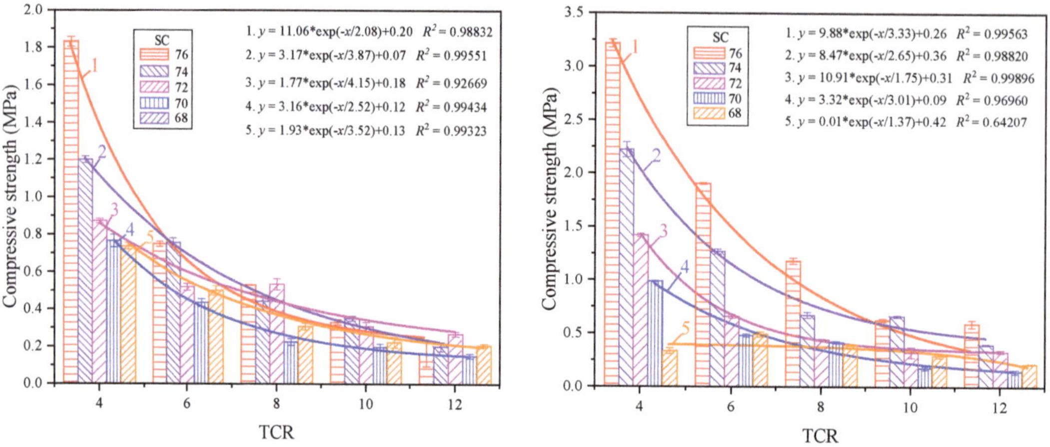

(**a**) (**b**)

Figure 2. Compressive strength of CSTB at different TCRs with curing age of 3 d (**a**) and 7 d (**b**).

3.1.2. Effect of SC on Compressive Strength of CSTB

Figure 3 shows the relationship between the early strength of CSTB and SC. It can be observed in Figure 3 that, when TCR is constant, the uniaxial compressive strength of CSTB increases with the increase in SC, and the growth rate gradually increases with the rise in SC. When TCR = 4, with SC increased from 68% to 70%, the CSTB's 3-day strength increased from 0.738 to 0.767 MPa, with an increased rate of 3.9%. Additionally, 7-day strength increased from 0.750 to 0.987 MPa, with an increased rate of 31.6%. When SC increases from 74% to 76%, the 3-day strength increases from 1.200 to 1.832 MPa with a growth rate of 52.7%, and the 7-day strength increases from 2.222 to 3.216 MPa with a growth rate of 44.7%. This can be attributed to the fact that when TCR is constant, SC can affect the cement segregation in CSTB. The reason for the segregation is that there is a difference in the settling speed of large and small particles in water. With the increase in SC, the smaller the cement segregation, the more uniform the particle distribution, and the lower the cement loss. Therefore, the relatively low SC backfill material has a higher content of high SC cement, and the strength of CSTB is relatively high. At the same time, with the decrease in SC, the more significant the difference in the sedimentation velocity of large and small particles, the more discrete the distribution, and the lower the intensity of low SC [36].

Under the same TCR, the CSTB intensities of different SCs were fitted. When the curing age was 3 days, except for the fitting curve of TCR = 4, the appropriate effect of the strengths of different SCs under the other TCRs was not noticeable. This is because, with the increase in TCR, the 3-day strength decreases with the decrease in SC. However, when TCR increases to a certain extent, the influence of SC on the 3-day power decreases, and the intensity change is not apparent enough, resulting in a poor fitting effect [37]. With the increase in curing time, the influence of SC on the strength of CSTB gradually increases. When the age reaches 7 days, the fitting of strength shows the exponential correlation shown in Equation (2):

$$y = a + be^{((x-c)/d)} \quad (2)$$

where y is the compressive strength, MPa; a, b and c are constants depending on TCR and SC factors, and x is SC.

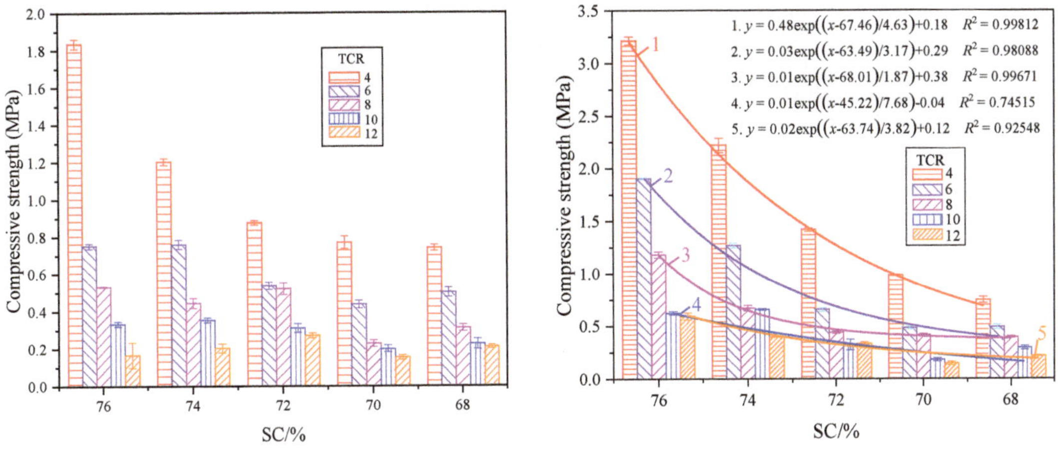

(a) (b)

Figure 3. Compressive strength of CSTB at different SCs with curing age of 3 d (**a**) and 7 d (**b**).

3.1.3. Coupling Effect of TCR and SC on Compressive Strength of CSTB

According to Abrams' law, the water–cement ratio (W/C) is considered the most critical factor affecting concrete strength. The relationship between compressive strength and W/C is not affected by sand type [38–40]. TCR and SC determine the W/C, so the early power of CSTB is closely related to TCR and SC, and its strength increases with the decrease in TCR and the increase in SC, and the lower the TCR, the greater the influence of SC on CSTB. Under a specific SC, the intensity of CSTB changes exponentially with the decrease in TCR. The lower the TCR, the more pronounced the intensity increase; at the same time, under a certain TCR, the CSTB intensity also has an exponential correlation with the rise in SC.

From the above analysis, it can be seen that the strength of CSTB is related to TCR and SC, that the two have different effects on the strength, and that the sensitivity of the strength to TCR and SC is different. In Figure 2, it can be observed that, when the SC is constant, the TCR decreases from 12 to 4 in turn, and the strength increases significantly at each stage of reduction. In Figure 3, at 3 days, except for the strength of CSTB increasing obviously with SC when TCR = 4, the strength of CSTB under other TCR does not increase obviously with SC and even cannot reach the fitting effect. However, with the increase in curing age, the strength of CSTB increases obviously with SC compared with that at 3 days, but it is still not as significant as the effect of TCR on the strength. It can be considered that the sensitivity of CSTB compressive strength is TCR > SC. This is consistent with the research results of some scholars [37,41,42], while the conclusions of other scholars are the opposite [30,43]. This difference can be attributed to the following two points. Firstly, the SC concentration gradient set in this paper is slight, and the CSTB intensity does not change significantly with SC. Secondly, this is related to the particle size distribution of backfill materials and chemical composition. For example, the research of Fu et al. [30] on the tailings of iron ore (setting SC as 65%, 68%, 70%, 73%, 75%, and TCR as 4, 5, 6, 8, 10) shows that the CSTB strength of the whole tailings of this iron ore is more sensitive to SC than TCR. However, the research of Yang et al. [42] on another iron ore (setting SC as 65%, 68%, 70%, 73%, and TCR as 4, 6, 8, 10) has reached the opposite conclusion.

3.2. Hydration Exothermic Analysis

Cement hydration is an exothermic process. A series of physical and chemical changes occur after CSTB is mixed with water, and a large amount of heat is released. The amount of heat released is related to the cement's hydration process [44,45].

The study shows that the hydration stage of CPB can be divided into five stages: (1) dissolution stage, (2) induction stage, (3) acceleration stage, (4) deceleration stage, and (5) slow reaction stage [46–48]. The dissolution stage starts from the contact between the backfill material and water. At this time, the hydration reaction is rapid, and a large amount of heat is released. Due to metastable barrier [49] or slow dissolution [50], after the initial fast reaction, the hydration reaction slows down, and the exothermic hydration decrease enters the induction stage. After the induction stage, the hydration rate is accelerated and enters the acceleration stage. The exothermic heat in the acceleration stage is closely related to the nucleation of the generated hydrated calcium silicate gel (C-S-H) [51–53]. With the progress of the hydration reaction, the hydration products increase, the small redactable particles, available space and redactable water decrease, and the hydration enter the deceleration stage. After that, due to the consumption of cementitious materials, hydration enters the slow reaction stage.

It is worth noting that, compared with the exothermic hydration curve of ordinary cement-based materials, the exothermic hydration curve of CSTB has a significant feature that a new hydration exothermic peak appears on its hydration exothermic curve (Figure 4). The following two reasons may cause this new exothermic peak. First, the hydration reaction of CSTB can be divided into two parts. One part is the hydration reaction of tricalcium silicate (C3S) contained in cement (the second exothermic peak of ordinary cement-based materials). The other part is the hydration reaction of tailings, the third exothermic peak, which is similar to the exothermic hydration curve produced by adding blast furnace slag [54] or iron tailings [22] to cement. Second, the new exothermic peak is related to the hydration reaction of C3A. In this process, sulfate ion concentration plays a key role. C-S-H initially adsorbs the sulfate ion. When solid sulfate is consumed, the slope of thermal evolution increases and C3A hydration leads to a new exothermic peak [55–58].

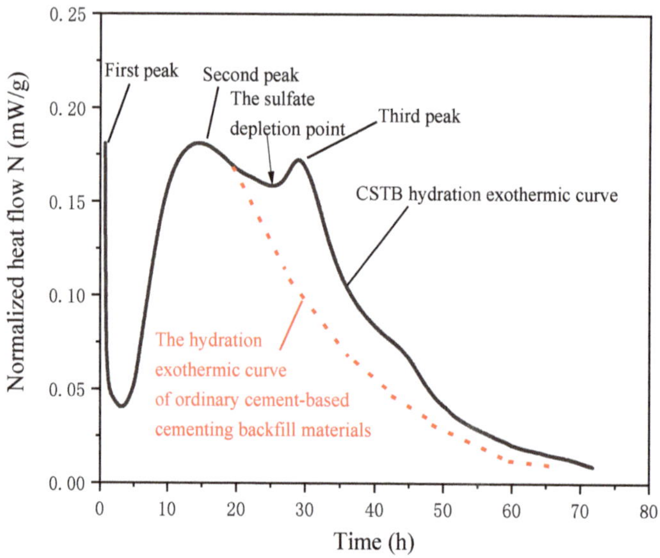

Figure 4. Differ exothermic hydration curves between ordinary cement-based materials [59] and CSTB.

3.2.1. Effect of TCR on the Heat Exotherm of CSTB Hydration

Figure 5 is the hydration reaction exothermic rate curve of different TCRs under the same SC and the total exothermic amount curve of SC = 72%. Table 2 indicates the specific changes in the hydration heat evolution of different TCRs under SC = 72%.

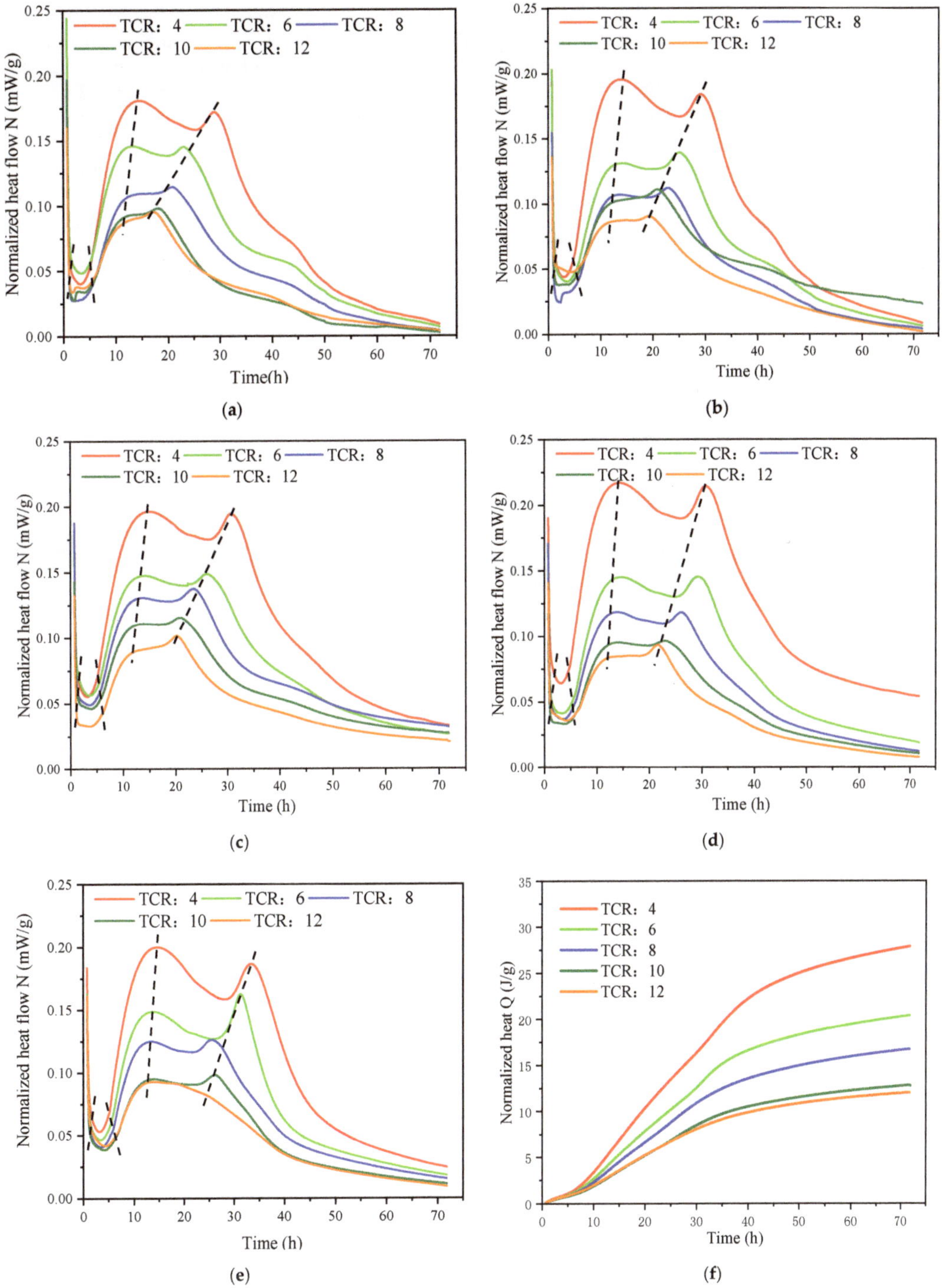

Figure 5. Hydration heat release rate curve: (**a**) SC = 68%, (**b**) SC = 70%, (**c**) SC = 72%, (**d**) SC = 74%, (**e**) SC = 76%, and total heat release: (**f**) SC = 72% of CSTB under different TCRs.

Table 3 demonstrates that CSTB with low TCR entered the induction phase somewhat later, but that the duration of this phase was shorter. For example, when TCR = 12, the induction phase was entered 1.1 h after the beginning of the test, and the duration of this phase was 6.6 h. When TCR was reduced to 4, the induction period was delayed by 0.3 h, but the time was shortened by 1.8 h. In this case, the exothermic hydration curve in Figure 5 c shows that the exothermic angle of low TCR is sharper at this stage. This situation can be explained as follows: when the dissolution reaction proceeds to a certain extent, the clinker particles are wrapped by the hydration film, and the formation of the hydration film hinders the hydration reaction, thus reducing the hydration reaction rate of CSTB, and the hydration reaction enters the induction period. With time, the wrapped clinker particles continue to dissolve and participate in the response slowly, the concentration of hydration products C-S-H or calcium hydroxide (CH) increases, and the pressure difference between the inside and outside of the hydrated film increases. The hydrated film breaks when the concentration rises to a critical point, and the hydration reaction enters an accelerated phase. The material with high TCR contains less cement, the water–cement ratio increases, and the increase in available water reduces the concentration of Ca^{2+} in the pore solution. This leads to a longer time being taken for Ca^{2+} to reach saturation and a longer duration of the induction period [46,60].

Table 3. Variation in hydration exothermic rate for different TCRs (h).

TCRs	Induction Stage	Acceleration Stage	Second Hydration Peak	Third Hydration Peak	Deceleration Stage
4	1.4	5.1	15.1	30.8	33.6
6	1.6	5.8	14.1	26.1	28.9
8	1.3	5.8	13.7	23.6	27.7
10	1.2	6.5	14.0	21.0	26.5
12	1.1	6.6	None	20.3	24.2

In the acceleration stage, it can be observed that the hydration heat release rate curve of high TCR, seen in Figure 5a–e, is gentler. The slower the hydration heat release growth is, the lower the height of the exothermic peak is, and the second exothermic peak completely disappears as TCR rises to 12. Table 3 shows that the second and third exothermic peaks of high TCR are advanced, and the duration of this phase is reduced. The hydration rate in the accelerated degree is related to the heterogeneous nucleation and growth of C-S-H on mineral surfaces, which has been supported by many scholars [61–63]. Under the same SC, the cement content of CSTB with high TCR is reduced, and the amount of C-S-H generated is relatively reduced. The hydration heat release rate increases more slowly as the hydration heat release curve is flatter and the hydration heat release rate is lower. Hence, the second heat release peak-to-peak value of high TCR is lower [60].

In the acceleration stage, the second hydration exothermic peak of high TCR appears earlier. The acceleration stage's duration is shortened, which is mainly related to the crystal nucleation and crystal growth of C-S-H and CH. With the increase in TCR, tailing particles increase and can be used as the nucleii for the precipitation and development of initial hydration products such as C-S-H. Therefore, the rise in TCR accelerates the hydration reaction and shortens the duration of the acceleration phase [64]. In addition, the increase in TCR leads to the rise in the water–cement ratio (W/C), the expansion of free water content per unit cement particle, and the decrease in cement that can participate in the hydration reaction, all of which will accelerate the hydration reaction of cement. Therefore, the second exothermic peak appears in advance [24,26]. This is because the tailing content increases with the increase in TCR, resulting in fewer sulfate ions being dissolved in water and the concentration of sulfate ions in the solution decreases. The time point of sulfate depletion is advanced, and C3A is re-dissolved and hydrated, and so the occurrence time of the third exothermic peak is advanced. In addition, with the advance in the third exothermic peak, the second exothermic peak overlapped with the third exothermic peak, which was more evident at high TCR. When TCR = 12, the second exothermic peak could not be observed. This is the same as what was found by Han et al. [27] and Yang [65], who made the observation that the hydration exothermic peak changes with the increase in iron tailing content.

In the deceleration and slow reaction stages, the heat release rate of the backfill material with high TCR decreases more slowly than that of that backfill material with low TCR, and its heat release curve is relatively flat. The hydration rate of backfill materials with high TCR is still lower than that of backfill materials with low TCR. The diffusion process determines the deceleration of the hydration reaction and the rate of the slow reaction stage and is affected by the size of reaction particles, reaction water, and reaction space. The hydration rate of the low-TCR product is relatively higher than that of the high-TCR product because the cement content is higher. There are more reactive particles in the later stage [66].

The total heat release curve, under different TCRs with SC of 72%, is shown in Figure 5f. It can be observed in the figure that the total amount of hydration heat release increases with the decrease in TCR. At 72 h, the difference reaches the maximum, and the whole heat release from TCR = 12 to TCR = 10, TCR = 8, TCR = 6, and TCR = 4 increases by 3.1 J/g, 3.3 J/g, 2.2 J/g, and 7.5 J/g, respectively. This shows that with the decrease in TCR, the water–cement ratio decreases correspondingly, which is conducive to the hydration reaction. As a result, the hydration reaction speed is faster, and more hydration products that can provide strength for CSTB can be produced simultaneously.

3.2.2. Effect of SC on the Heat Exotherm of CSTB Hydration

As shown in Figure 6, SC's hydration heat release rate increased from 68% to 72% under the same TCR and the total heat release amount of CSTB when TCR = 4. It can be observed in Figure 6a–e that the exothermic hydration curves of different SCs are similar under a given TCR—taking TCR = 4 as an example for analysis. The data in Table 4 show that SC = 68% enters the induction phase after 1.5 h and enters the acceleration phase after 4.9 h. Although the SC increased to 72%, in the hydration heat release curve, the time to join the acceleration phase is only 0.1 h ahead of schedule, and the duration of the acceleration phase is only 0.3 h ahead. In addition, it can be seen from the hydration heat release curve that the process of the acceleration stage is only 1.7 h. The appearance time of the second exothermic peak was delayed by 0.5 h. The third exothermic peak was delayed by 1.7 h. The changes in total heat release in different periods in Table 5 show that the complete heat release Q did not change significantly in the early stage. For example, when T = 18 h, Q only increased by about 13.0% as SC rose from 68% to 72%. The above experimental results show that, compared with the effect of TCR on CSTB, SC has a limited impact in the early stage. This is because when TCR is constant (cement content is consistent), the change in sulfate ion concentration, caused by the shift in SC, is small. Therefore, although SC increases from 68% to 72%, there was only a small difference in the time of sulfate depletion point.

Table 4. Time of hydration exothermic phase of different SCs.

SCs	Induction Stage (h)	Acceleration Stage (h)	Second Hydration Peak (h)	Third Hydration Peak (h)	Deceleration Stage (h)
68%	1.5	4.9	31.7	3.4	26.8
70%	1.5	4.7	31.9	3.2	27.2
72%	1.4	5.1	33.6	3.7	28.5

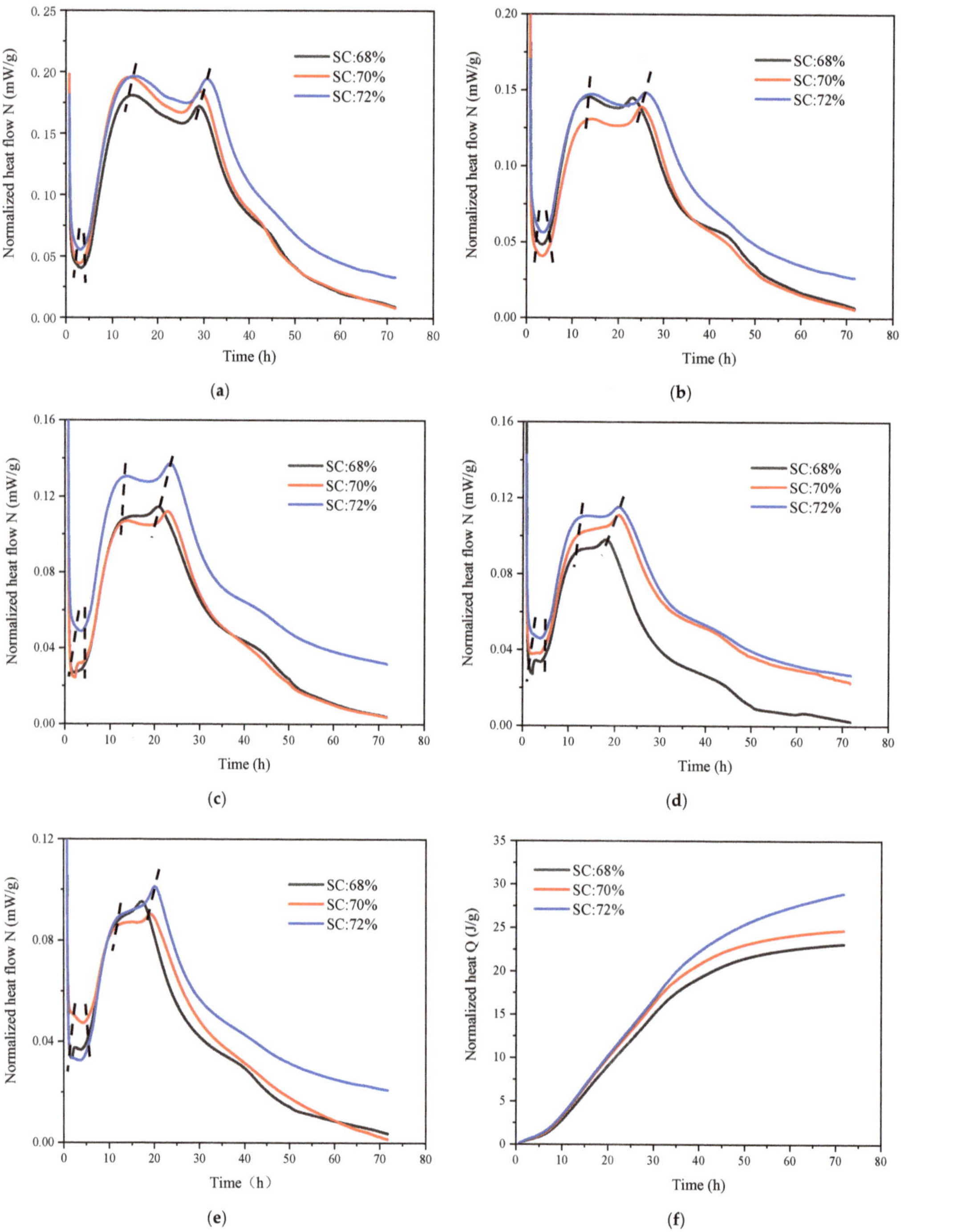

Figure 6. Hydration heat release rate curve: (**a**) TCR = 4, (**b**) TCR = 6, (**c**) TCR = 8, (**d**) TCR = 10, (**e**) TCR = 12 and total heat release: (**f**) TCR = 4 of CSTB under different SCs.

Table 5. Total heat release Q (J/g) under different SCs.

SCs	18 h	36 h	54 h	72 h
68%	7.7	17.8	21.9	23.1
70%	8.6	19.3	23.5	24.7
72%	8.7	20.4	26.2	28.9

Meanwhile, when the cement content is constant, the increase in SC increases W/C to a certain extent. However, its impact is not as great as the influence of TCR on W/C. Therefore, the difference in hydration exothermic curve under different SC is not as significant as that under TCR. However, with time, the influence of SC will be more evident due to the different reaction rates in the slow reaction stage. For example, when t = 72 h, SC = 68% increases to 72%, and Q increases by 25.1%.

3.2.3. Effect of TCR and SC Coupling on Hydration Heat Release of CSTB

The change in TCR and SC will cause the shift of W/C, which is proportional to TCR and inversely proportional to SC. W/C is an essential parameter for studying the cementation of cement-based materials. It can affect hydration rate, microstructure [67], and porosity [68]. Table 6 lists the changes in W/C by different TCR and SC, and Figure 7 further shows the changes in the total hydration heat of CSTB when W/C changes. It can be observed from Table 6 that the ratio of W/C increases by about 0.8 with each gradient of TCR, and decreases by only 0.2–0.3 with each rise in SC. It can be seen from Figure 7 that the total amount of heat released is inversely proportional to W/C. W/C changes faster with the increase in TCR, and the total amount of heat released decreases more obviously. W/C changes little with the rise in SC, and the change in entire heat release was insignificant. Therefore, the effect of TCR on the hydration process is more significant than that of SC.

Table 6. Variation in W/C with TCR and SC under given conditions.

TCRs/SCs	TCR = 4	TCR = 6	TCR = 8	TCR = 10	TCR = 12	SC = 68%	SC = 70%	SC = 72%
W/C	1.9	2.7	3.5	4.3	5.1	2.4	2.1	1.9

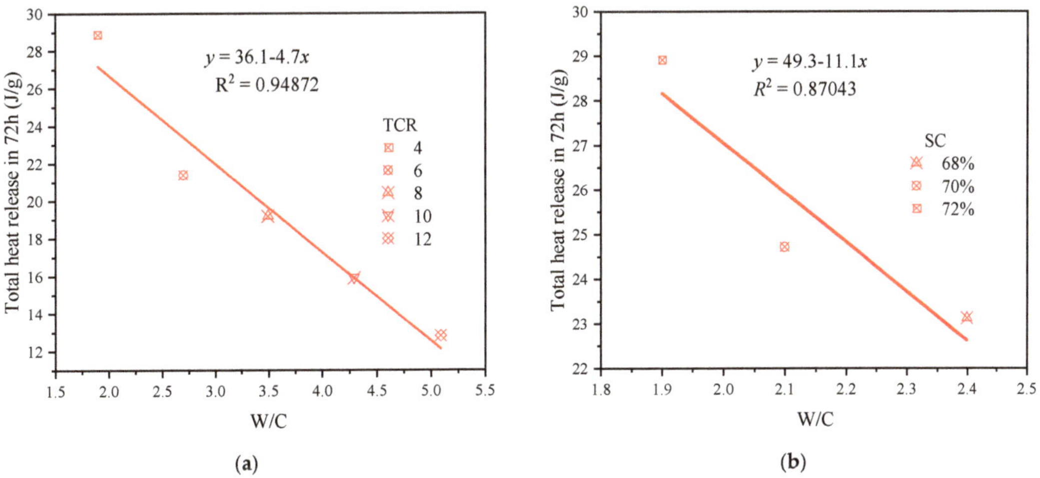

Figure 7. The effect of W/C on the total heat release Q under certain conditions: (**a**) SC = 72%; (**b**) TCR = 4.

Besides the effect of TCR on the hydration heat release of CSTB due to SC can be attributed to W/C, there may be other reasons. For example, the change in TCR will have an impact on cement and tailings in CSTB. The decrease in TCR reduces the content of tailings and leads to the reduction in nucleation sites of C-S-H/CH. However, compared with the increase in cement content, the decrease in nucleation sites may have a limited impact. The nucleation sites provided by tailings may have reached saturation when TCR = 4. Therefore, with the decline in TCR, there is no reduction in hydration heat release [66]. The increase in SC leads to the decrease in W/C and increases the content of cement and tailings per unit volume. The spacing between particles in CSTB is smaller, the dispersion effect on cement and tailings is weakened, and the hydration reaction speed increases. Therefore, the heat release rate and total heat release are increased [69,70].

Figure 8 displays the relationship between the total heat release of hydration in 72 h and the 3d compressive strength. The results show a positive correlation between the intensity and the total heat release and the relationship equations obtained by fitting, which is consistent with the conclusions of other scholars [47,64,71].

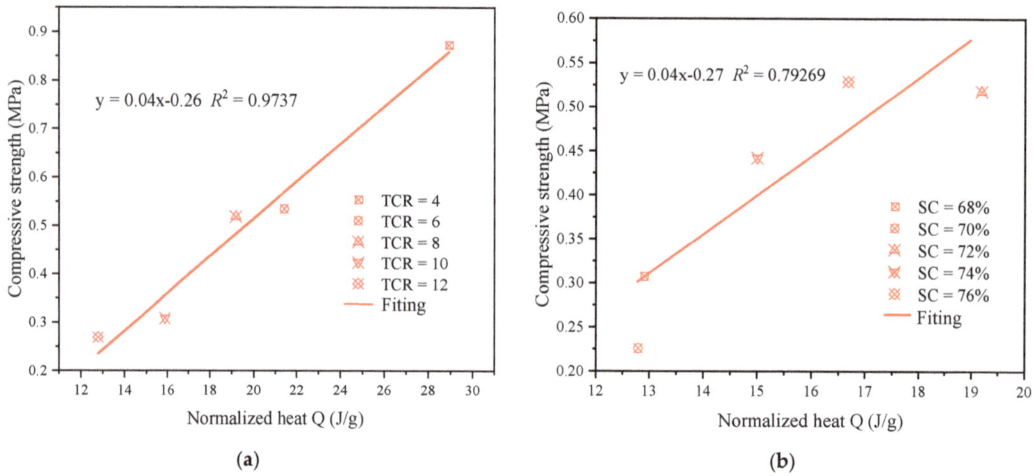

Figure 8. Hydration heat versus the compressive strength values at 3 days of CSTB: (**a**) SC = 72%; (**b**) TCR = 8.

3.3. *Hydration Products and Microstructure Analysis*

The strength development of CSTB is closely related to the hydration products formed [72]. After CSTB is mixed with water, calcium silicate in cement is hydrated to create hydration products such as CH, C-S-H, and AFt. The XRD and SEM images of CSTB can analyze the microstructure of CSTB, thus revealing the visible strength change in CSTB from mineral composition and microscopic morphology.

3.3.1. Effect of TCR on Hydration Products and Microstructure of CSTB

Figures 9 and 10 show the XRD and SEM of CSTB samples at different curing ages when SC = 72%. It can be observed in the XRD pattern that the 3-day and 7-day ages mainly contain peaks corresponding to CH, C-S-H, and AFt, among which CH and C-S-H are due to the hydration reaction of calcium silicate, as shown in Equations (3) and (4):

$$3CaO \cdot SiO_2 + H_20 \rightarrow xCaO \cdot SiO_2 \cdot yH_2O(C\text{-}S\text{-}H) + Ca(OH)_2 \quad (3)$$

$$2CaO \cdot SiO_2 + H_20 \rightarrow xCaO \cdot SiO_2 \cdot yH_2O(C\text{-}S\text{-}H) + Ca(OH)_2 \quad (4)$$

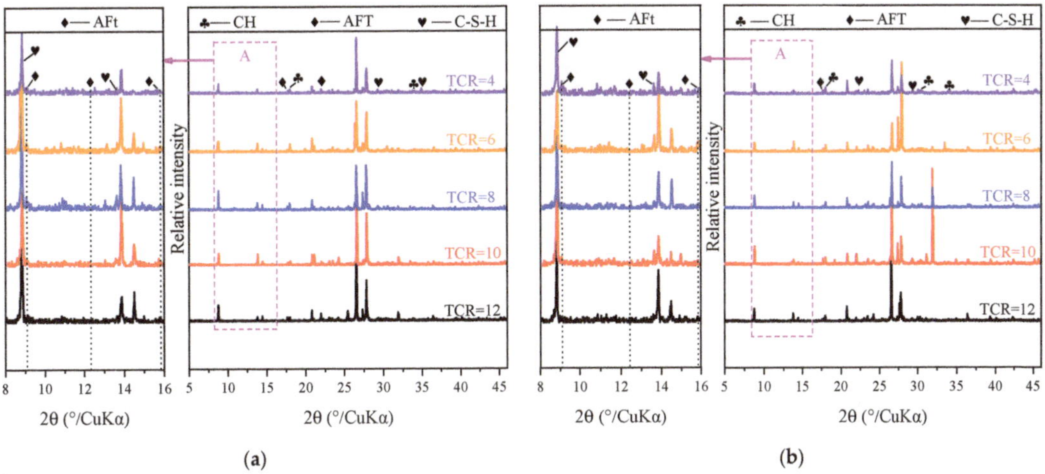

Figure 9. XRD patterns of CSTB in 3d (**a**) and 7d (**b**) under SC = 72%.

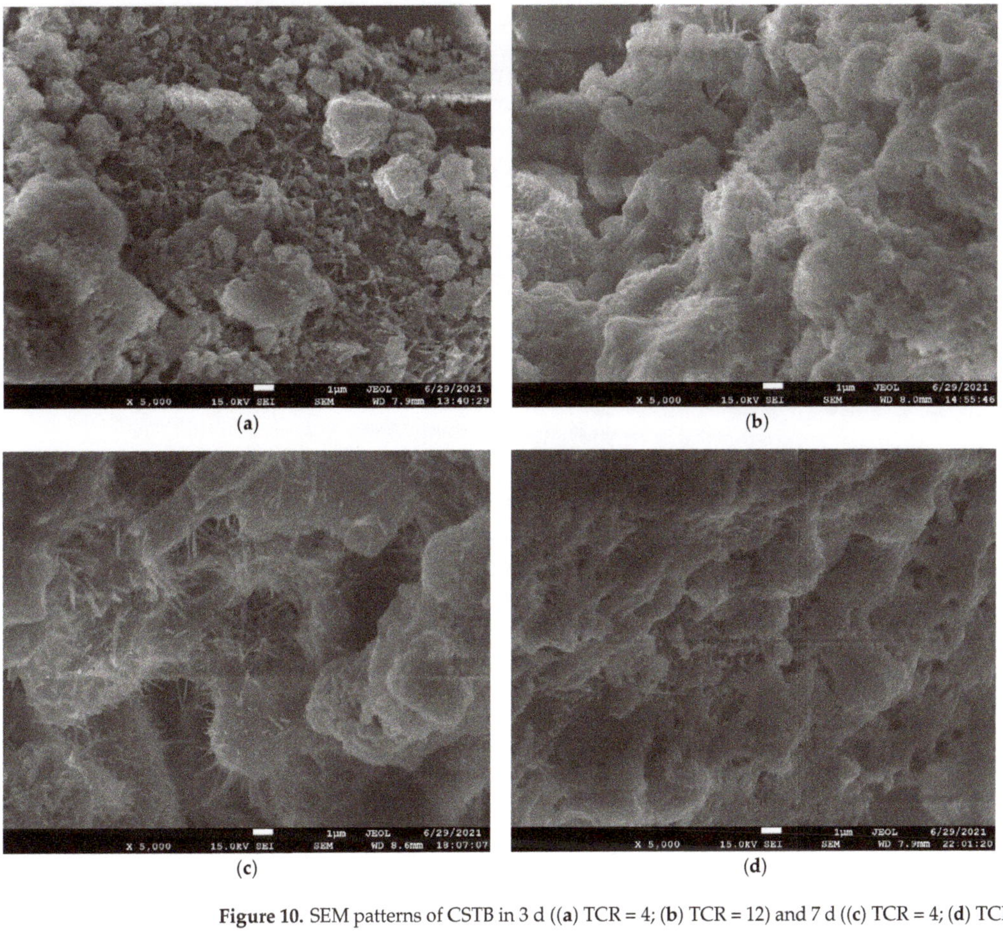

Figure 10. SEM patterns of CSTB in 3 d ((**a**) TCR = 4; (**b**) TCR = 12) and 7 d ((**c**) TCR = 4; (**d**) TCR = 12) under SC = 72%.

Since the generated CH provides an alkaline environment for CSTB, CH reacts with SiO_2 in tailings in the alkaline environment, as shown in Equation (5). At the same time, with the participation of water, CH reacts with Al_2O_3 in tailings and $CaSO_4$ in cement to form AFt through Equation (6) [73]:

$$xCa(OH)_2 + SiO_2 + (y - x)H_2O \rightarrow xCaO \cdot SiO_2 \cdot yH_2O(C\text{-}S\text{-}H) \quad (5)$$

$$Al_2O_3 + 3Ca(OH)_2 + 3CaSO_4 \cdot 2H_2O + 23H_2O \rightarrow 3CaO \cdot Al_2O_3 \cdot 3CaSO_4 \cdot 32H_2O(AFt) \quad (6)$$

The peaks corresponding to the hydration products can be observed in Figure 9, but the peak intensities are not wholly consistent. For example, at the diffraction angle of 34°2-theta, CH's maximum relative peak intensity is TCR = 4, and the minimum peak intensity is TCR = 12. Even the peak cannot be observed. This indicates that more CH is generated at low-TCR conditions than high-TCR conditions. This phenomenon exists, not only for CH, but also for other hydration products.

The hydration products generated and the pores between tailings particles can be observed from the SEM of Figure 10 (a and c are TCR = 4; b and d are TCR = 12). In the case of TCR = 12, it can be seen that there are a lot of pores between tailings particles. When TCR = 4, AFt can be perceived on the surface of CSTB particles and connected. Hydration products wrap tailings particles, and pores are effectively filled. Only a few pores can be noted. At the same time, it can be observed that when the curing time reaches 7 days, the structure is dense.

The reason for the above phenomenon is that with the decrease in TCR, more cement can participate in the hydration reaction, and as such the hydration reaction speed is faster. In the same curing time, more hydration products are produced. That is, the relative intensity value of each hydration product in the XRD diffraction pattern is more significant, and more hydration products can be observed in the SEM image. After the hydration product is generated, it can effectively fill the pores between the tailings particles and play a role in connecting the tailings particles. Therefore, the pores observed at the lower TCR are relatively small, and the internal structure is more compact [74]. The compressive strength of CSTB is closely related to its internal pores and hydration products. Therefore, the compressive strength of CSTB with low TCR is higher than that of CSTB with high TCR. The strength changes and hydration heat analysis results mentioned above have been further confirmed.

3.3.2. Effect of SC on Hydration Products and Microstructure of CSTB

As shown in Figures 11 and 12 (a and c are SC = 68%; b and d are SC = 76%), it can be observed that the hydration products of CSTB increase with the increase in SC. For example, SC increases from 68% to 76%, and C-S-H peak intensity, corresponding to 31.7° 2-theta, gradually increases. AFt corresponding to 18.8 ° 2-theta and CH corresponding to 34° 2-theta also follow the same law. The higher the relative intensity of the peak, the more the hydration product is produced, and the hydration product increases with the increase in SC.

By observing SEM images, it can be observed that when SC is 68%, the formation of needlelike AFt is not seen when its age is 3 days, and only a tiny amount of needlelike AFt and flocculent gelled product C-S-H can be observed when the age increases to 7 days. When SC increased to 76%, the number of hydration products that could be observed in 3 days was higher than that in 7 days when SC was 68%, which was the same as the conclusion obtained by XRD.

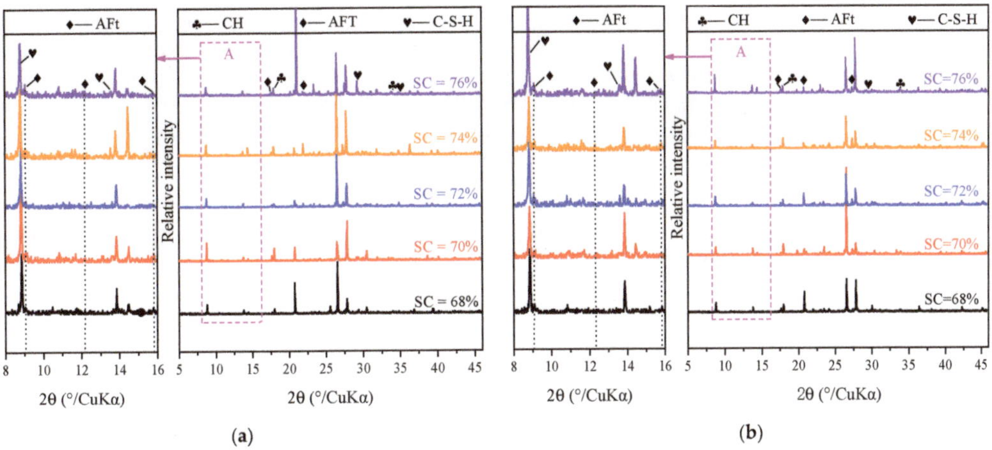

Figure 11. XRD patterns of CSTB in 3 d (**a**) and 7 d (**b**) under TCR = 4.

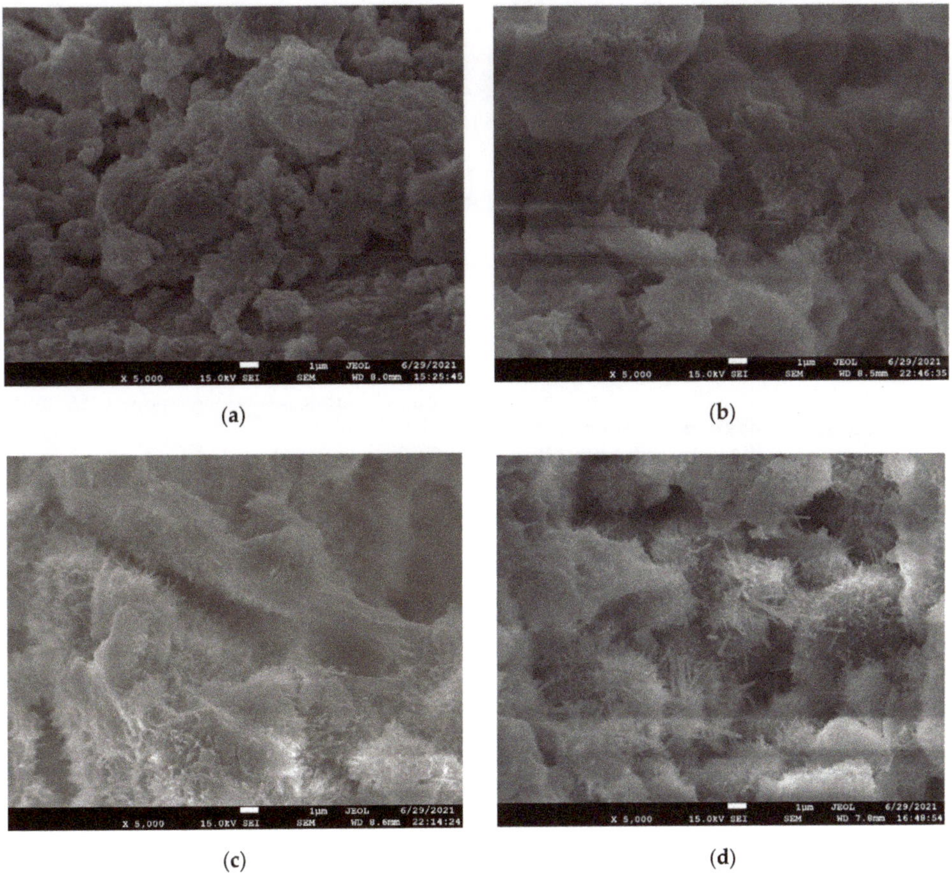

Figure 12. SEM patterns of CSTB in 3 d ((**a**) SC = 68%; (**b**) SC = 76%) and 7 d ((**c**) SC = 68%; (**d**) SC = 76%) under TCR = 4.

The above phenomenon can be attributed to tailings, cement, and water reaction after mixing with the increase in SC. The water–cement ratio decreases, and the relative concentration of cement increases. The rise in concentration is conducive to producing more hydration products. The XRD diffraction pattern's relative intensity values of C-S-H and CH are larger. These products will fill the pores of the particles in the backfill body, increase the compactness of the tailing's backfill body and reduce the pore spacing, so the CSTB strength with higher SC will also be higher.

3.3.3. Effect of Coupling Action of TCR and SC on Hydration Products and Microstructure of CSTB

TCR and SC do not affect the composition of hydration products of CSTB, and both will produce hydration products such as CH, C-S-H, and AFt that can provide strength for CSTB. The difference lies in the number of hydration products built. Hydration products increase with the decrease in TCR, and at the same time, they grow with the increase in SC. TCR and SC jointly determine the number of hydration products of CSTB.

At the same time, TCR and SC also impact the microstructure of CSTB, and the formation of hydration products can fill the internal pores of CSTB. TCR and SC can affect the appearance of hydration products and their structure. When SC is constant, the microstructure of CSTB with lower TCR is denser. When TCR is consistent, the microstructure of CSTB with higher SC is more compact.

4. Conclusions

In this paper, the influence of TCR and SC on the early mechanical strength, hydration characteristics, and microstructure of CSTB and the relationship between them are studied, and the following conclusions are reached:

The early intensity of CSTB increased with the decrease in TCR and the increase in SC. Still, the early power of CSTB had different sensitivity to TCR and SC and was more susceptible to TCR. The results show that the early strength of CSTB is exponentially related to TCR or SC, except that TCR = 4 at the curing age of 3 days.

With the increase in TCR, the peak values of the second and third hydration exothermic peaks gradually decreased and appeared earlier. The second exothermic peak disappeared when TCR = 12, the exothermic hydration rate slowly fell, the duration of the induction phase was prolonged, the acceleration phase was shortened, and the total heat release decreased.

The second and third hydration exothermic peaks increased with SC, and the occurrence time was delayed. The exothermic hydration rate gradually increased, but TCR did not significantly affect the difference. The total exothermic amount began to show a significant difference 18 h after the start of the test.

The main hydration products of CSTB are CH, C-S-H, and AFt. With the decrease in TCR, there are more hydration products, the corresponding XRD diffraction peak intensity is higher, the densification degree of the matrix in SEM image is enhanced, and the internal pores are reduced. The hydration products also increased with the increase in SC, which was consistent with the decrease in TCR.

The early strength, hydration heat, hydration products, and microstructure of CSTB are also affected by each other. The exothermic hydration rate of CSTB materials with low TCR and high SC is faster. The total amount of exothermic hydration increases, indicating that more hydration reactions are carried out and will generate more hydration products. These hydrates can better fill the gap between the CSTB particles, make the internal structure of CSTB more compact, and finally show higher CSTB strength.

Author Contributions: Software, C.Z.; Formal analysis, S.D.; Investigation, L.X.; Data curation, Y.L.; Writing – review & editing, S.D. and P.Y.; Funding acquisition, L.L.; Supervision, L.L.; Validation, L.L. All authors have read and agreed to the published version of the manuscript.

Funding: This research was supported by the National Natural Science Foundation of China: Nos. 51674188, 51874229, 51504182, 51974225, 51904224,51904225, 51704229.

Institutional Review Board Statement: Not applicable.

Informed Consent Statement: Informed consent was obtained from all subjects involved in the study.

Data Availability Statement: The data used to support the findings of this study are included in the article.

Conflicts of Interest: The authors declare no conflict of interest.

References

1. Sahoo, M.; Atkinson, J.T.N. Magnesium-lithium-alloys—Constitution and fabrication for use in batteries. *J. Mater. Sci.* **1982**, *17*, 3564–3574. [CrossRef]
2. Feng, K.; Li, D.; He, C. Progress in Superlight Mg-Li Alloys for Aerospace Industry. *Spec. Cast. Nonferrous Alloy.* **2017**, *37*, 140–144. [CrossRef]
3. Martin, G.; Rentsch, L.; Höck, M.; Bertau, M. Lithium Market Research—Global Supply, Future Demand and Price Development. *Energy Storage Mater.* **2016**, *6*, 171. [CrossRef]
4. China nonferrous metals news. The price of lithium carbonate exceeds 100,000 yuan/t SQM. The global demand for lithium is expected to grow by about 40% this year. *China Nonferrous Metall.* **2021**, *50*, 48.
5. Chen, J.; Xie, H.; Liu, Y. Research progress on mineral processing of spodumene. *J. Min. Metall.* **2022**, *31*, 3.
6. Vm, A.; Drb, B. The potential of lithium in alkali feldspars, quartz, and muscovite as a geochemical indicator in the exploration for lithium-rich granitic pegmatites: A case study from the spodumene-rich Moblan pegmatite, Quebec, Canada—ScienceDirect. *J. Geochem. Explor.* **2015**, *205*, 106336. [CrossRef]
7. Chen, D.; Hu, X.; Shi, L. Synthesis and characterization of zeolite X from lithium slag. *Appl. Clay Sci.* **2012**, *59–60*, 148–151. [CrossRef]
8. Burritt, R.; Christ, K. Water risk in mining: Analysis of the Samarco dam failure. *J. Clean. Prod.* **2018**, *178*, 196–205. [CrossRef]
9. Wang, X. *Experimental Study on Resource Utilization of Lithium Extraction Tailings*; China University of Geosciences: Beijing, China, 2021.
10. Yang, J.; Xu, L.; Wu, H. Preparation and properties of porous ceramics from spodumene flotation tailings by low-temperature sintering. *Trans. Nonferrous Met. Soc. China* **2021**, *31*, 2797–2811. [CrossRef]
11. Lemougna, P.; Yliniemi, J.; Adediran, A. Synthesis and characterization of porous ceramics from spodumene tailings and waste glass wool. *Ceram. Int.* **2021**, *47*, 33286–33297. [CrossRef]
12. Jie, Y.; Lxa, B.; Hwa, C. Microstructure and mechanical properties of metakaolin-based geopolymer composites containing high volume of spodumene tailings. *Appl. Clay Sci.* **2022**, *218*, 106412. [CrossRef]
13. Zanelli, C.; Conte, S.; Molinari, C. Waste recycling in ceramic tiles: A technological outlook. *Resour. Conserv. Recycl.* **2020**, *168*, 105289. [CrossRef]
14. Pnl, A.; Aa, A.; Jy, A. Thermal stability of one-part metakaolin geopolymer composites containing high volume of spodumene tailings and glass wool. *Cem. Concr. Comp.* **2020**, *114*, 103792. [CrossRef]
15. Wu, H.; Yang, J.; Xue, K. Effects of spodumene flotation tailings as aggregates on mechanical properties of cement mortar. *Colloid Surf. A* **2022**, *640*, 128346. [CrossRef]
16. Zhu, M.; Liu, L.; Wang, S. Short- and long-walls backbackfill pillarless coal mining method. *J. Min. Saf. Eng.* **2022**. [CrossRef]
17. Zhu, M.; Cheng, J.; Zhang, Z. Quality control of microseismic P-phase arrival picks in coal mine based on machine learning. *Comput. Geosci-UK* **2021**, *156*, 104862. [CrossRef]
18. Fridjonsson, E.; Hasan, A.; Fourie, A.; Johns, M. Pore structure in a gold mine cemented paste backfill. *Miner Eng.* **2013**, *53*, 144–151. [CrossRef]
19. Behera, S.; Mishra, D.P.; Ghosh, C.N. Characterization of lead-zinc mill tailings, fly ash and their mixtures for paste backfill in underground metalliferous mines. *Environ. Geol.* **2019**, *78*, 394.1–394.13.
20. Qiu, J.; Xiang, J.; Zhang, W. Effect of microbial-cemented on mechanical properties of iron tailings backfill and its mechanism analysis. *Constr. Build. Mater.* **2022**, *318*, 126001. [CrossRef]
21. Li, Y. Investigation on the Early Hydration and Microstructure of Cement-Based Materials. *J. Wuhan Univ. Technol.* **2012**.
22. Zhu, M.; Xie, G.; Liu, L.; Yang, P.; Qu, H.; Zhang, C. Influence of Mechanical Grinding on Particle Characteristics of Coal Gasification Slag. *Materials* **2022**, *15*, 6033. [CrossRef] [PubMed]
23. Shao, X.; Tian, C.; Li, C.; Fang, Z.; Zhao, B.; Xu, B.; Ning, J.; Li, L.; Tang, R. The Experimental Investigation on Mechanics and Damage Characteristics of the Aeolian Sand Paste-like Backfill Materials Based on Acoustic Emission. *Materials* **2022**, *15*, 7235. [CrossRef] [PubMed]
24. Yin, B.; Kang, T.; Kang, J.; Chen, Y.; Wu, L.; Du, M. Investigation of the Hydration Kinetics and Microstructure Formation Mechanism of Fresh Fly Ash Cemented Filling Materials Based on Hydration Heat and Volume Resistivity Characteristics. *Appl. Clay Sci.* **2018**, *166*, 146–158. [CrossRef]
25. Zhang, J.; Ke, G.; Liu, Y. Early Hydration Heat of Calcium Sulfoaluminate Cement with Influences of Supplementary Cementitious Materials and Water to Binder Ratio. *Materials* **2021**, *14*, 642. [CrossRef]
26. Han, F.; Li, L.; Song, S.; Liu, J. Early-Age Hydration Characteristics of Composite Binder Containing Iron Tailing Powder. *Powder Technol.* **2017**, *315*, 322–331. [CrossRef]

27. *ASTM C1702-14*; Standard Test Method for Measurement of Heat of Hydration of Hydraulic Cementitious Materials Using Isothermal Conduction Calorimetry. ASTM International: West Conshohocken, PA, USA, 2017.
28. Dong, Q.; Liang, B.; Jiang, L.; Jia, L.; Wang, K. Effects on short-term strength of cemented paste backfill under different sulfate contents and cement-tailings ratios. *J. Environ. Sci.-China* **2016**, *10*, 5886–5892.
29. Li, W.; Fall, M. Sulphate effect on the early age strength and self-desiccation of cemented paste backfill. *Constr. Build. Mater.* **2016**, *106*, 296–304. [CrossRef]
30. Wang, P.; Feng, S.; Liu, X. Research approaches of cement hydration degree and their development. *J. Build. Mater.* **2005**, *8*, 646.
31. Scrivener, K.; Juilland, P.; Monteiro, P.J.M. Advances in understanding hydration of Portland cement. *Cem. Concr. Res.* **2015**, *78*, 38–56. [CrossRef]
32. Shuai, C.; Erol, Y.; Song, W. Evaluation of Viscosity, Strength and Microstructural Properties of Cemented Tailings Backfill. *Minerals* **2018**, *8*, 352. [CrossRef]
33. Zhang, Q.; Wang, Z.; Rong, S. Early strength characteristics and microcosmic in fluence mechanism analysis of cemented backfill body of full tailings in deep mine. *Nonferrous Met. Eng.* **2019**, *9*, 97.
34. Wang, X.; Hu, Y.; Wang, S. Orthogonal Test of Optimization of Ultrafine Whole-tailings Backfill Materials. *Gold Sci. Technol.* **2015**, *23*, 45.
35. Zhao, L.; Wang, N. Study on influencing factors of strength of cemented tailings backfill. *Min. Technol.* **2021**, *21*, 91. [CrossRef]
36. Yang, F.; Nie, Y.; Wang, X. Experimental study on early strength characteristics of whole tailings cemented backfill. *Min. Res. Dev.* **2018**, *38*, 58. [CrossRef]
37. Rao, G. Generalization of Abrams' Law for Cement Mortars. *Cem. Concr. Res.* **2001**, *31*, 495–502. [CrossRef]
38. Currie, D.; Sinha, B. Survey of Scottish Sands and Their Characteristics Which Affect Mortar Strength. 1981. Available online: https://xueshu.baidu.com/usercenter/paper/show?paperid=c118266079b9b823b665cc9b9ca0fd71&site=xueshu_se&hitarticle=1 (accessed on 5 December 2022).
39. Nagaraj, T.; Banu, Z. Generalization of Abrams' law. *Cem. Concr. Res.* **1996**, *26*, 933–942. [CrossRef]
40. Liu, S.; Sun, Y.; Li, G. Analysis of early strength performance of full tailings backfill in a gold mine. *Min. Res. Dev.* **2020**, *40*, 125. [CrossRef]
41. Yang, L.; Qiu, J.; Fan, K. Analysis on strength characteristics of super-fine unclassified tailings cemented backfills. *Bull. Chin. Ceram. Soc.* **2017**, *36*, 249. [CrossRef]
42. Wang, L.; Yan, G.; Fu, J. Study on Influencing factors and proportion optimization of full tailings backfill material in a gold mine. *Min. Res. Dev.* **2022**, *42*, 11. [CrossRef]
43. Fu, J.; Du, C.; Song, W. Strength sensitivity and failure mechanism of full tailings cemented backfills. *J. Univ. Sci. Technol. B* **2014**, *36*, 1149. [CrossRef]
44. Zhang, Z. *Study on the Hydration Kinetics Model of Cement-Slag Composite Cementitious Materials*; Tsinghua University: Beijing, China, 2018.
45. Bullard, J.; Jennings, H.; Livingston, R.; Nonat, A.; Scherer, G.W.; Schweitzer, J.S.; Scrivener, K.L.; Thomas, J.J. Mechanisms of Cement Hydration. *Cem. Concr. Res.* **2011**, *41*, 1208–1223. [CrossRef]
46. Liu, L.; Yang, P.; Zhang, B.; Huan, C.; Guo, L.; Yang, Q.; Song, K. Study on Hydration Reaction and Structure Evolution of Cemented Paste Backfill in Early-Age Based on Resistivity and Hydration Heat. *Constr. Build. Mater.* **2021**, *272*, 121827. [CrossRef]
47. Shi, H.; Huang, X. Research on hydration heat of Portland cement and its progress. *Cement.* **2009**, *4*, 4–10.
48. Pratt, H. An Experimental Argument for the Existence of a Protective Membrane Surrounding Portland Cement during the Induction Period. *Cem. Concr. Res.* **1979**, *9*, 501–506. [CrossRef]
49. Damidot, D.; Bellmann, F.; Mser, B.; Sovoidnich, T. Calculation of the Dissolution Rate of Tricalcium Silicate in Several Electrolyte Compositions. *Cem. Wapno Beton.* **2007**, *74*, 57–67.
50. Ings, J.; Brown, P.; Frohnsdorff, G. Early Hydration of Large Single Crystals of Tricalcium Silicate. *Cem. Concr. Res.* **1983**, *13*, 843–848. [CrossRef]
51. Bullard, J.; Flatt, R. New Insights Into the Effect of Calcium Hydroxide Precipitation on the Kinetics of Tricalcium Silicate Hydration. *J. Am. Ceram. Soc.* **2010**, *93*, 1894–1903. [CrossRef]
52. Garrault, S. Study of C-S-H Growth on C3S Surface during Its Early Hydration. *Mater. Struct.* **2005**, *38*, 435–442. [CrossRef]
53. Gruyaert, E.; Robeyst, N.; Belie, N. Study of the Hydration of Portland Cement Blended with Blast-Furnace Slag by Calorimetry and Thermogravimetry. *J. Therm. Anal. Calorim.* **2010**, *102*, 941–951. [CrossRef]
54. Wu, X.; Roy, D.; Langton, C.A. Early Stage Hydration of Slag-Cement. *Cem. Concr. Res.* **1983**, *13*, 277–286. [CrossRef]
55. Bensted, J. Hydration of Portland Cement. *Adv. Cem. Technol.* **1983**, 307–347. [CrossRef]
56. Gallucci, E.; Mathur, P.; Scrivener, K. Microstructural development of early age hydration shells around cement grains. *Cem. Concr. Res.* **2010**, *40*, 4–13. [CrossRef]
57. Quennoz, A.; Scrivener, K.L. Interactions between alite and C3A-gypsum hydrations in model cements. *Cem. Concr. Res.* **2013**, *44*, 46–54. [CrossRef]
58. Study on the curing mechanism of cemented backfill materials prepared from sodium sulfate modified coal gasification slag. *J. Build. Eng.* **2022**, *62*, 105318. [CrossRef]
59. Frías, M.; de Rojas, M.; Cabrera, J. The Effect That the Pozzolanic Reaction of Metakaolin Has on the Heat Evolution in Metakaolin-Cement Mortars. *Cem. Concr. Res.* **2000**, *30*, 209–216. [CrossRef]

60. Thomas, J. A New Approach to Modeling the Nucleation and Growth Kinetics of Tricalcium Silicate Hydration. *J. Am. Ceram. Soc.* **2007**, *90*, 3282–3288. [CrossRef]
61. Bishnoi, S.; Scrivener, K. Ic: A New Platform for Modelling the Hydration of Cements. *Cem. Concr. Res.* **2009**, *39*, 266–274. [CrossRef]
62. Thomas, J.; Jennings, H.; Chen, J. Influence of Nucleation Seeding on the Hydration Mechanisms of Tricalcium Silicate and Cement. *J. Phys. Chem. C* **2009**, *113*, 4327–4334. [CrossRef]
63. Li, Z.; Lu, D.; Gao, X. Analysis of correlation between hydration heat release and compressive strength for blended cement pastes. *Constr. Build. Mater.* **2020**, *260*, 120436. [CrossRef]
64. Baran, T.; Pichniarczyk, P. Correlation factor between heat of hydration and compressive strength of common cement. *Constr. Build. Mater.* **2017**, *150*, 321–332. [CrossRef]
65. Li, G.; Sheng, H.; Song, P.; Zhu, J. Study and Optimization on Rheological Properties of High-concentration Tailings Slurry with Different Gradations in a Mine. *Min. Res. Dev.* **2021**, *41*, 55–59. [CrossRef]
66. Huang, Z.; Yuan, Y.; Chen, Z.; Wen, Z. Microstructure of Autoclaved Aerated Concrete Hydration Products in Different Water-to-Binder Ratio and Different Autoclaved System. *Adv. Mater. Res.* **2013**, *602–604*, 1004–1009. [CrossRef]
67. Piasta, W.; Zarzycki, B. The Effect of Cement Paste Volume and w/c Ratio on Shrinkage Strain, Water Absorption and Compressive Strength of High Performance Concrete. *Constr. Build. Mater.* **2017**, *140*, 395–402. [CrossRef]
68. Bentz, D.; Peltz, M.; Winpigler, J. Early-Age Properties of Cement-Based Materials. II: Influence of Water-to-Cement Ratio. *J. Mater. Civil. Eng.* **2009**, *21*, 512–517. [CrossRef]
69. Bentz, D.; Aietcin, P. The Hidden Meaning of Water- Cement Ratio. *Concr. Int.* **2008**, *30*, 51–54.
70. Tian, X.; Fall, M. Non-Isothermal Evolution of Mechanical Properties, Pore Structure and Self-Desiccation of Cemented Paste Backfill. *Constr. Build. Mater.* **2021**, *297*, 123657. [CrossRef]
71. Liu, L.; Yang, P.; Qi, C.; Zhang, B.; Guo, L.; Song, K. An Experimental Study on the Early-Age Hydration Kinetics of Cemented Paste Backfill. *Constr. Build. Mater.* **2019**, *212*, 283–294. [CrossRef]
72. Li, Z.; Zhang, Y.; Zhao, H.; Chen, H.; He, R. Structure Characteristics and Composition of Hydration Products of Coal Gasification Slag Mixed Cement and Lime. *Constr. Build. Mater.* **2019**, *213*, 265–274. [CrossRef]
73. Xu, W.; Tian, M.; Li, Q. Time-Dependent Rheological Properties and Mechanical Performance of Fresh Cemented Tailings Backfill Containing Flocculants. *Miner. Eng.* **2020**, *145*, 106064. [CrossRef]
74. Wang, H.; Qiao, L. Coupled Effect of Cement-to-Tailings Ratio and Solid Content on the Early Age Strength of Cemented Coarse Tailings Backfill. *Geotech. Geol. Eng.* **2019**, *37*, 2425–2435. [CrossRef]

materials

MDPI

Article

Engineering, Mechanical and Dynamic Properties of Basalt Fiber Reinforced Concrete

Han Wu, Xia Qin, Xu Huang and Sakdirat Kaewunruen *

Department of Civil Engineering, School of Engineering, University of Birmingham, Edgbaston B15 2TT, UK
* Correspondence: s.kaewunruen@bham.ac.uk

Abstract: This study investigates the engineering and mechanical properties of basalt fiber-reinforced (FRF) concrete, giving special attention to residual flexural strength and dynamic modal parameters. These properties, which have not been thoroughly investigated elsewhere, are a precursor to structural design applications for dynamic compliant structures (i.e., bridges, offshore platforms, railways, and airport pavement). Accordingly, the standard notched flexural tests have been carried out to assess the basalt fiber-reinforced concrete's residual flexural strength with an additional 0.125%, 0.25%, 0.375%, and 0.5% of volume fraction of basalt fiber. In addition, dynamic modal tests were then conducted to determine the dynamic modulus of elasticity (MOE) and damping of the FRF concrete beams. The results indicate that concrete's toughness and crack resistance performance are significantly improved with added fiber in basalt fiber reinforced concrete, and the optimum fiber content is 0.25%. It also exhibits the highest increment of compressive strength of 4.48% and a dynamic MOE of 13.83%. New insights reveal that although the residual flexural performance gradually improved with the addition of basalt fiber, the damping ratio had an insignificant change.

Keywords: basalt fiber reinforced concrete; residual flexural strength; modal test; damping ratio; dynamic modulus of elasticity

Citation: Wu, H.; Qin, X.; Huang, X.; Kaewunruen, S. Engineering, Mechanical and Dynamic Properties of Basalt Fiber Reinforced Concrete. *Materials* **2023**, *16*, 623. https://doi.org/10.3390/ma16020623

Academic Editors: Angelo Marcello Tarantino, Hrvoje Smoljanović, Ivan Balić and Nikolina Zivaljic

Received: 1 December 2022
Revised: 30 December 2022
Accepted: 4 January 2023
Published: 9 January 2023

1. Introduction

Fiber-reinforced concrete (FRC) has gained momentum in industrial applications in recent decades, especially for bespoke products and components using a low-to-moderate strength of concrete. With the pioneering publications of steel fiber reinforced in concrete by Romualdi and Batson [1] and Romualdi and Mandel [2], the world has entered a new era of fiber reinforced concrete. The research in fiber reinforced concrete (FRC) has a profound history.

Generally, the improvement of fiber reinforced concrete from ordinary concrete includes flexural strength, spitting strength, compressive strength, elastic modulus, crack resistance, self-healing ability, abrasion resistance, shrinkage and expansion resistance, and thermal characteristics. Meanwhile, with the development of high-performance materials technology, there are more types of fiber, such as glass polypropylene and carbon fiber, that have been applied in the research field of fiber reinforced concrete. The advanced fiber not only contributes to the improvement of mechanical properties, but is more environmental-friendly and even more inexpensive than the traditional steel fiber [2–14].

Recently, basalt fiber, as one of advanced fibers applied in the field of fiber reinforced concrete, has drawn attention from concrete engineers due to its excellent work in the improvement of mechanical properties and its contribution to sustainable development [15]. However, most previous studies mainly focused on the mechanical properties of basalt fiber reinforced concrete, and the dynamic properties of concrete have not been identified anywhere.

1.1. The Mechanical Properties of Basalt Fiber Reinforced Concrete

Basalt is known as a natural, solid, dense, and dark brown to black volcanic igneous rock formed about 100 to 300 km under the Earth's surface. It can turn the surface into molten lava with a melting temperature ranging from 1500 to 1700 °C [15,16].

The condition of basalt is significantly influenced by the temperature of the quenching process, which results in highly complete crystallization. As the two significant minerals, plagiocene and pyroxene account for over 80% of basalts. By examining the chemical makeup, Fiore V et al. [17] made it feasible to see that SiO_2 and Al_2O_3 are the two primary components.

Basalt fiber is also regarded as a new highly-recommended reinforcement material in concrete compared to the relatively common types, including carbon, aramid, and glass fibers. As the quenching process is environmentally friendly with high security [17,18], such an innovative material has several advantages, such as high thermal and chemical resistance and excellent modulus [11,17,19,20]. Meanwhile, basalt fiber shows a higher tensile strength than E-glass fibers. According to Branston J. et al. [21], it is many times stronger than steel fibers and possesses a few comprehensive qualities, including acid and alkaline resistance [22–24]. Therefore, the mechanical properties of basalt fiber reinforced concrete have a comprehensive improvement in flexural strength, spitting strength, modulus of rupture, fracture energy, and abrasion resistance [4,21,25–30]. Among the measurements, compression, tension, flexural, and splitting strength have become the main factors in understanding and accessing the fundamental engineering and mechanical properties. After a critical review, the most widely used basalt fiber has a length of 10 to 65 mm with a diameter of 13–20 μm. In addition, the density of basalt is about 1.3 to 2.75 g/cm^3, the tensile strength is between 2600 to 4840 MPa, and the range of the elastic modulus is 80–115 GPa [31].

Although basalt fiber can be relatively more expensive than some other fibers, it is cheaper than carbon fiber and S-glass fiber. Meanwhile, basalt fiber is more stable in a radioactive environment than synthetic and organic fibers. Since the appearance of related research on the composition of basalt fibers in the 1980s, exhaustive studies of the features and durability of concrete reinforced with basalt fiber have been conducted in the past five years which shows the material's comparatively good resistance to salt, water, corrosion, and severe degradation in an alkaline environment [32,33].

Moreover, it is globally accepted that chopped basalt fibers can be added to concrete to offset the defects. Therefore, in the past two decades, blending fibers into concrete has been widely applied in some projects [34–38], such as dams, airport sidewalks, and railway sleepers.

1.2. The Mechanical Properties of Basalt Fiber Reinforced Concrete

The dosage of basalt fiber and aspect ratio could deeply influence the mechanical behavior of basalt fiber reinforced concrete [39]. Therefore, many researchers have investigated the mechanical properties of basalt fiber reinforced concrete under static conditions.

By studying the impact of the length and volume fraction of basalt fiber on the compressive and flexural strength test under three different lengths and three different dosages, Iyer et al. [40] have further claimed that the optimal length, diameter, and volume fraction under compression and moment of resistance are 36 mm, 16 μm, and 18 kg/m^3, respectively. According to the pervious study, the performance improvement with the addition of basalt in normal strength concrete or high strength concrete are similar with each other [41]. This shows that the optimal dosage of basalt fiber is 2% with the addition of metakaolin, ranging from 1% to 3% for the compression strength, modulus of elasticity, and split tensile strength.

Following the examination by Highet al. [42] of the usage of dry and procured discrete basalt fiber under the compression and modulus of rupture, Girgin and Yıldırım [43] have also studied the replacement of glass fibers with basalt fibers, showing that basalt fibers are more operational compared to glass fibers under flexural performance. In addition, Hannawi et al.'s [44] examination of the microstructure and mechanical properties of ultrahigh performance concrete with different types of fiber up to 1% shows that basalt fibers can increase the ductility of the specimen, even with the lateral strain, which is the result of its greater aspect ratio than steel fiber, PP-PE fiber, PVA fiber and Barchip fiber.

Furthermore, according to Algin's [45] study on fresh and hardened concrete with chopped basalt fiber with a length of 12 mm at 0.2%, and 0.4%. 0.6%, and 0.8% of the total volume of concrete mix, the optimization analysis shows that a dosage of 0.356% with a w/c ratio of 0.47 produces the best results.

Table 1 summarises different studies on the impact of dosage on the flexural strength of basalt fiber reinforced concrete in the past decade.

Table 1. Research status of the optimal dosage of basalt fiber for flexural performance.

Scholar	Test Methodology	Flexural Properties	Dosage (%)	V_{opt} (%)	Increment (%)
Kabay [3]	Three-Point Bending Test on Notched Beams	Flexural Stress	0.07, 0.14	0.14	13
Jiang [19]	Australian Standard, As1012.11-1985	Strength-Effectiveness and Flexural stress	0.05, 0.1, 0.3	0.3	25.51
Jun [46]	Three-Point Bending Test	Flexural Stress	0.1, 0.15, 0.2, 0.25, 0.3, 0.35	0.3	12.3
Arslan [25]	Three-Point Bending Test on Notched Beams	Flexural Stress	0.021, 0.042, 0.084, 0.126	0.084	25.37
Branston [21]	Astm C1609	Residual Flexural Stress	0.15, 0.31, 0.46	0.46	180
Jalasutram [27]	Astm C1609	Residual Flexural Stress	0.5, 0.1, 0.15, 0.2	0.2	80
Katkhuda [47]	Three-Point Bending Test	Flexural Stress	0.1, 0.3, 0.5, 1, 1.5	1.5	74
Zhou [28]	Three-Point Bending Test	Flexural Stress	0.1, 0.2, 0.3, 0.4, 0.5, 0.6	0.3~0.4	40

1.3. Dynamic Properties of Concrete Properties

Limited studies have been conducted on the dynamic properties of basalt fiber reinforced concrete under dynamic loading conditions such as blast loading, impact response, damping ratio, dynamic modulus of elasticity, and dynamic compression characteristics.

Several studies have recognised moisture content as a primary element of damping capacity. For example, according to Jordan [48], the moisture content in the pore structure will influence the addition of a viscous damping component. Meanwhile, the effect of ageing has also been investigated. In fact, many studies [49–51] have suggested that the damping property of concrete decreases as the concrete ages, proving that the interfacial relations of the microstructure within the concrete matrix are the primary property of damping, unlike the compressive strength [48] property.

Meanwhile, the effect of frequency [52] on damping properties has been investigated. In fact, changes existed in the damping properties of the concrete beam under the three main vibrational modes: longitudinal, flexural, and torsional. As a brittle material with various levels of micro-cracks during dynamic bending, the cracks of concrete may cause the matrix to rub on the fiber surface, resulting in energy loss during vibration and a larger damping ratio for fiber reinforced concrete [53,54].

Although studies on the impact of adding steel fiber on the dynamic properties of fiber reinforced concrete show that the damping property of wet-cured fiber reinforced concrete is over 50% greater than plain concrete, the dynamic modulus of elasticity stays relatively constant. Furthermore, ageing and drying can significantly decrease the logarithmic decrement on both fiber reinforced concrete and plain concrete [55]. In addition, as shown in related research [56] on the relation between the damping ratio and other concrete matrix properties the damping ratio may be leveraged as an indication of changing concrete mechanical properties, such as flexural strength, elastic modulus, and ductility. These damping tests are usually advantageous because of their general non-destructive characteristics, meaning that the structure, element, or specimen is unharmed afterwards.

According to Zhang et al. [57],the impact of basalt fiber at various strain rates on the mechanical property of basalt fiber reinforced concrete under a dynamic load, the volume fraction of basalt fiber can vary from 1.325 to 6.625 kg/m^3. Meanwhile, the concrete specimen is also found to behave in a notably ductile manner due to the increased toughness with the strain rate. In addition, basalt fiber has been tested to optimize the micropore, which is also suitable for high-grade concrete.

After exploring the dynamic properties of basalt fiber reinforced concrete with various fiber volume fractions (1~3%) under different temperatures, Ren et al. [58] have concluded that the behavior of concrete specimens changes weakly under the impact loading due to temperature changes.

Furthermore, according to the micro studies of Fu et al. [59], basalt fibers perform better than polypropylene fibers in stopping fractures due to their increased stiffness. As a result of the more vital relationships between the matrixes, there is no notable change in the surface of the fiber. Meanwhile, some studies, such as those of Fu et al. and Feng et al. [59,60] on basalt fiber reinforced concrete under blast loading show that basalt rebars performed better than steel rebars on many occasions, especially at higher reinforcing ratios. As basalt fiber is far more resistant to blast forces than steel rebars, it can resist explosion energy far better than steel rebars. In this way, the damage degree has been threshold spall at the reinforcement ratio of 1.8%, indicating that micro cracks have occurred followed by hollow sounds and considerable bulges in the concrete, scattering a few fragments on the floor.

The amplitude and frequency content of the excitation source [61] can greatly impact the vibrations in a structure generated by such dynamic loads. Hence, the dynamic properties of concrete materials play a vital role in reducing noise and vibration.

However, the dynamic properties of basalt fiber reinforced concrete have not been identified in the literature, and limited studies have been conducted on combining the residual flexural performance and dynamic properties of fiber reinforced concrete. Therefore, this unprecedented study will primarily explore the performance of basalt fiber-reinforced concrete on residual mechanical properties with the variation of fiber dosage and its dynamic modulus of elasticity test. Moreover, the micro-crack in the real construction work can hardly be identified and some advanced non-destructive testing will be expensive. However, the micro-cracks will be sensitive to vibrations, and the natural frequency is a key parameter to describe the vibrations in concrete. In that case, using the model test to get the natural frequency calculating the dynamic modulus of basalt fiber reinforced concrete in real construction, then comparing the results with the dynamic modulus proposed in this study, is beneficial for engineers to justify the location and number of micro cracks. In this context, this study is the world's first to identify the dynamic properties of basalt fiber reinforced concrete and the potential relationship between the dynamic and mechanical properties by a series of experimental studies. The new dynamic property tests of basalt fiber reinforced concrete are essential to further clarify the performance of basalt fiber reinforced concrete in shock absorption and sound insulation, and is beneficial to the development of new non-destructive testing.

2. Materials and Methods

2.1. Material

The concrete specimens in this experiment are all made with Portland cement, well-graded aggregates, water, and basalt fiber [62]. Portland cement (Normal Cem II A-L 42.5 N) [63] was adopted from O'Brien Cement Ltd. in Waterford, Ireland, up to European cement standard I.S. EN197-1:2011. The coarse and fine aggregate materials are sand, gravel, granite, gritstone, crushed rock with quartz, and limestone, all manufactured by Tarmac Trupak of Plumtree Farm Industrial Estate in Birmingham, UK.

The alkaline-resistant basalt fiber is adopted from Deutsche Basalt Faser GmbH (DBF), in which the basalt is chopped to the desired length under a high temperature. This type of fiber also known as integral fibers, such chopped alkaline-resistant fibers have been tested to show no damage when exposed to an alkaline environment at 40 °C and a pH value of 13.5 for 365 days according to the description of Institute of Construction Materials of the Technical University of Dresden. Furthermore, they also present better adhesion performance in concrete due to their special coating. In addition, as shown in Table 2, which shows the mechanical properties of basalt fiber, alkaline-resistant basalt fiber has many advantages compared to other types of fiber. For example, these fibers are not moisture-absorbent compared to mineral wool. Meanwhile, they are not toxic according to DIN 102. They are equipped with a high density at a low insulation thickness of 8 mm and low thermal conductivity. Most importantly, all these characteristics have made them more environmentally friendly than others.

Table 2. Parameter of basalt fiber in the study.

Basalt Type	Short-chopped Filament	Length [mm]	25
Single Diameter [μ]	13~17	Temperature Resistance [°C]	750
Density [kg/m^3]	2800	Thermal Conductivity [W/mK]	0.031
Areal Density [kg/m^2]	1	Tensile Strength [MPa]	4840
Width × Run Length [mm]	1 × 10	Elastic Modulus (GPa)	89
Absorption of Humidity (%)	≤0.1	Elongation (%)	3.15

2.2. Concrete Mixes

Five concrete mixes have been designed for this study with a water-to-cement (w/c) ratio of 0.55. For all the mixes, the proportions of each concrete mix are 1: 3.1: 2.25 by the mass of cement, fine aggregate, and coarse aggregate, with the details of each concrete mix shown in Table 3.

Table 3. Proportion of concrete mix design for 1 m$^{3.}$

Materials	Cement	Gravel	Sand	Water
Weight (kg/m^3)	317.3	713.925	983.63	174.515

The coarse aggregate has been prepared under full washing in case of sand's adhesion on the surface. In addition, both coarse and fine aggregates are dried at the temperature of 105 °C for 48 h.

As demonstrated in a review of previous literature, limited studies on the mechanical properties of basalt fiber-reinforced concrete have shown that the optimal dosage of the basalt fiber roughly ranges within 0.5% of the total concrete volume. Hence, four different dosages are used for each type of basalt fiber reinforcement, ranging from a low dosage to the maximum mixable dosage. In addition, the chopped basalt fiber is added at 0.125%, 0.25%, 0.375%, and 0.5% of content by the total volume of the concrete mix. Finally, the concrete mixes were labelled as M_0 (plain concrete), M_1 (0.125%), M_2 (0.25%), M_3 (0.375%), and M_4 (0.5%).

According to the standard test method for fundamental transverse, longitudinal, and torsional resonant frequencies of concrete specimens (ASTM C215-19), to test the hardened concrete, two sizes of concrete specimen have been prepared for compression testing (Cube: 150 mm × 150 mm × 150 mm). This study employs the modal test and the three-points load bending test (rectangular prism: 150 mm × 150 mm × 600 mm), while three beams and five cubes have been made for each mix to ensure enough sample size and avoid errors. Hence, each concrete specimen is labelled as M_XC/B_X. For example, M_1C_2 means that the concrete specimen is the second cube containing 0.125% of basalt fiber reinforcement. Similarly, M_2B_3 shows that the concrete specimen is the third beam containing 0.5% of basalt fiber reinforcement.

2.3. Experimental Setup

Some relevant tests in the laboratory have been conducted, including the slump test (BS EN 12350-2:2019), the compression test (BS EN 12390-3:2019), the residual flexural strength test (BS EN 14651:2005), and the dynamic modal test (ASTM C215-19).

In these tests, the components have been placed in the remixing container in the following order: aggregates, cement, basalt fiber, and water. When the cone and base plate are dampened and the excessive moisture with the moist cloth is removed, the concrete matrix is filled to the slump retention at a specific time with three layers compacted with 25 strokes of the compacting rod, respectively. The cone is then raised for 2 to 5 s to avoid the lateral or torsional motion formed on the concrete. By measuring the heights of the cone and the highest point of the slumped test specimen, the slump h can be calculated and recorded, as shown in Figure 1.

Figure 1. Slump test of the M_3 concrete mix.

When the mixing of the concrete matrix is done, it is cast in a mold with the sizes of $150 \times 150 \times 150$ mm^3 (cube) and $150 \times 150 \times 600$ mm^3 (rectangular prism). Moreover, the 5-mm-wide notch is placed on one side of the cuboid mold for a three-point bending test of a single-side notched beam on the specimens. Finally, the specimens are demoulded after 48 h and cured under the water for at least 28 days.

The compressive strength of the cube specimens has then been tested under a universal testing machine from Norton Hydraulics (Midland) in Birmingham, UK at the rate of 0.5 MPa/s. In fact, the flexural performance of the beam specimen can be determined by the load-displacement curve obtained by simply testing the supported beam specimens and load-CMOD (Crack Mouth Opening Displacement) curve obtained by the Crack Opening Displacement (COD) gauges.

Furthermore, the resonant frequencies of the dynamic properties of specimens can be obtained from the detected driving force with a hammer, which is shown in Figure 2. Meanwhile, the amplifiers are adjusted in the driving and pickup circuits to provide a noticeable indication. As the driver is positioned, the driving force can be perpendicular to the specimen surfaces when the specimens are supported for free vibration in the transverse mode. In this way, the specimen will be forced to vibrate at different frequencies, while the different indications of the amplified pickup outputs are recorded.

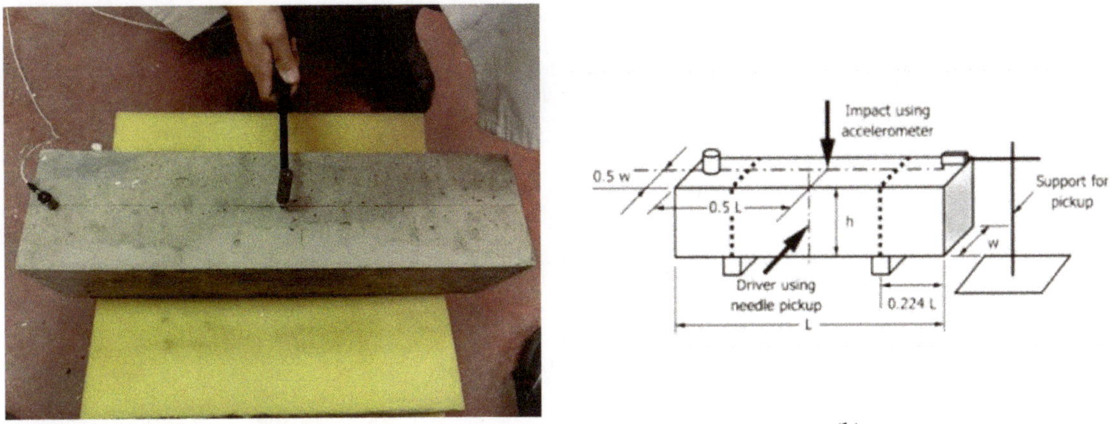

(a) (b)

Figure 2. Modal test of the concrete specimen (ASTM C215-19). (**a**) The experiment setting; (**b**) details of the equipment setting.

Lastly, the modal test is normally set ahead of the three-point loading test because it causes no damage to the specimen. In addition, the mass and dimensions of the specimen are measured to calculate the dynamic MOE (modulus of elasticity) from Equation (1) in ASTM C215-19.

$$\text{Dynamic } E = CMn^2 \tag{1}$$

M = mass of the specimen, kg,
n = fundamental transverse frequency, Hz,
C = 0.9464 (L^3T/dt^3), m^{-1} for a prism,
d = diameter of cylinder, m,
t = dimension of prism cross section in the direction the prism is driven or impacted, m,
b = dimension of prism cross section perpendicular to t, m,
T = correction factor that depends on the radio of the radius of gyration, (This experiment was scored as 1.39).

3. Results and Discussion

3.1. Fresh Concrete Workability

According to Table 4, which lists the slump of the basalt fiber-reinforced concrete, the slump has slightly dropped with the increasing dosage of basalt fiber up to 0.5%, which peaked at 43 mm in the plain concrete matrix M_0. However, when adding basalt fiber with the volume fraction of 0.125%, 0.25%, 0.375% and 0.5%, the slumps of the concrete matrix has fell to 40 mm, 33 mm, 26 mm, and 18 mm, respectively. Therefore, the addition of fibers in concrete can result in a reduction in the workability of the concrete. In this way, the dispersed basalt fiber in the concrete, which might form a net-shaped structure, makes this phenomenon possible because it has prevented the mixture's segregation and flow. According to the degree of workability suggested in the standard, the range of slump is 25–50 mm, which means that the workability of concrete is low and the slump between 0–25 mm is very low. It can be seen that the addition of basalt fiber led to a decrease in the workability of concrete, and the level of workability dropped to very low workability with a fiber content of 0.5%. Due to the fiber's high content and surface area, it may absorb more cement paste to wrap around the fiber itself, making the mixture more viscous while reducing the slump [64].

Table 4. The result of the compression and slump test.

Mix	Slump (mm)	Compressive Strength (MPa) 28 Days						
		1	2	3	4	5	* $f_c \pm$ SD (MPa)	Increment (%)
Plain	43	42.02	40.25	43.08	42.06	42.06	41.91 ± 1.03	42.06
0.125%	40	41.22	42.41	44.26	42.99	42.99	42.68 ± 1.10	42.99
0.25%	33	41.78	44.01	43.97	43.98	43.98	43.56 ± 1.10	43.98
0.375%	26	43.17	42.21	43.56	42.51	42.51	42.88 ± 0.53	42.51
0.5%	18	42.98	42.87	43.11	42.92	42.92	42.68 ± 0.66	42.92

* f_c is the compressive strength and SD is the standard deviation.

3.2. *Mechanical Property Test*

3.2.1. Compression Test

According to Table 4, which presents the results of the concrete cube specimen's compression after 28 days, the compressive strength of all the mixes does not show a conspicuous change as the increment varies in the small zone while forming a slight trend. Firstly, the results of five specimens from the same patch has shown stable properties as the standard deviation value ranges within 1.00. Secondly, the trend has grown to 43.684 MPa with a 0.25% basalt fiber addition, before going down flatly to 41.5 MPa with a 0.5% basalt fiber addition. In addition, when it comes to M_4, the value of the 0.5% volume fraction appears to be flat on the M_1 (0.125%). However, compared to plain concrete, certain improvements can still be achieved.

According to Figure 3, the plain concrete specimen seems to have developed more longitudinal cracks than specimens with added basalt fiber, because the cracks on the longitudinal cross-section of M_2 are shorter than plain concrete specimens after the compressive loading. In other words, the distribution of cracks is considered more dispersed with the increasing fiber in the concrete specimen. Figure 4 portrays the influence of fiber addition on the compressive strength of basalt fiber reinforced concrete.

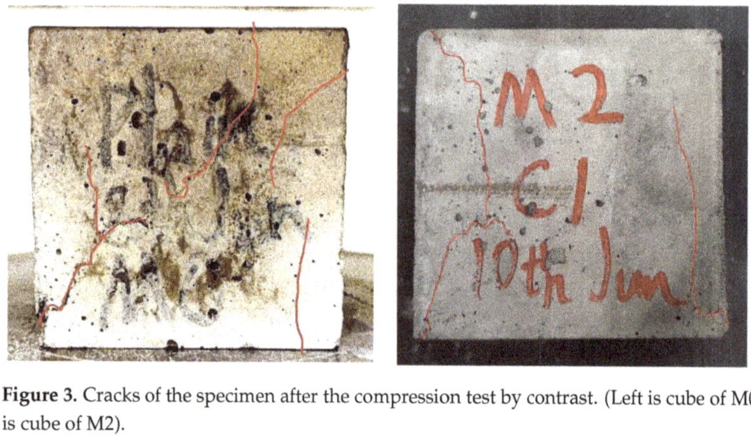

Figure 3. Cracks of the specimen after the compression test by contrast. (Left is cube of M0, and right is cube of M2).

Besides, a previous Branston study [22] shows that there might be a monotonical decrease with the increment of fiber. The applied basalt fiber is 36 mm and 50 mm. Compared to its irregular result, the shorter length of the basalt fiber seems to show more details to test the impact of the fiber dosage on the compressive performance of the basalt fiber-reinforced concrete. In addition, the fiber-bridging force existing in fiber-reinforced concrete can delay the macro crack propagation while restricting the development and opening of the macro cracks [28]. Moreover, such a behaviour also helps to prevent the concrete from lateral expansion by holding back the fracture formation. Therefore, although adding more basalt fibers to concrete cannot reinforce its peak compressive resistance, a gradual decrease in the load capacity and the enhancement of structural integrity can be seen. In this way, the long and chopped fibre can cause conjunction during the mixing, resulting in blockages and clumps in concrete.

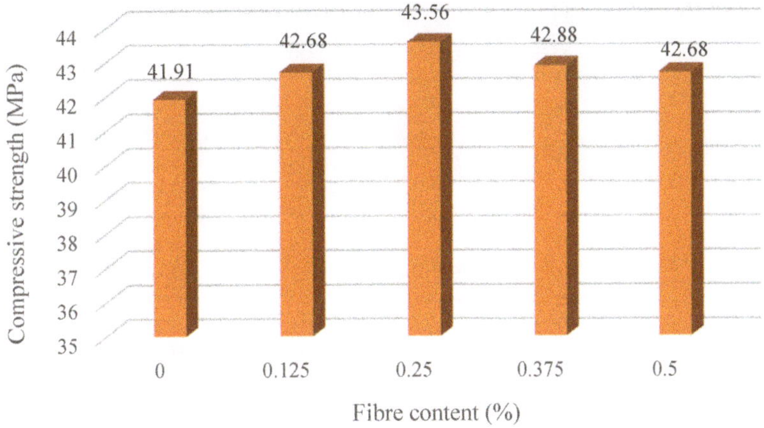

Figure 4. Compressive strength under different mixes.

3.2.2. Flexural Performance

Table 5 lists the experimental results of the basalt fiber-reinforced concrete beam specimens. The values of the proportionality (LOP) presented as $F_{fc,t}$, the span length shown as L, and the residual flexural tensile strengths referred to as $F_{R,1}$, $F_{R,2}$, $F_{R,3}$, and $F_{R,4}$ listed in Table 5 are the averages of the three beam specimens per test variable. Therefore, the residual flexural strength has been calculated with the following Equation (2), which is defined in the BS EN 14651: 12.

$$f_{R,j} = \frac{3P_jL}{2bh_{sp}^2} \tag{2}$$

L is the span length, mm;
P_j is the applied load, N;
b is the width of the specimen, mm;
h_{sp} is the distance between the tip of the notch and the top of the specimen, mm.

Where $f_{R,j}$ and P_j are the residual flexural tensile strength and the applied load, respectively, which correspond to CMOD_j. In this way, $F_{R,1}$, $F_{R,2}$, $F_{R,3}$, and $F_{R,4}$ are determined by the corresponding CMOD values of 0.5 mm, 1.5 mm, 2.5 mm, and 3.5 mm, respectively.

Table 5. The result of residual flexural strength.

Flexural Strength (MPa) 28 Days					
Mix	**Plain**	**0.125%**	**0.25%**	**0.375%**	**0.5%**
Peak Load (kN)	9.078	9.455	10.949	11.684	13.331
Increment (%)		4.15%	20.61%	28.70%	46.85%
Displacement (mm)	0.003	0.055	0.042	0.032	0.028
fct, L	3.486	3.631	4.204	4.487	5.119
FR, 1	0.127	0.865	0.923	0.956	1.132
FR, 2	0.061	0.114	0.238	0.319	0.356
FR, 3	0.000	0.063	0.141	0.166	0.188
FR, 4	0.000	0.000	0.103	0.104	0.113

The results of the three specimens of each mix were obtained under the there-point load bending test. Each median result of the three specimens has been taken as a reference for a more concise analysis on flexural performance considering the results of each mix varied in a small proportion, as shown in Table 5.

The continuous growth of flexural strength can be observed to exist with the increasing dosage of basalt fiber in the concrete specimens. Meanwhile, the LOP (f_{ct}, L) of the plain concrete specimen stays at 3.49 Mpa with the peak load of 9.078 kN. With the increasing volume fraction of basalt fiber from 0.125% to 0.5%, the increment amount ranges from approximately 4.15% to 46.85%. In addition, the displacement at the peak load is shorter and earlier than plain concrete when more fiber volume fractions are added. Therefore, the increment has shown a uniform growth curve, with M_4 (0.5%) having the peak load of 13.331 kN and the fastest speed to reach it.

From Figure 5, it can be seen that the residual flexural tensile strength is linearly proportional to the CMOD before it reaches LOP, generally forming a relative elastic zone. However, the curve is abruptly shifted once it reaches the LOP. Meanwhile, the peak load value shows a trend of increase with basalt fibers in concrete, indicating that basalt fiber-reinforced concrete can improve the initiation of tensile cracks. On the other hand, once the first peak or cracking strength occurs, the flexural tensile strength sees a sudden decrease. Therefore, the magnitude of the strength reinforcement is mainly dependent on the content ratio of the fibers.

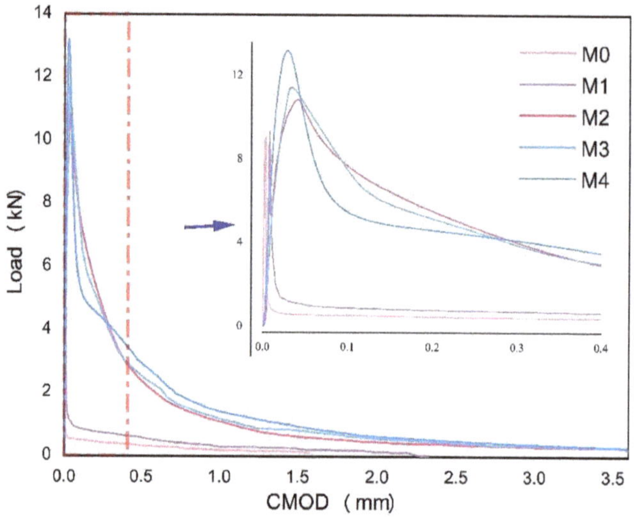

Figure 5. The Load-CMOD curve of the bending test (3.5 mm and 1.0 mm).

Comparatively, when the curves dropped after the peak, the curves of each mix can be divided into two groups regarding the residual flexural tensile performance, with M_0 and M_1 showing a more drastic drop when reaching their peaks. As shown in Figure 6, while the applied load values of M_0 and M_1 are 0.21 and 0.29 kN, respectively, those of M_2, M_3, and M_4 are 5.1, 7.0, and 7.4 kN, respectively. Moreover, the two groups have shown different slopes when the CMOD comes to 0.1 mm. The curves of mixes containing less than 0.25% volume fraction of basalt fiber remains at approximately 1 kN. On the other hand, a load of mixes with over 0.25% volume fraction of basalt fiber shows a sign of dropping at the relatively higher level but with a slower slope.

According to Figure 7, several cracks appeared on the plain concrete specimen when the load increased, causing it to suddenly break into two pieces on the machine. Hence, the load stopped at a CMOD of 1.7 mm. In this way, the M_1 specimens can usually last longer on the machine than the plain specimens. In addition, the M_1 is likely to be broken into two pieces when it is removed from the machine. When the volume fraction of basalt fiber is over 0.25%, the load-CMOD curve will have a smooth downward trend until it is over 3.5 mm of the test range. Moreover, the residual flexural tensile strengths of M_2, M_3, and M_4 are higher when more fiber is added. As shown in Table 5, residual flexural tensile strength will increase slightly between M_2 and M_3 with the trend being more conspicuous from M_3 to M_4. In those cases, the addition of fiber contribution to the increment of residual flexural strength from 0.125% to 0.5% of basalt fiber. Figure 7 shows that there are abundant ruptured basalt fibers on the split cross-section of the beam specimens. The concretes [3] perform the best with high ultimate loads, greater deflections, and higher fracture energy values at the post-peak zone as the basalt fiber content rises. Therefore, the bridging force of the fibers can significantly impact the property of plain concrete from brittleness to ductility.

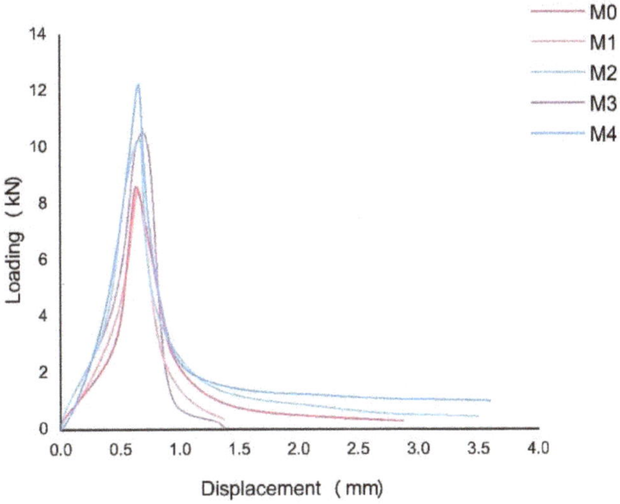

Figure 6. The load-displacement curve of the bending test.

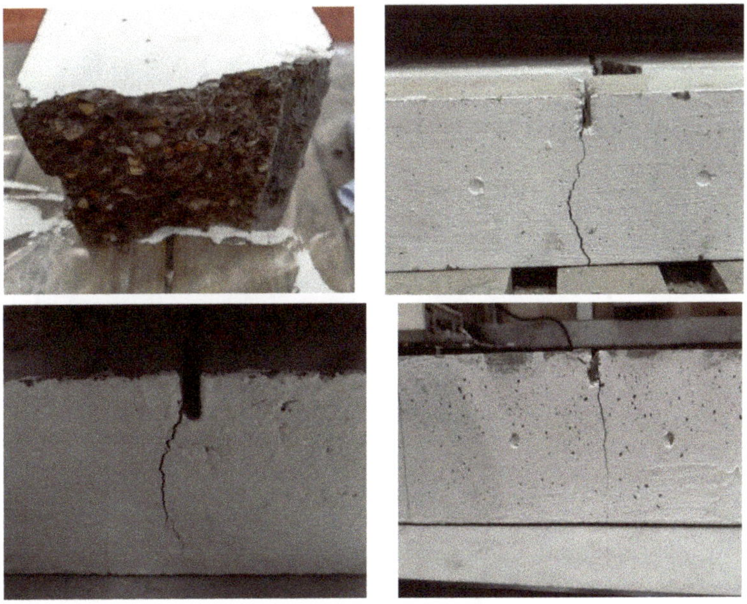

Figure 7. The post-crack of the concrete specimen after the bending test by contrast (left upper is M0, right upper is M1, left bottom is M2 and right bottom is M3).

3.3. *Modal Testing*

3.3.1. Damping Ratio

The dynamic tests have been carried out above a strong floor with very little noise and vibration in order to avoid the interference of external vibrations. The specimen is placed above two layers of sponge cushion with a total thickness of 10 cm, which aims to provide a free-free condition. To ensure the accuracy of the experiments, they have all been performed by one operator and used a similar excitation force. Ten taps are performed for each experiment, and the FRF curve (see Figure 8) is calculated based on the software. According to Table 6, which shows the natural frequency and damping ratio in transverse mode, the standard deviation value of the damping ratio of each three beams varies from a small range, especially on M_2 and M_3, which have shown a high level of consistency. In fact, the fiber-reinforced specimen has a similar damping ratio with plain concrete, which is approximately 0.59%. Due to its variation, changes can be deduced with the addition of basalt fiber. Furthermore, the 0.5% volume fraction of basalt fiber has no obvious impact on the damping ratio of the basalt fiber-reinforced concrete.

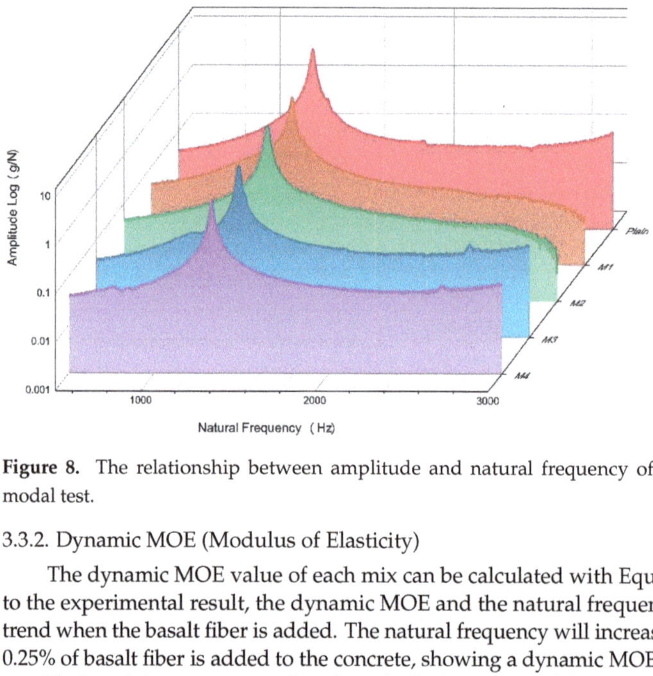

Figure 8. The relationship between amplitude and natural frequency of each mix during the modal test.

3.3.2. Dynamic MOE (Modulus of Elasticity)

The dynamic MOE value of each mix can be calculated with Equation (1). According to the experimental result, the dynamic MOE and the natural frequency present a similar trend when the basalt fiber is added. The natural frequency will increase to 1345.6 Hz when 0.25% of basalt fiber is added to the concrete, showing a dynamic MOE of 31.44 GPa. When considering plain concrete as a benchmark, the increment of dynamic MOE is 13.83% at the peak (0.25%). The dynamic MOE then sees a continuous decline with a slower slope than the rising slope. In comparison, when adding 0.5% of basalt fiber, the dynamic MOE is 29.86 GPa, which is lower than the M2 (0.25%) value. However, compared to plain concrete, it still has an 8.1% increment.

From the results of the modal test, as shown in Table 7, and the compressive test, as presented in Table 4 and Figure 4, the compressive strength and dynamic MOE have shown a similar trend when the volume fraction of basalt fiber is increased continuously, with M_2 (0.25%) showing the highest compression and dynamic MOE value. In addition, according to the load-displacement curve of the residual flexural tensile strength, M_2 (0.25%) is the mix that has reached its peak load, which, in another method, can represent the relationship between the strength and strain of the concrete specimens.

Table 6. The result of the modal test.

Concrete	Mix	Natural Frequency (Hz)			Damping Ratio		
		Sample Value	Average	SD	Sample Value	Average	SD
Plain	M_0B_1	1267.2	1265.3	1.36	0.50%	0.56%	0.052
	M_0B_2	1264.1			0.55%		
	M_0B_3	1264.6			0.62%		
0.125%	M_1B_1	1301.9	1298.9	2.10	0.64%	0.60%	0.033
	M_1B_2	1297.4			0.57%		
	M_1B_3	1297.5			0.57%		
0.25%	M_2B_1	1347.4	1345.6	1.34	0.60%	0.60%	0.003
	M_2B_2	1345.2			0.59%		
	M_2B_3	1344.2			0.60%		
0.375%	M_3B_1	1327.9	1323.967	2.79	0.58%	0.59%	0.008
	M_3B_2	1321.8			0.60%		
	M_3B_3	1322.2			0.59%		
0.5%	M_4B_1	1321.2	1322.167	0.97	0.61%	0.58%	0.021
	M_4B_2	1321.8			0.56%		
	M_4B_3	1323.5			0.57%		

Table 7. Dynamic MOE of each mix in the modal test.

	Density of Beams (kg/m^3)	Natural Frequency [Hz ± SD]	Damping Ratio [% ± SD]	Dynamic—MOE [GPa ± SD]
Plain	2277.04	1265.3 ± 1.36	0.56 ± 0.052	27.62 ± 0.06
0.125%	2264.44	1298.8 ± 2.10	0.60 ± 0.033	29.06 ± 0.09
0.25%	2275.56	1345.6 ± 1.34	0.60 ± 0.003	31.44 ± 0.06
0.375%	2282.22	1324.0 ± 2.79	0.59 ± 0.008	30.05 ± 0.13
0.5%	2276.30	1322.2 ± 0.97	0.58 ± 0.0021	29.86 ± 0.04

From the result of the modal test (Table 7) and compressive test (Table 4), the compressive strength and dynamic MOE showed a similar trend when continuously adding the volume fraction of basalt fiber. M_2 (0.25%) has the highest compression and dynamic MOE value. In addition, the load-displacement curve of residual flexural tensile strength also showed that M_2 (0.25%) is the fastest mix to reach its peak load, which, in another method, could display the relationship between the stress and strain of the concrete specimens.

Research [65] shows that the dynamic modulus of elasticity could be correlated to the compressive strength of high-strength self-compacting concrete. Lu et al. [64] stated that both dynamic MOE and static MOE are relevant to compressive resistance. An empirical equation was concluded for estimation purposes. Hence, concrete with higher stiffness/MOE normally has a higher compressive performance.

Furthermore, a previous study [65] also shows that the dynamic modulus of elasticity can be correlated to compressive high-strength self-compacting concrete. For example, according to Lu et al. [66] and Li et al. [67–70], both dynamic and static MOE are relevant to compressive resistance. Therefore, an empirical equation can be settled for estimating, showing that concrete with higher stiffness/MOE usually has a higher compressive performance.

4. Conclusions

Based on the results from a series of experiments, conclusions can be drawn to develop new insights into basalt fiber reinforced concrete for industrial applications.

The workability of fresh concrete elicits signs of decline with the increasing dosage of basalt fiber, with the slump decreasing from 44 mm to 18 mm from plain concrete to concrete, with a volume fraction of 0.5% for basalt fiber. Among concrete with volume fractions of 0.125%, 0.25%, 0.5%, 0.375%, and 0.5% of basalt fiber in plain concrete, the compression of concrete with 0.25% of basalt fiber dosage has the highest performance (43.98 MPa), which is 4.48% higher than that of plain concrete. Meanwhile, the flexural strength (LOP) has greatly improved, showing a monotonical trend of increase when the volume fraction of basalt fiber is increased from 0.125% to 0.5%. The peak flexural strength is 4.15%, 20.61%, 28.70%, and 46.85%, respectively. Furthermore, the residual flexural tensile strength shows a smooth descending CMOD-load curve after the peak when the basalt fiber content is above 0.25%. In addition, the damping ratio remains around 0.60, which ranges from plain concrete to a volume fraction of 0.5% for basalt fiber. Given the results of the experimental study, these new conclusions can be drawn:

- The addition of basalt reduces the workability of concrete significantly
- The compressive strength of basalt fiber concrete does not increase with the addition of fiber, and the suggested fiber content is 0.25%
- The flexural strength increases sharply with the addition of basalt fiber when keeping the fiber content lower than 0.5%
- The MOE of basalt fiber concrete is affected by the compressive strength, and there is no significant impact on the damping ratio of concrete with the addition of basalt fiber.

This study is the first to demonstrate the essential parameters of mechanical and dynamic properties of basalt fiber concrete through experimental studies. Based on these new insights, engineers can evaluate the performance of their basalt fiber concrete structures, especially in aggressive environments. In addition, the dynamic moduli and damping coefficients reported in this study can be applied for dynamic structural designs and the non-destructive testing of basalt fiber reinforced concrete structures.

Author Contributions: S.K. developed the concept and investigation; X.Q. and H.W. did the data collection, and data analyzed; H.W., X.H. and X.H. contributed the manuscript; S.K., X.Q. and X.H. reviewed the paper. All authors wrote the paper. All authors have read and agreed to the published version of the manuscript.

Funding: The research was funded by the European Commission for the H2020-RISE Project No. 691135 and H2020-S2R Project No. 730849. The APC is sponsored by the MDPI's Invited Paper Initiative.

Institutional Review Board Statement: Not applicable.

Informed Consent Statement: Not applicable.

Data Availability Statement: Data can be made available upon reasonable request to the corresponding author.

Acknowledgments: The authors are sincerely grateful to European Commission for the financial sponsorship of the H2020-RISE Project No. 691135 "RISEN: Rail Infrastructure Systems Engineering Network," which enables a global research network that tackles the grand challenge in railway infrastructure resilience and advanced sensing in extreme environments. In addition, this project is partially supported by European Commission's Shift2Rail, H2020-S2R Project No. 730849 "S-Code: Switch and Crossing Optimal Design and Evaluation".

Conflicts of Interest: The authors declare that they have no conflicts of interest.

References

1. Romualdi, J.P.; Batson, G.B. Mechanics of Crack Arrest in Concrete. *J. Eng. Mech. Div.* **1963**, *89*, 147–168. [CrossRef]
2. Romualdi, J.P.; Mandel, J.A. Tensile Strength of Concrete Affected by Uniformly Distributed and Closely Spaced Short Lengths of Wire Reinforcement. *JP* **1964**, *61*, 657–672. [CrossRef]
3. Kabay, N. Abrasion Resistance and Fracture Energy of Concretes with Basalt Fiber. *Constr. Build. Mater.* **2014**, *50*, 95–101. [CrossRef]
4. Ramadevi, D.K. Experimental study on strength properties of concrete with different aspect ratios of basalt. *Int. J. Civ. Eng. Technol.* **2017**, *8*, 629–637.
5. Alberti, M.G.; Enfedaque, A.; Gálvez, J.C. Fracture Mechanics of Polyolefin Fibre Reinforced Concrete: Study of the Influence of the Concrete Properties, Casting Procedures, the Fibre Length and Specimen Size. *Eng. Fract. Mech.* **2016**, *154*, 225–244. [CrossRef]

6. Memon, I.A.; Jhatial, A.A.; Sohu, S.; Lakhiar, M.T.; Khaskheli, Z.H. Influence of Fibre Length on the Behaviour of Polypropylene Fibre Reinforced Cement Concrete. *Civ. Eng. J.* **2018**, *4*, 2124–2131. [CrossRef]
7. Hu, X.; Guo, Y.; Lv, J.; Mao, J. The Mechanical Properties and Chloride Resistance of Concrete Reinforced with Hybrid Polypropylene and Basalt Fibres. *Materials* **2019**, *12*, 2371. [CrossRef]
8. de Sousa Menezes, M.; Queiroz, E.C.; Soares, P.V.; Faria-e-Silva, A.L.; Soares, C.J.; Martins, L.R.M. Fiber Post Etching with Hydrogen Peroxide: Effect of Concentration and Application Time. *J. Endod.* **2011**, *37*, 398–402. [CrossRef]
9. Gondokusumo, G.S.; Venkateshwaran, A.; Tan, K.H.; Liew, J.Y.R. Unified Equations to Predict Residual Flexural Tensile Strength of Lightweight Steel Fiber-Reinforced Concrete. *Struct. Concr.* **2021**, *22*, 2202–2222. [CrossRef]
10. Borhan, T.M. Properties of Glass Concrete Reinforced with Short Basalt Fibre. *Mater. Des.* **2012**, *42*, 265–271. [CrossRef]
11. Yang, Y.; Wu, C.; Liu, Z.; Wang, H.; Ren, Q. Mechanical Anisotropy of Ultra-High Performance Fibre-Reinforced Concrete for 3D Printing. *Cem. Concr. Compos.* **2022**, *125*, 104310. [CrossRef]
12. Simonova, H.; Frantik, P.; Kersner, Z.; Schmid, P.; Rovnanik, P. Components of the Fracture Response of Alkali-Activated Slag Composites with Steel Microfibers. *Appl. Sci.-Basel* **2019**, *9*, 1754. [CrossRef]
13. Abdollahnejad, Z.; Mastali, M.; Falah, M.; Shaad, K.M.; Luukkonen, T.; Illikainen, M. Durability of the Reinforced One-Part Alkali-Activated Slag Mortars with Different Fibers. *Waste Biomass Valorization* **2021**, *12*, 487–501. [CrossRef]
14. Zhang, B.; Zhu, H.; Dong, Z.; Wang, Q. Enhancement of Bond Performance of FRP Bars with Seawater Coral Aggregate Concrete by Utilizing Ecoefficient Slag-Based Alkali-Activated Materials. *J. Compos. Constr.* **2022**, *26*, 04021059. [CrossRef]
15. Deng, W.; Zhao, J. Structure Characteristics and Mechanical Properties of Fiber Reinforced Concrete. *Adv. Build. Mater. Pts 1-3* **2011**, *168–170*, 1556–1560. [CrossRef]
16. Militky, J.; Kovacic, V. Ultimate Mechanical Properties of Basalt Filaments. *Text. Res. J.* **1996**, *66*, 225–229. [CrossRef]
17. Militký, J.; Kovačič, V.; Rubnerová, J. Influence of Thermal Treatment on Tensile Failure of Basalt Fibers. *Eng. Fract. Mech.* **2002**, *69*, 1025–1033. [CrossRef]
18. Fiore, V.; Scalici, T.; Di Bella, G.; Valenza, A. A Review on Basalt Fibre and Its Composites. *Compos. Part B Eng.* **2015**, *74*, 74–94. [CrossRef]
19. Sim, J.; Park, C.; Moon, D.-Y. Characteristics of Basalt Fiber as a Strengthening Material for Concrete Structure. *Compos. Part B Eng.* **2005**, *36*, 504–512. [CrossRef]
20. Jiang, C.; Fan, K.; Wu, F.; Chen, D. Experimental Study on the Mechanical Properties and Microstructure of Chopped Basalt Fibre Reinforced Concrete. *Mater. Des.* **2014**, *58*, 187–193. [CrossRef]
21. Wei, B.; Cao, H.; Song, S. Surface Modification and Characterization of Basalt Fibers with Hybrid Sizings. *Compos. Part A Appl. Sci. Manuf.* **2011**, *42*, 22–29. [CrossRef]
22. Branston, J.; Das, S.; Kenno, S.Y.; Taylor, C. Mechanical Behaviour of Basalt Fibre Reinforced Concrete. *Constr. Build. Mater.* **2016**, *124*, 878–886. [CrossRef]
23. Tang, C.; Jiang, H.; Zhang, X.; Li, G.; Cui, J. Corrosion Behavior and Mechanism of Basalt Fibers in Sodium Hydroxide Solution. *Materials* **2018**, *11*, 1381. [CrossRef]
24. Lee, J.; Song, J.; Kim, H. Chemical Stability of Basalt Fiber in Alkaline Solution. *Fibers Polym.* **2014**, *15*, 2329–2334. [CrossRef]
25. Rabinovich, F.N.; Zueva, V.N.; Makeeva, L.V. Stability of Basalt Fibers in a Medium of Hydrating Cement. *Glass Ceram.* **2001**, *58*, 431–434. [CrossRef]
26. Arslan, M.E. Effects of Basalt and Glass Chopped Fibers Addition on Fracture Energy and Mechanical Properties of Ordinary Concrete: CMOD Measurement. *Constr. Build. Mater.* **2016**, *114*, 383–391. [CrossRef]
27. El-Gelani, A.M.; High, C.M.; Rizkalla, S.H.; Abdalla, E.A. Effects of Basalt Fibres on Mechanical Properties of Concrete. *MATEC Web Conf.* **2018**, *149*, 01028. [CrossRef]
28. Jalasutram, S.; Sahoo, D.R.; Matsagar, V. Experimental Investigation of the Mechanical Properties of Basalt Fiber-Reinforced Concrete. *Struct. Concr.* **2017**, *18*, 292–302. [CrossRef]
29. Zhou, H.; Jia, B.; Huang, H.; Mou, Y. Experimental Study on Basic Mechanical Properties of Basalt Fiber Reinforced Concrete. *Materials* **2020**, *13*, 1362. [CrossRef]
30. Jiao, H.; Wu, Y.; Chen, X.; Yang, Y. Flexural Toughness of Basalt Fibre-Reinforced Shotcrete and Industrial-Scale Testing. *Adv. Mater. Sci. Eng.* **2019**, *2019*, e6568057. [CrossRef]
31. Bheel, N. Basalt Fibre-Reinforced Concrete: Review of Fresh and Mechanical Properties. *J. Build Rehabil.* **2021**, *6*, 12. [CrossRef]
32. Scheffler, C.; Förster, T.; Mäder, E.; Heinrich, G.; Hempel, S.; Mechtcherine, V. Aging of Alkali-Resistant Glass and Basalt Fibers in Alkaline Solutions: Evaluation of the Failure Stress by Weibull Distribution Function. *J. Non-Cryst. Solids* **2009**, *355*, 2588–2595. [CrossRef]
33. Wu, G.; Wang, X.; Wu, Z.; Dong, Z.; Zhang, G. Durability of Basalt Fibers and Composites in Corrosive Environments. *J. Compos. Mater.* **2015**, *49*, 873–887. [CrossRef]
34. Alsaif, A.; Koutas, L.; Bernal, S.A.; Guadagnini, M.; Pilakoutas, K. Mechanical Performance of Steel Fibre Reinforced Rubberised Concrete for Flexible Concrete Pavements. *Constr. Build. Mater.* **2018**, *172*, 533–543. [CrossRef]
35. Mahoutian, M.; Shekarchi, M.; Libre, N.A. Application of Steel Fiber Reinforced Lightweight Aggregate Concrete in Underground Mining. *J. Min. Sci.* **2011**, *47*, 606–617. [CrossRef]
36. Briffaut, M.; Benboudjema, F.; D'Aloia, L. Effect of Fibres on Early Age Cracking of Concrete Tunnel Lining. Part I: Laboratory Ring Test. *Tunn. Undergr. Space Technol.* **2016**, *59*, 215–220. [CrossRef]

37. Zhang, J.; Wang, Z.; Ju, X. Application of Ductile Fiber Reinforced Cementitious Composite in Jointless Concrete Pavements. *Compos. Part B Eng.* **2013**, *50*, 224–231. [CrossRef]
38. Li, X.; Xue, W.; Fu, C.; Yao, Z.; Liu, X. Mechanical Properties of High-Performance Steel-Fibre-Reinforced Concrete and Its Application in Underground Mine Engineering. *Materials* **2019**, *12*, 2470. [CrossRef] [PubMed]
39. John, V.J.; Dharmar, B. Influence of Basalt Fibers on the Mechanical Behavior of Concrete—A Review. *Struct. Concr.* **2021**, *22*, 491–502. [CrossRef]
40. Iyer, P.; Kenno, S.Y.; Das, S. Mechanical Properties of Fiber-Reinforced Concrete Made with Basalt Filament Fibers. *J. Mater. Civ. Eng.* **2015**, *27*, 04015015. [CrossRef]
41. Ayub, T.; Shafiq, N.; Nuruddin, M.F. Mechanical Properties of High-Performance Concrete Reinforced with Basalt Fibers. *Procedia Eng.* **2014**, *77*, 131–139. [CrossRef]
42. High, C.; Seliem, H.M.; El-Safty, A.; Rizkalla, S.H. Use of Basalt Fibers for Concrete Structures. *Constr. Build. Mater.* **2015**, *96*, 37–46. [CrossRef]
43. Girgin, Z.C.; Yıldırım, M.T. Usability of Basalt Fibres in Fibre Reinforced Cement Composites. *Mater. Struct.* **2015**, *49*, 3309–3319. [CrossRef]
44. Hannawi, K.; Bian, H.; Prince-Agbodjan, W.; Raghavan, B. Effect of Different Types of Fibers on the Microstructure and the Mechanical Behavior of Ultra-High Performance Fiber-Reinforced Concretes. *Compos. Part B Eng.* **2016**, *86*, 214–220. [CrossRef]
45. Algin, Z.; Mermerdaş, K.; Khalid, L.W. Mechanical Performance of Basalt Fibre Reinforced Concretes. *J. Inst. Sci. Technol.* **2020**, *10*, 1093–1106. [CrossRef]
46. Jun, W.; Yue, M.A.; Ye, Z.; Wei, C. Experimental research and analysis on mechanical properties of chopped basalt fiber reinforced concrete. *Gclx* **2014**, *31*, 99–102, 114. [CrossRef]
47. Katkhuda, H.; Shatarat, N. Improving the Mechanical Properties of Recycled Concrete Aggregate Using Chopped Basalt Fibers and Acid Treatment. *Constr. Build. Mater.* **2017**, *140*, 328–335. [CrossRef]
48. Jordan, R.W. The Effect of Stress, Frequency, Curing, Mix and Age upon the Damping of Concrete. *Mag. Concr. Res.* **1980**, *32*, 195–205. [CrossRef]
49. Yan, L.; Jenkins, C.H.; Pendleton, R.L. Polyolefin Fiber-Reinforced Concrete Composites: Part, I. Damping and Frequency Characteristics. *Cem. Concr. Res.* **2000**, *30*, 391–401. [CrossRef]
50. Lokajicek, T.; Petružálek, M.; Svitek, T.; Kuchařová, A.; Šachlová, Š.; Přikryl, R. Dynamic Elastic Properties of Hardened Experimental Mortar Bars Affected by Accelerated Alkali–Silica Reactivity Test: A Laboratory Approach. *Bull. Eng. Geol. Environ.* **2021**, *80*, 8921–8933. [CrossRef]
51. Amick, H.; Monteiro, P.J.M. Experimental Determination of Modal Damping in Concrete Beams. *ACI Mater. J.* **2006**, *103*, 153.
52. Jones, R. The Effect of Frequency on the Dynamic Modulus and Damping Coefficient of Concrete. *Mag. Concr. Res.* **1957**, *9*, 69–72. [CrossRef]
53. Fu, X.; Chung, D.D.L. Vibration Damping Admixtures for Cement. *Cem. Concr. Res.* **1996**, *26*, 69–75. [CrossRef]
54. Nelson, D.; Hancock, J. Interfacial Slip and Damping in Fibre Reinforced Composites. *J. Mater. Sci.* **1978**, *13*, 2429–2440. [CrossRef]
55. Swamy, R.N.; Mangat, P.S. Influence of Fiber Geometry on the Properties of Steel Fiber Reinforced Concrete. *Cem. Concr. Res.* **1974**, *4*, 451–465. [CrossRef]
56. Prasad, S. The Energy Absorption Capability of Composite Materials and Structures: Influence of Impact Loading. *J. Mod. Mech. Eng. Technol.* **2018**, *5*, 32–46. [CrossRef]
57. Zhang, H.; Wang, B.; Xie, A.; Qi, Y. Experimental Study on Dynamic Mechanical Properties and Constitutive Model of Basalt Fiber Reinforced Concrete. *Constr. Build. Mater.* **2017**, *152*, 154–167. [CrossRef]
58. Ren, W.; Xu, J.; Su, H. Dynamic Compressive Behavior of Basalt Fiber Reinforced Concrete after Exposure to Elevated Temperatures. *Fire Mater.* **2016**, *40*, 738–755. [CrossRef]
59. Fu, Q.; Niu, D.; Zhang, J.; Huang, D.; Wang, Y.; Hong, M.; Zhang, L. Dynamic Compressive Mechanical Behaviour and Modelling of Basalt–Polypropylene Fibre-Reinforced Concrete. *Arch. Civ. Mech. Eng.* **2018**, *18*, 914–927. [CrossRef]
60. Feng, J.; Zhou, Y.; Wang, P.; Wang, B.; Zhou, J.; Chen, H.; Fan, H.; Jin, F. Experimental Research on Blast-Resistance of One-Way Concrete Slabs Reinforced by BFRP Bars under Close-in Explosion. *Eng. Struct.* **2017**, *150*, 550–561. [CrossRef]
61. Remennikov, A.M.; Kaewunruen, S. Determination of dynamic properties of rail pads using an instrumented hammer impact technique. *Acoust Aust.* **2005**, *33*, 63–67.
62. Basalt Rebar for Reinforcement—Deutsche Basalt Faser GmbH. Deutsche Basalt Faser (DBF Company). Available online: https://www.deutsche-basalt-faser.de/en/products/basalt-rebar-for-reinforcement/ (accessed on 5 January 2023).
63. O'Brien Cement Company. Available online: http://obriencement.ie/ (accessed on 5 January 2023).
64. Chen, B.; Liu, J. Contribution of Hybrid Fibers on the Properties of the High-Strength Lightweight Concrete Having Good Workability. *Cem. Concr. Res.* **2005**, *35*, 913–917. [CrossRef]
65. Priyanka, P.C.; Manoj, R.V. Correlation of Compressive Strength and Dynamic Modulus of Elasticity for High Strength SCC Mixes. *Int. J. Eng. Tech. Res. (IJETR)* **2015**, *3*, 42–46.
66. Lu, X.; Sun, Q.; Feng, W.; Tian, J. Evaluation of Dynamic Modulus of Elasticity of Concrete Using Impact-Echo Method. *Constr. Build. Mater.* **2013**, *47*, 231–239. [CrossRef]
67. Li, D.; Kaewunruen, S.; You, R. Remaining Fatigue Life Predictions of Railway Prestressed Concrete Sleepers Considering Time-Dependent Surface Abrasion. *Sustainability* **2022**, *14*, 11237. [CrossRef]

68. Li, D.; You, R.; Kaewunruen, S. Mechanisms and Evolution of Cracks in Prestressed Concrete Sleepers Exposed to Time-Dependent Actions. *Appl. Sci.* **2022**, *12*, 5511. [CrossRef]
69. Li, D.; You, R.; Kaewunruen, S. Crack Propagation Assessment of Time-Dependent Concrete Degradation of Prestressed Concrete Sleepers. *Sustainability* **2022**, *14*, 3217. [CrossRef]
70. Li, D.; Kaewunruen, S.; You, R.; Liu, P. Fatigue life modelling of railway prestressed concrete sleepers. *Structures* **2022**, *41*, 643–656. [CrossRef]

materials

MDPI

Article

A Design Process for Preventing Brittle Failure in Strengthening RC Slabs with Hybrid FRP-HPC Retrofit Systems

Huy Q. Nguyen [1], Taek Hee Han [2], Jun Kil Park [2] and Jung J. Kim [1,*]

1 Department of Civil Engineering, Kyungnam University, Changwon-si 51767, Republic of Korea; nguyenquochuy@muce.edu.vn

2 Coastal Development and Ocean Energy Research Center, Korea Institute of Ocean Science and Technology, 385 Haeyang-ro, Yeongdo-gu, Busan 49111, Republic of Korea; taekheehan@kiost.ac.kr (T.H.H.); jkpark@kiost.ac.kr (J.K.P.)

* Correspondence: jungkim@kyungnam.ac.kr; Tel.: +82-552-496-421; Fax: +82-505-999-2165

Abstract: The retrofitting of existing RC slabs with an innovative system comprising FRP and HPC has been demonstrated to be effective in strengthening and overcoming the logistical challenges of installation. Nonetheless, the excessive improvement of flexural strength over shear strength would cause the sudden failure of rehabilitated flexural members. The literature has previously recommended failure limits to determine the additional moment strength compared with the shear strength to prevent brittle shear failure of strengthened, continuous RC slabs. This study suggests a design process for preventing shear failure and inducing the ductile-failure mode to improve the safety and applicability of retrofitted RC slabs based on the proposed failure limits. The effectiveness of the procedure in brittle-failure prevention for the end and interior spans of retrofitted RC slabs is illustrated via a case study. The outcomes showed that the retrofit system with 0.53-mm-thick-CFRP prevented brittle failure and significantly enhanced the design-factored load and ultimate failure load by up to 2.07 times and 2.13 times, respectively.

Keywords: CFRP; high-performance concrete; brittle failure; reinforced concrete slab; retrofit

Citation: Nguyen, H.Q.; Han, T.H.; Park, J.K.; Kim, J.J. A Design Process for Preventing Brittle Failure in Strengthening RC Slabs with Hybrid FRP-HPC Retrofit Systems. *Materials* **2023**, *16*, 755. https://doi.org/10.3390/ma16020755

Academic Editors: Hrvoje Smoljanović, Ivan Balić and Nikolina Zivaljic

Received: 16 December 2022
Revised: 6 January 2023
Accepted: 9 January 2023
Published: 12 January 2023

1. Introduction

Rehabilitating and improving reinforced concrete (RC) structures with fiber-reinforced polymers (FRP) have recently seen enormous advances in methods and techniques thanks to widespread interest in the engineering community [1,2]. It has emerged as one of the preferred solutions to minimize environmental deterioration and save costs for upgrading existing RC infrastructure compared to constructing entirely new structures [3–6]. A wide range of literature confirmed the advantage of FRP as carbon FRP (CFRP), glass FRP (GFRP), and basalt FRP (BFRP) in repairing and enhancing RC structures due to its high tensile strength, lightweight, non-corrosion resistance, and flexible dimensions [7–12]. CFRP has superior tensile strength, whereas the others have lower material costs and poorer mechanical properties [13,14]. Despite this, relying solely on these FRP materials' high tensile strength will cause logistical obstacles and technical difficulties. For instance, getting a well-prepared underneath surface for RC slabs or bridge decks can be challenging and expensive since traffic under overpasses cannot be interrupted, or electrical wiring, suppression systems, and ventilation ducts cannot be blocked [15,16].

In contrast, FRP installation on the top surface of slabs or bridge decks has not encountered many problems. Thus, Mosallam et al. [15] developed an innovative retrofit system made of CFRP laminate and high-performance concrete (HPC) installed on top of slabs for enhancing strength and ductility, as shown in Figure 1. The retrofitting mechanism of strengthened slabs using retrofit systems based on ACI 440.2R [17] was proposed and validated with experimental data. Nevertheless, one of the primary reasons retrofitted slabs failed to achieve their ultimate carrying capacity was a premature debonding failure between CFRP and overlay interface [18]. A specialized adhesive layer permits interfacial stress to be transferred from slab to CFRP or vice versa more effectively than from CFRP to overlay concrete, resulting in a higher potential debonding failure between overlay and CFRP [19,20]. Moon et al. [21] also suggested using shear connectors to enhance the bond strength and flexural carrying capacity of retrofitted slabs. While these attempts were valuable, the brittle shear failure and premature debonding failure of the FRP-to-concrete interface have not been sufficiently noted to yield efficient and safe structures for actual designs.

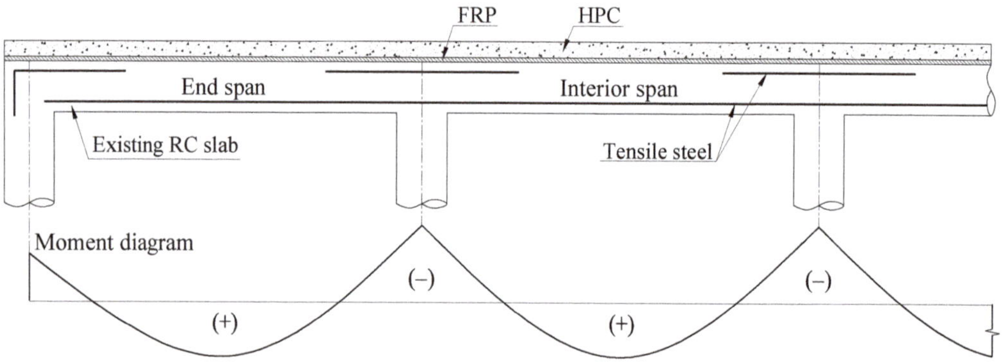

Figure 1. The existing RC slab with an HPC-FRP retrofit system.

Besides, the ductility of flexural members restored by FRP is also a matter of particular concern because it can warn of ultimate failure and reduce the impact of dynamic loads [22–24]. In contrast with steel-reinforced concrete structures, strengthened RC structures tend to fail more brittle because of the intrinsic bond conditions between FRP and concrete [25–30]. Additionally, a sudden failure of rehabilitated flexural-continuous members would originate from an excessive enhancement in flexural strength over the shear strength. The forewarning of potential failures and preventing sudden collapses requires considering structural ductility as a critical design factor [31–33]. Thus, design limits to prevent shear failure and induce ductile failure of retrofitted continuous slabs were established for the end span by Nguyen et al. [34] and interior span by Kim et al. [35]. The method is used to limit the moment-carrying capacities of the strengthened slab to less than or equal to its shear-carrying capacities, which can prevent shear failure. Failure limit regions are also defined according to moment-carrying capacities to limit the added moment at the support compared to mid-span or vice versa. Even so, the difference in design limits, span lengths, and slab structures between the interior and exterior spans promotes an attempt to develop a single design guide to match the practice design purposes. The design process based on the proposed failure limits and ACI committee 440 will help to simplify a complicated and understudied problem in structural strengthening. Fortunately, it can be achieved by simply adapting the retrofit system without a violent impact on existing RC structures. CFRP and the overlay HPC significantly affect additional flexural and shear strength at the support, whereas only the HPC overlay is responsible for these properties at mid-span [36]. As a result, it is possible to adjust retrofit systems related to CFRP and overlay HPC separately or simultaneously, depending on the design purpose.

This study describes the failure limits for the end and interior span of the continuous RC slabs relevant to their bearing-carrying capacities. The limit equations for determining the design-factored load and the ultimate failure load corresponding to the failure mode are discussed. The design process for preventing brittle failure in strengthening RC slabs with hybrid FRP-HPC retrofit systems based on novel failure limit classifications is recommended. The case study is also used to illustrate the efficiency of the proposed design process. Based on the obtained result, modifications related to retrofit systems in sudden failure prevention for the end and interior spans of retrofitted RC slabs are discussed and suggested.

2. Overview Theory and Proposed Process

2.1. Failure Limits Overview

According to ACI 318M [37], members of continuous construction should be designed to withstand maximum-factored-load effects as defined by the elastic analysis theory. Moments and shears can be achieved by elastic analysis using the strength design method. The reasonably conservative moments and shears of flexural members subjected to the uniformly distributed load can be estimated if they have more than two spans with a longer span no greater than 1.2 times the shorter span. It can be conducted based on the applied load, length of clear span, and approximate coefficients, as shown in Figure 2.

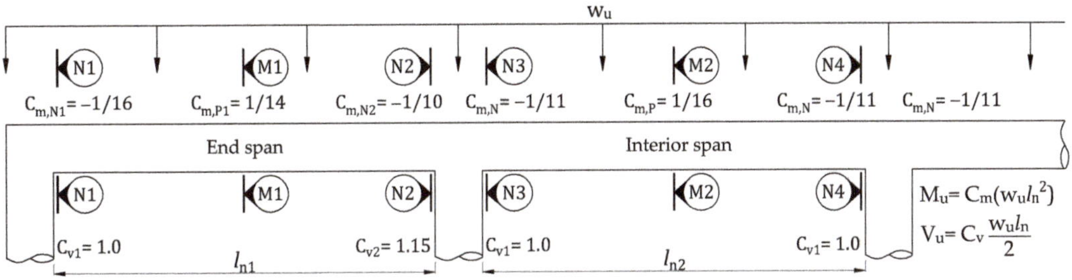

Figure 2. Approximate coefficients for estimating moments and shears of flexural members on continuous slabs with column supports in accordance with ACI 318M.

Nguyen et al. [34] and Kim et al. [35] established failure limits following shear limits for flexural members of frames to prevent brittle failure based on the theory of elastic analysis, as shown in Figure 3. The RC slab is considered a ductile failure when collapse occurs after forming all plastic hinges at high-internal-forces locations such as the support and mid-span. By contrast, it is supposed in brittle failure. Different failure modes according to the order of forming plastic hinges for the end and interior span of continuous RC slabs are summarized in Table 1. The proposed approach has contributed to simplifying a complex issue that prevents the sudden failure of strengthened continuous structures using external bonded FRP materials. Previous studies have demonstrated a positive effect on strengthened slabs using FRP-HPC retrofit systems [15,18,21]. We can derive the failure limits for each region of the end span from the expressions below:

$$M_{n,N} = \frac{2C_{m,N2}}{C_{v2}} V_n l_{n1} \tag{1}$$

$$M_{n,P} = \frac{2C_{m,P1}}{C_{v2}} V_n l_{n1} \tag{2}$$

$$M_{N1} = \frac{2C_{m,N1}}{C_{v2}} V_n l_{n1} \tag{3}$$

$$\frac{M_{N1}}{M_{n,P}} = \frac{C_{m,N1}}{C_{m,P1}} \tag{4}$$

$$\frac{M_{n,N}}{M_{n,P}} = \frac{C_{m,N2}}{C_{m,P1}} \tag{5}$$

$$M_{n,P} + M_{n,N}\left(\frac{C_{v2}/8 + C_{m,N1} - C_{m,P1} - C_{v2}C_{m,N1}}{C_{m,N2}} + C_{v2} - 1\right) = \frac{1}{4} V_n l_{n1} \tag{6}$$

$$M_{n,P}(2C_{v2} - 1) + M_{n,N}\left(\frac{C_{v2}/4 + C_{m,P1} - C_{m,N1} - 2C_{v2}C_{m,P1}}{C_{m,N2}} + 1\right) = \frac{1}{2} V_n l_{n1} \tag{7}$$

$$M_{n,P}\left(\frac{C_{v2}/8 - C_{m,N2}}{C_{m,P1}}\right) + M_{n,N} = \frac{1}{4}V_n l_{n1} \quad (8)$$

$$M_{n,P}\left(\frac{C_{v2}/4 + C_{m,N2} - C_{m,N1} - 2C_{v2}C_{m,N2}}{C_{m,P1}}\right) + 2C_{v2}M_{n,N} = \frac{1}{2}V_n l_{n1} \quad (9)$$

Additionally, the formulas define the failure limits for each region of the interior span as follows,

$$M_{n,N} = \frac{2C_{m,N}}{C_{v1}}V_n l_{n2} \quad (10)$$

$$M_{n,P} = \frac{2C_{m,P}}{C_{v1}}V_n l_{n2} \quad (11)$$

$$\frac{M_{n,N}}{M_{n,P}} = \frac{C_{m,N}}{C_{m,P}} \quad (12)$$

$$M_{n,N}\left(\frac{C_{v1}/8 - C_{m,P}}{C_{m,N}}\right) + M_{n,P} = \frac{1}{4}V_n l_{n2} \quad (13)$$

$$M_{n,N} + M_{n,P}\left(\frac{C_{v1}/8 - C_{m,N}}{C_{m,P}}\right) = \frac{1}{4}V_n l_{n2} \quad (14)$$

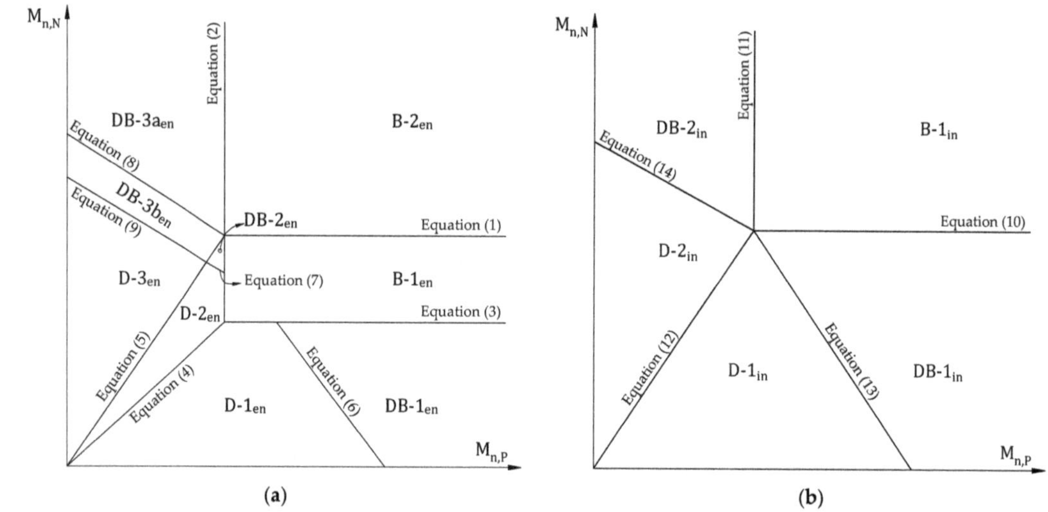

Figure 3. A classification of continuous RC slab failure modes based on moment and shear capacities: (**a**) the end span; (**b**) the interior span.

Table 1. Failure modes of continuous RC slabs according to the order of plastic hinge formation.

Span Type	Failure Modes	First Plastic Hinge	Second Plastic Hinge	Third Plastic Hinge	Shear Failure	Failure Type
End span	D-1$_{en}$	N2	N1	M1	-	Ductile failure
	D-2$_{en}$	N2	M1	N1	-	Ductile failure
	D-3$_{en}$	M1	N2	N1	-	Ductile failure
	DB-1$_{en}$	N2	N1	-	N2	Brittle failure
	DB-2$_{en}$	N2	M1	-	N2	Brittle failure
	DB-3a$_{en}$	M1	-	-	N2	Brittle failure
	DB-3b$_{en}$	M1	N2	-	N2	Brittle failure
	B-1$_{en}$	N2	-	-	N2	Brittle failure
	B-2$_{en}$	-	-	-	N2	Brittle failure

Table 1. *Cont.*

Span Type	Failure Modes	First Plastic Hinge	Second Plastic Hinge	Third Plastic Hinge	Shear Failure	Failure Type
Interior span	D-1$_{in}$	N3, N4	M2		-	Ductile failure
	D-2$_{in}$	M2	N3, N4		-	Ductile failure
	DB-1$_{in}$	N3, N4	-		N3, N4	Brittle failure
	DB-2$_{in}$	M2	-		N3, N4	Brittle failure
	B-1$_{in}$	-	-		N3, N4	Brittle failure

Based on the order to form plastic hinges, the ultimate failure loads considering plastic redistribution of the slab can be determined using the superposition method. The ultimate failure loads for the end and interior spans of continuous slabs will be calculated in accordance with Table 2.

Table 2. Calculating ultimate failure loads of continuous RC slabs.

Failure Modes	Failure Load	
D-1$_{en}$	$w_f = \phi_f \frac{8}{l_{n1}^2}\left(M_{n,P} + M_{n,N}\frac{(1/8 - C_{m,P1})}{C_{m,N2}}\right)$	(15)
D-2$_{en}$	$w_f = \phi_f \frac{4}{l_{n1}^2}\left(M_{n,P} + M_{n,N}\frac{(1/4 + C_{m,N2} - C_{m,N1} - C_{m,P1})}{C_{m,N2}}\right)$	(16)
D-3$_{en}$	$w_f = \phi_f \frac{4}{l_{n1}^2}\left(M_{n,P}\frac{(1/4 - C_{m,N1})}{C_{m,P1}} + M_{n,N}\right)$	(17)
DB-1$_{en}$, DB-2$_{en}$, DB-3a$_{en}$, DB-3b$_{en}$, B-1$_{en}$, B-2$_{en}$	$w_f = \phi_v \frac{2V_n}{C_{v2} l_{n1}}$	(18)
D-1$_{in}$	$w_f = \phi_f \frac{8}{l_{n2}^2}\left(M_{n,P} + M_{n,N}\frac{(1/8 - C_{m,P})}{C_{m,N}}\right)$	(19)
D-2$_{in}$	$w_f = \phi_f \frac{8}{l_{n2}^2}\left(M_{n,P}\frac{(1/8 - C_{m,N})}{C_{m,P}} + M_{n,N}\right)$	(20)
DB-1$_{in}$, DB-2$_{in}$, B-1$_{in}$	$w_f = \phi_v \frac{2V_n}{C_{v1} l_{n2}}$	(21)

2.2. Retrofitting Mechanism and Design Process Preventing Brittle Failure

The retrofitting mechanism for the strengthened slab originated from the sectional compressive force in HPC and the sectional tensile force in FRP and steel according to ACI 440.2R. For the positive moment section, it is suggested to have enough thickness to pull the neutral axis into the HPC overlay, as shown in Figure 4a. Besides, the minimum HPC thickness is recommended due to its high self-weight. For the negative moment section, the retrofitting mechanism can be predicted in a typical way, as shown in Figure 4b.

In this study, either evaluating the effects of long-term service loads or harsh environmental conditions is beyond the scope. The FRP sheet's thickness, which can increase bending moments significantly of the strengthened slab, is considered a modifiable variable. The design process is based on ACI 440.2R [17] and proposed failure limits [34,35]. Before applying the design process, it is necessary to determine the input parameters related to the reference slab (i.e., l, h, b, A_s, d, f'_c, f_y, E_s) and retrofit system (i.e., f'_H, f^*_{fu}, E_{fe}, C_E) to perform the relevant preliminary calculations. The flowchart of design for preventing brittle failure is described in Figure 5 by the following steps:

(1) Assume FRP thickness (t_F).

(2) The overlay strength (f'_H) should be greater than the limits in the following equations to ensure the neutral axis within the overlay and FRP in the tensile zone [15]; otherwise, re-assume FRP thickness.

$$f'_H \geq \frac{\varepsilon_{cu} E_{fe}}{1.445}\left(\frac{t_F}{t_H}\right)^2 + \frac{f_y(A_s/b)}{0.7225 t_H} \tag{22}$$

$$f'_H \geq 0.15 f'_c + \frac{\varepsilon_{cu} E_{fe}}{1.7}\left(\frac{t_F}{t_H}\right)^2 + \frac{f_y(A_s/b)}{0.85 t_H} \tag{23}$$

(3) Compute the design strain of FRP (ε_{fd}) at support.

$$\varepsilon_{fd} = 0.41\sqrt{\frac{f'_c}{n E_{fe} t_F}} \leq 0.9\varepsilon_{fu} \tag{24}$$

(4) Assume the neutral axis depth (c).
(5) Compute FRP stress (f_{fe}) corresponding to FRP strain (ε_{fe}) and concrete strain at failure (ε_c) by applying similar triangles based on strain compatibility.

$$f_{fe} = \varepsilon_{fe} E_{fe} \leq f_{fu} \tag{25}$$

For the support section:

$$\varepsilon_{fe,N} = \varepsilon_{cu}\left(\frac{h - c_N}{c_N}\right) - \varepsilon_{bi} \leq \varepsilon_{fd} \tag{26}$$

$$\varepsilon_{c,N} = (\varepsilon_{fe,N} + \varepsilon_{bi})\left(\frac{c_N}{h - c_N}\right) \tag{27}$$

For the mid-span section:

$$\varepsilon_{fe,P} = \varepsilon_{cu}\left(\frac{t_H - c_P}{c_P}\right) \leq \varepsilon_{fd} \tag{28}$$

$$\varepsilon_{c,P} = \varepsilon_{fe,P}\left(\frac{c_P}{t_H - c_P}\right) \tag{29}$$

(6) Compute reinforced steel stress (f_s) and strain (ε_s).

$$f_s = \varepsilon_s E_s \leq f_y \tag{30}$$

For the support section ($\varepsilon_{s,N}$):

$$\varepsilon_{s,N} = (\varepsilon_{fe,N} + \varepsilon_{bi})\left(\frac{d - c_N}{h - c_N}\right) \tag{31}$$

For the mid-span section ($\varepsilon_{s,P}$):

$$\varepsilon_{s,P} = \varepsilon_{c,P}\left(\frac{d + t_H + t_F - c_P}{c_P}\right) \tag{32}$$

(7) Check the equilibrium condition by comparing c defined in Equation (33) with the value in step 4. If it is satisfied, go to step 9; otherwise, return to step 4.

$$c = \frac{A_s f_s + A_F f_{fe}}{\alpha_1 f'_c \, \beta_1 b} \tag{33}$$

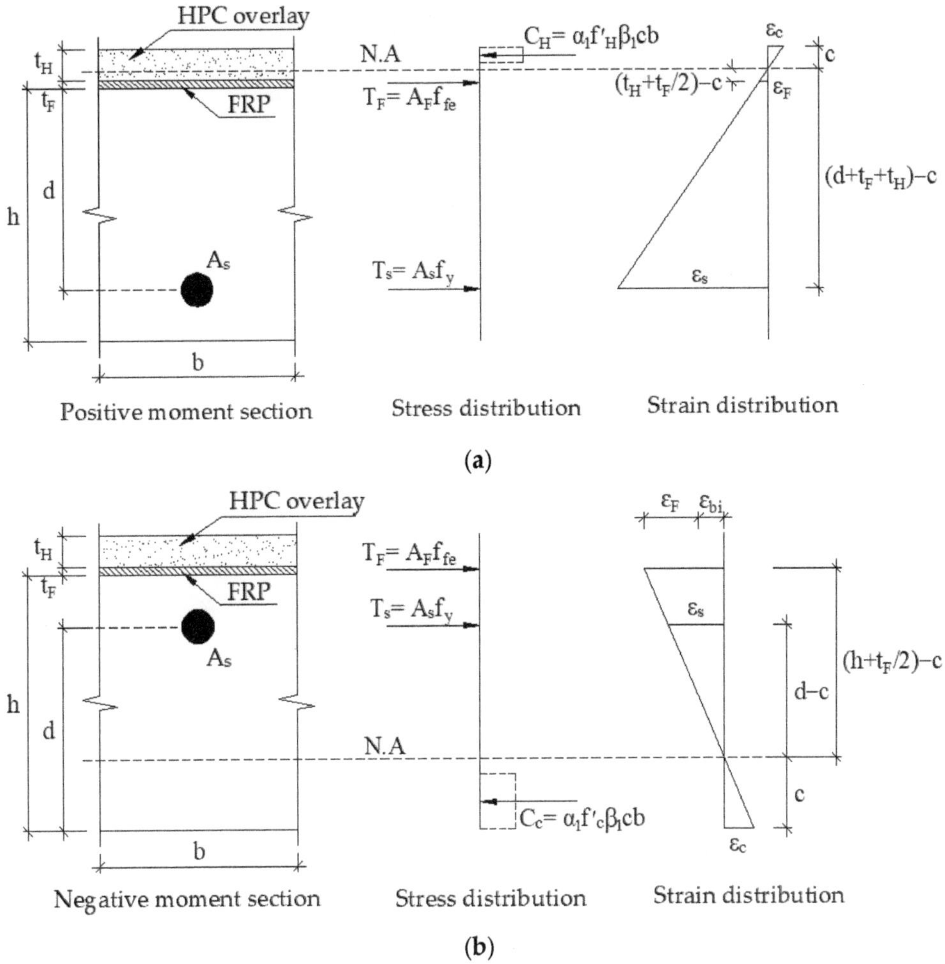

Figure 4. Retrofitting mechanism for a retrofitted slab (**a**) positive-moment section; (**b**) negative moment section.

Coefficients related to stress block (α_1 and β_1) are calculated as recommended in ACI 318M once concrete strain (ε_c) is equal to or greater than the ultimate value (ε_{cu}); otherwise, these values shall be estimated following the Whitney stress block, as reported by the ACI 440 committee, where $\varepsilon'_c = \dfrac{1.7f'_c}{E_c}$; $\beta_1 = \dfrac{4\varepsilon'_c - \varepsilon_c}{6\varepsilon'_c - 2\varepsilon_c}$; $\alpha_1 = \dfrac{3\varepsilon'_c\varepsilon_c - \varepsilon_c^2}{3\beta_1\varepsilon'^2_c}$

(8) Compute strength in flexure ($\phi_f M_n$) and shear ($\phi_v V_n$)

$$\phi_f M_n = \phi_f(M_{ns} + \psi_f M_{nf}) \tag{34}$$

$$\phi_v V_n = \phi_v\left(d\sqrt{f'_c} + t_H\sqrt{f'_H}\right)\frac{b}{6} \tag{35}$$

For the support section, the contribution of steel ($M_{ns,N}$) and FRP ($M_{nf,N}$), as

$$M_{ns,N} = A_s f_{s,N}\left(d - \frac{\beta_{1,N}c_N}{2}\right) \tag{36}$$

$$M_{nf,N} = A_F f_{fe,N}\left(h - \frac{\beta_{1,N} c_N}{2}\right) \quad (37)$$

For the mid-span section, the contribution of steel ($M_{ns,P}$) and FRP ($M_{nf,P}$), as

$$M_{ns,P} = A_s f_{s,P}\left(d + t_H + t_F - \frac{\beta_{1,P} c_P}{2}\right) \quad (38)$$

$$M_{nf,P} = A_F f_{fe,P}\left(t_H - \frac{\beta_{1,P} c_P}{2}\right) \quad (39)$$

(9) Define the design factored load as specified in Figure 2.

$$w_u = \min(w_{u,M}; w_{u,V}) = \min\left(\frac{\phi_f M_n}{C_m l_n^2}; \frac{\phi_v V_n}{C_v l_n}\right) \quad (40)$$

(10) Define the failure mode and failure load (w_f) in accordance with the failure limits. If the failure mode is ductile, the design process preventing brittle failure can be achieved; otherwise, re-assume FRP thickness.

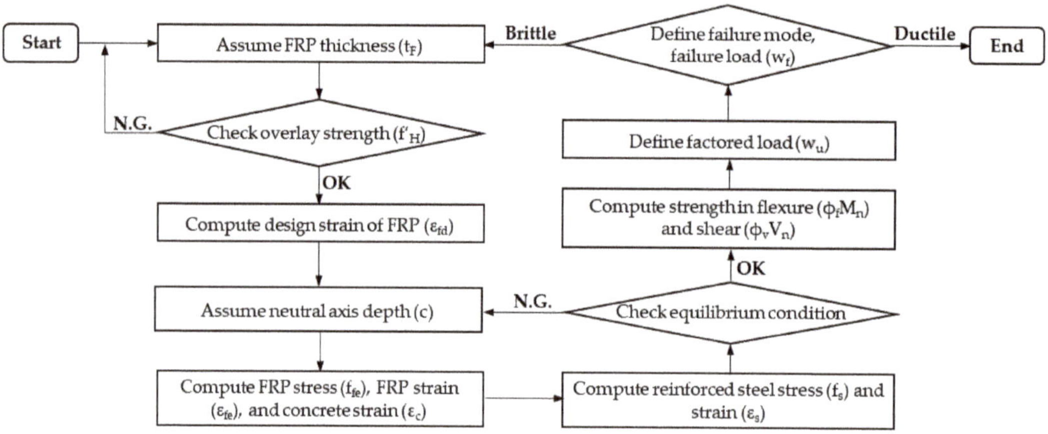

Figure 5. Flowchart of design for preventing brittle failure.

3. Design Example

In this case, a calculation is performed on a continuous RC slab with three spans, where the length of the end span of 2.6 m and interior span of 2.4 m are used. A uniformly distributed load is applied to the strengthened slab using the HPC-CFRP retrofit system. The moments and shears coefficients for the end and interior span are described in Figure 2. According to ACI 440 committee [17], the reduction factors of CFRP for strength contribution (ψ_f) of 0.85 and environment of 0.95 (C_E) are used. Besides, the flexural and shear strength reduction factors (ϕ_f and ϕ_v) are 0.9 and 0.75, respectively. Material properties for the end and interior span of the continuous RC slab are given in Table 3. The reference slab's reinforcement detail and dimensions are shown in Figure 6. The adjustable CFRP thickness with initial value and material properties of the retrofit system are provided in Table 4. For retrofit systems, shear anchors and HPC are suggested to avoid potential shear failures in overlay and premature debonding of CFRP. Previous work demonstrated that shear connectors could increase the bond strength of retrofit systems. The integrity of the retrofitted slab until reaching the ultimate bearing capacity is assumed. Table 5 shows preliminary analyses for the reference slab and retrofit system before implementing the proposed design process.

Table 3. Dimensions and material properties of the existing RC slab.

Type	l (mm)	h (mm)	b (mm)	f'_c (MPa)	A_s (mm^2)	d (mm)	f_y (MPa)	E_s (GPa)
End span	2600	145	900	32	426	110	410	200
Interior span	2400	-	-	-	-	-	-	-

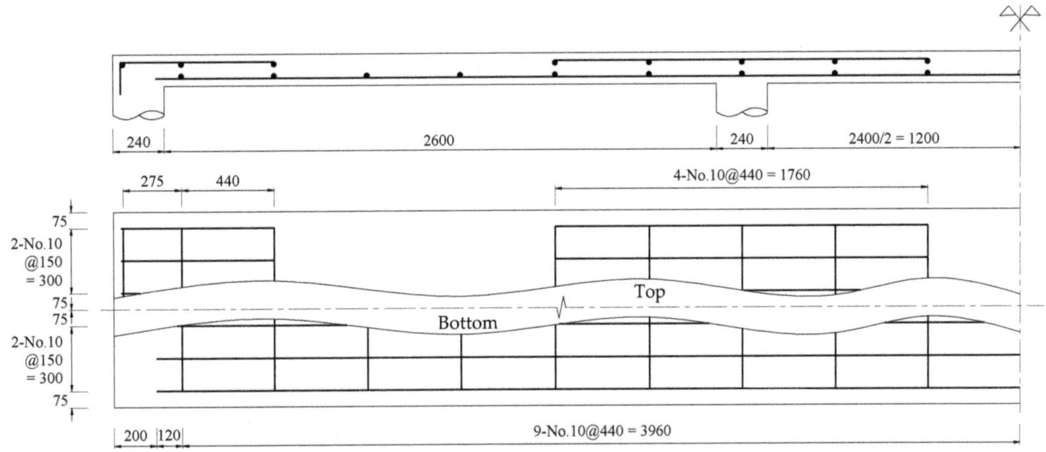

Figure 6. The reference slab's reinforcement detail and dimensions (in mm).

Table 4. The initial CFRP thickness and retrofit system's material properties.

HPC Overlay		CFRP		
t_H (mm)	f'_H (MPa)	t_F (mm)	f^*_{fu} (MPa)	E_{fe} (GPa)
30	75	1	600	40

Table 5. Initial analyses of the retrofit system and reference slab.

Analysis	Reference Slab
Sectional capacity	$\phi_f M_{n,N}$ = 16.73 kNm; $\phi_f M_{n,P}$ = 16.73 kNm; $\phi_v V_n$ = 70 kN
Design factored load	For end span: w_u = min(39.6; 24.7; 34.6; 53.8; 46.8) = 24.7 kN/m For interior span: w_u = min(32; 46.5; 58.3) = 32 kN/m
Define failure mode and failure load	For end span: D-2$_{en}$ according to Figure 7a; Equation (16), w_f = 31.3 kN/m For interior span: D-1$_{in}$ according to Figure 7b; Equation (19), w_f = 39.2 kN/m
Self-weight	$w_c = \gamma_c bh = 24(0.9)(0.145) = 3.13$ kN/m
Elastic modulus	$E_c = 4700\sqrt{f'_c} = 4700\sqrt{32} = 26{,}600$ MPa
At support, kd	kd = 24.65 mm
The crack moment at support	$I_{cr,N} = 2.78 \times 10^7$ mm^4
The ultimate strength and strain of CFRP	$f_{fu} = C_E f^*_{fu} = 570$ MPa; $\varepsilon_{fu} = \frac{f_{fu}}{E_{fe}} = 0.0143$
Moment due to dead load	At N2 section: $M_{D,N2} = \frac{1}{10}(3.13)(2.6^2) = 2.12$ kNm At N3 and N4 sections: $M_{D,N3} = M_{D,N4} = \frac{1}{11}(3.13)(2.4^2) = 1.64$ kNm
The existing state of strain	At N2 section: $\varepsilon_{bi} = \frac{2.12 \times 10^6(145 - 24.65)}{2.78 \times 10^7(26600)} = 0.00034$ At N3 and N4 sections: $\varepsilon_{bi} = \frac{1.64 \times 10^6(145 - 24.65)}{2.78 \times 10^7(26600)} = 0.00027$

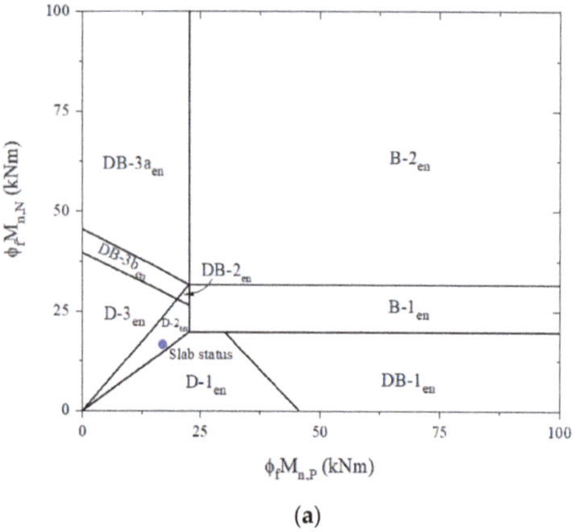

Figure 7. Predict failure mode based on moment-carrying capacities of the reference slab: (**a**) the end span; (**b**) the interior span.

4. Results and Discussion

For the reference slab, it can be seen from Table 5 that the design-factored load (w_u) and the ultimate failure load (w_f) for the end span are determined to be 24.7 kN/m and 31.3 kN/m, respectively. Figure 5a reveals that the failure mode is named D-2_{en}. Furthermore, there is a prediction of 32 kN/m design-factored load and 39.2 kN/m ultimate failure load for the interior span with failure mode D-2_{in}, as shown in Figure 7b. Before applying the design process, a preliminary analysis is performed at the adverse load-bearing section of support for the end span (section N2) and interior span (sections N3 and N4) of the reference slab and retrofit system.

For the retrofitted slab, the design process is carried out with an initial CFRP thickness of 1mm, as shown in Table 6. In step two, the compressive strength of the overlay is checked and shown that it is large enough to ensure a neutral axis in the overlay and FRP in the tensile zone. As a result of the analysis, the induction of tension in CFRP laminate at the midspan can be accomplished using a relatively low-compressive-strength concrete overlay. Nevertheless, it is preferred to use high-strength concrete to enhance the mid-span flexural strength of the slab and avoid potential shear failures of the overlay. The design strain of FRP is computed in step three before assuming the neutral axis depth in step four. The CFRP stress corresponding to its strain and concrete strain at failure is calculated in step five using similar triangles based on strain compatibility before the steel stress level is calculated in step six. According to step seven, an iterative process to achieve force equilibrium is performed as recommended by the ACI 440 committee before predicting design flexural and shear strengths in step eight. At steps nine and ten, the retrofitted slab was identified as a brittle failure for the end and interior span classified as DB-$3a_{en}$ and DB-2_{in}, respectively, as shown in Figure 8. There is an ultimate failure load at the end span of 66.38 kN/m and the interior span of 82.69 kN/m. It is not the desired outcome for the retrofitted slab to fail brittlely, regardless of the fact that its ultimate failure load is over 2.1 times that of the reference slab.

The failure mode of the strengthened slab is evaluated using the proposed process regarding brittle-failure prevention by adjusting CFRP thickness. For the end span of the strengthened slab, Table 6 shows the design process used to induce ductile failure. With a CFRP thickness of 0.53 mm, the retrofitted slab can fail in ductile failure mode D-3_{en}, as shown in Figure 9a. There is a design-factored load of 51.05 kN/m and an ultimate failure load of 62.59 kN/m, which are higher by 2.07 times and 2.01 times, respectively, than the existing RC slab.

Table 6. Analysis of the retrofitted slab considering brittle-failure prevention.

Process	End Span	Interior Span
1. Assume CFRP thicknesses	$t_F = 1$ mm	$t_F = 1$ mm
2. Check overlay strength	For both spans $f'_H \geq \dfrac{0.003(40000)}{1.445}\left(\dfrac{1}{30}\right)^2 + \dfrac{410(426/900)}{0.7225(30)} = 9.05$ MPa (OK) $f'_H \geq 0.15(32) + \dfrac{0.003(40000)}{1.7}\left(\dfrac{1}{30}\right)^2 + \dfrac{410(426/900)}{0.85(30)} = 12.49$ MPa (OK)	
3. Compute the design strain of FRP at support	$\varepsilon_{fd} = 0.41\sqrt{\dfrac{32}{1(40000)(1)}} = 0.0116 \leq 0.9\varepsilon_{fu} = 0.0128$	
4. Assume neutral axis depth	At the N2 section: $c_N = 27.61$ mm At mid-span section: $c_P = 10.26$ mm	At support sections (N3 and N4): $c_N = 27.64$ mm At mid-span section: $c_P = 10.26$ mm
5. Compute FRP stress (f_{fe}), FRP strain (ε_{fe}), and concrete strain (ε_c)	At the N2 section $\varepsilon_{fe,N} = 0.003\left(\dfrac{145-27.61}{27.61}\right) - 0.00034 = 0.0124 > \varepsilon_{fd}$ $\rightarrow \varepsilon_{fe,N} = \varepsilon_{fd} = 0.0116$ $\varepsilon_{c,N} = (0.0116 + 0.00034)\left(\dfrac{27.61}{145-27.61}\right) = 0.0028$ $f_{fe,N} = 0.0116(40{,}000) = 463.86$ MPa At mid-span section $\varepsilon_{fe,P} = 0.003\left(\dfrac{30-10.26}{10.26}\right) = 0.0058 \leq \varepsilon_{fd}$ $\varepsilon_{c,P} = 0.0058\left(\dfrac{10.26}{30-10.26}\right) = 0.003$ $f_{fe,P} = 0.0058(40{,}000) = 230.97$ MPa	At support sections $\varepsilon_{fe,N} = 0.003\left(\dfrac{145-27.64}{27.64}\right) - 0.00027 = 0.0125 > \varepsilon_{fd}$ $\rightarrow \varepsilon_{fe,N} = \varepsilon_{fd} = 0.0116$ $\varepsilon_{c,N} = (0.0116 + 0.00027)\left(\dfrac{27.64}{145-27.64}\right) = 0.0028$ $f_{fe,N} = 0.0116(40{,}000) = 463.86$ MPa At mid-span section (same as the end span case)
6. Compute reinforced steel stress (f_s) and strain (ε_s)	At the N2 section $\varepsilon_{s,N} = (0.0116 + 0.00034)\left(\dfrac{110-27.61}{145-27.61}\right) = 0.0084$ $f_{s,N} = 0.0084(200000) = 1680$ MPa $> f_y = 410$ MPa $\rightarrow f_{s,N} = f_y = 410$ MPa At mid-span section $\varepsilon_{s,P} = 0.003\left(\dfrac{110+30+1-10.26}{10.26}\right) = 0.0382$ $f_{s,P} = 0.0382(200{,}000) = 7640$ MPa $> f_y = 410$ MPa $\rightarrow f_{s,P} = f_y = 410$ MPa	At the support sections $\varepsilon_{s,N} = (0.0116 + 0.00027)\left(\dfrac{110-27.64}{145-27.64}\right) = 0.0083$ $f_{s,N} = 0.0083(200{,}000) = 1660$ MPa $> f_y = 410$ MPa $\rightarrow f_{s,N} = f_y = 410$ MPa At mid-span section (same as the end span case)
7. Check the equilibrium condition	At the N2 section, due to $\varepsilon_{c,N} < \varepsilon_{cu}$ $\varepsilon'_{c,N} = \dfrac{1.7(32)}{26600} = 0.002;$ $\beta_{1,N} = \dfrac{4(0.002) - 0.0028}{6(0.002) - 2(0.0028)} = 0.807$ $\alpha_{1,N} = \dfrac{3(0.002)(0.0028) - (0.0028)^2}{3(0.807)(0.002)^2} = 0.922$ $c_N = \dfrac{426(410) + 900(463.86)}{0.922(32)(0.807)(900)} = 27.61$ mm (OK) At mid-span section, due to $\varepsilon_{c,P} = \varepsilon_{cu}$ $\beta_{1,P} = \beta_1 = 0.65;\ \alpha_{1,P} = \alpha_1 = 0.85$ $c_P = \dfrac{426(410) + 900(230.97)}{0.85(75)(0.65)(900)} = 10.26$ mm (OK)	At support sections, due to $\varepsilon_{c,N} < \varepsilon_{cu}$ $\varepsilon'_{c,N} = \dfrac{1.7(32)}{26600} = 0.002;$ $\beta_{1,N} = \dfrac{4(0.002) - 0.0028}{6(0.002) - 2(0.0028)} = 0.806$ $\alpha_{1,N} = \dfrac{3(0.002)(0.0028) - (0.0028)^2}{3(0.806)(0.002)^2} = 0.923$ $c_N = \dfrac{426(410) + 900(463.86)}{0.923(32)(0.806)(900)} = 27.64$ mm (OK) At mid-span section (same as the end span case)
8. Compute strength in flexure and shear 8.1 Compute strength in flexure at the support section	$M_{ns,N} = \dfrac{426(410)}{10^6}\left(110 - \dfrac{0.807(27.61)}{2}\right) = 17.27$ kNm $M_{nf,N} = \dfrac{(900 \times 1)(463.86)}{10^6}\left(145 - \dfrac{0.807(27.61)}{2}\right) = 55.89$ kNm $\phi_f M_{n,N} = 0.9[17.27 + 0.85(55.88)] = 58.29$ kNm	$M_{ns,N} = \dfrac{426(410)}{10^6}\left(110 - \dfrac{0.806(27.64)}{2}\right) = 17.27$ kNm $M_{nf,N} = \dfrac{(900 \times 1)(463.86)}{10^6}\left(145 - \dfrac{0.806(27.64)}{2}\right) = 55.89$ kNm $\phi_f M_{n,N} = 0.9[17.27 + 0.85(55.88)] = 58.29$ kNm
8.2 Compute strength in flexure at the mid-span section	$M_{ns,P} = \dfrac{426(410)}{10^6}\left(110 + 30 + 1 - \dfrac{0.65(10.26)}{2}\right) = 24.04$ kNm $M_{nf,P} = \dfrac{(900 \times 1)(230.97)}{10^6}\left(30 - \dfrac{0.65(10.26)}{2}\right) = 5.54$ kNm $\phi_f M_{n,P} = 0.9[24.04 + 0.85(5.54)] = 25.88$ kNm	same as end span case
8.3 Compute strength in shear	$\phi_v V_n = \dfrac{0.75}{10^3}\left(110\sqrt{32} + 30\sqrt{75}\right)\dfrac{900}{6} = 99.23$ kN	same as end span case

Table 6. *Cont.*

Process	End Span	Interior Span
9. Define design factored load	$w_u = \min(53.6; 66.38) = 53.6$ kN/m $w_{u,M} = \min(53.6; 137.96; 86.23) = 53.6$ kN/m $w_{u,V} = \min(76.33; 66.38) = 66.38$ kN/m	$w_u = \min(71.46; 82.69) = 71.46$ kN/m $w_{u,M} = \min(71.46; 111.32) = 71.46$ kN/m $w_{u,V} = 82.68$ kN/m
10. Define failure mode and failure load	DB-3a$_{en}$ according to Figure 6a Equation (18), $w_f = \dfrac{2(99.23)}{1.15(2.6)} = 66.38$ kN/m	DB-2$_{in}$ according to Figure 6b Equation (21), $w_f = \dfrac{2(99.23)}{1(2.4)} = 82.68$ kN/m
Adjust iteratively CFRP thicknesses to achieve ductile failure mode	It can be achieved with $t_F = 0.53$ mm; $w_u = 51.05$ kN/m; failure mode D-3$_{en}$; $w_f = 62.95$ kN/m, as shown in Figure 9a.	It can be achieved with $t_F = 0.62$ mm; $w_u = 68.92$ kN/m; failure mode D-2$_{in}$; $w_f = 82.38$ kN/m. However, to be consistent with the end span, $t_F = 0.53$ mm; $w_u = 68.17$ kN/m; failure mode D-2$_{in}$; $w_f = 76.48$ kN/m, as shown in Figure 9b.

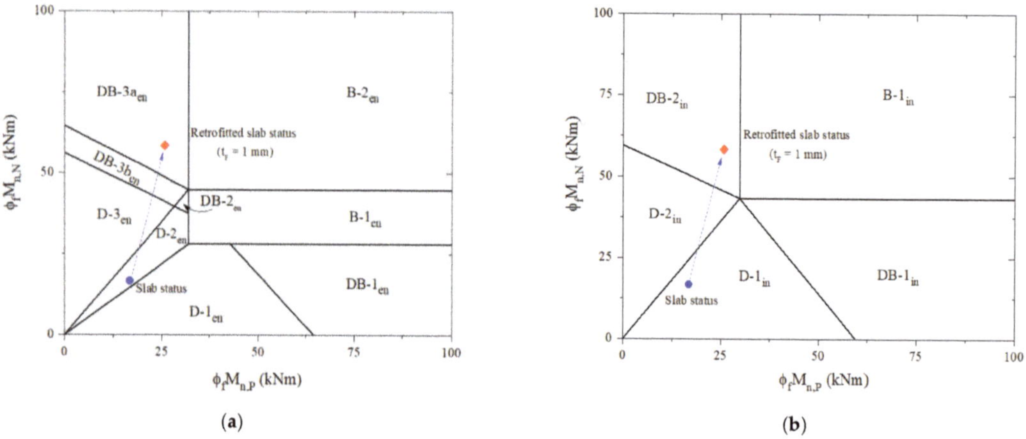

Figure 8. Predicting failure limits and failure mode for the retrofitted slab: (**a**) the end span; (**b**) the interior span.

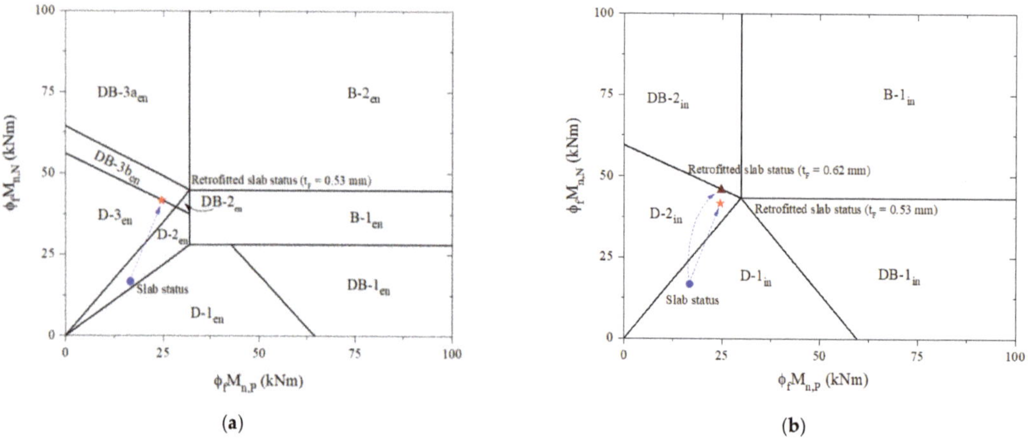

Figure 9. Predicting failure limits and failure mode considering brittle-failure prevention for the retrofitted slab: (**a**) the end span; (**b**) the interior span.

The interior span of the retrofitted slab is also similarly analyzed. The design procedure for preventing brittle failure can be achieved with 0.62-mm-thick CFRP laminate. Based on Table 6, the design-factored load is 68.92 kN/m. Figure 9b shows that the ductile failure mode D-2$_{in}$ is determined to correspond to the ultimate failure load of 82.38 kN/m. Once the thickness of the CFRP laminate applied for the interior span is thicker than the end span or vice versa, it should be adjusted to be consistent with practical design requirements. With 0.53-mm-thick CFRP, the design-factored load of the strengthened slab is estimated at 68.17 k/m, which is 2.13 times higher than that of the reference slab. In accordance with Figure 9b, the failure mode remains D-2_{in} relevant to the ultimate failure load for the retrofitted slab of 76.48 kN/m, increasing by 1.97 times over the reference slab.

The proposed design process allows for the optimization of bearing capacity and increased safety required in strengthening slabs using CFRP-HPC retrofit systems. Compared with the 1-mm-thick CFRP strengthening solution, it can reduce the amount of CFRP by up to 47% without considerable changes in w_u and w_f by up to 5% and 8%, respectively, as shown in Table 6. The failure mode of the end span or the interior span will determine the failure mode of RC slabs depending on the length ratio of the end-to-interior span, slab structure, and load characteristics. Thus, it is necessary to have detailed analyses for each type of span in continuous RC slabs. The retrofit results for the end span and interior span of the strengthened RC slab are shown in Table 7.

Table 7. Summarize the retrofit results of the strengthened RC slab.

Span	Failure Mode	w_u (kN/m)		w_f (kN/m)		t_F (mm)
Existing end span	D-2_{en}	24.70	[100%]	31.30	[100%]	-
Retrofitted end span	D-3_{en}	51.05	[207%]	62.95	[201%]	0.53
Existing interior span	D-1_{in}	32.00	[100%]	39.20	[100%]	-
Retrofitted interior span	D-2_{in}	68.92	[215%]	82.83	[211%]	0.62
Retrofitted interior span (for consistency)	D-2_{in}	68.17	[213%]	76.48	[195%]	0.53

5. Conclusions

The present study describes the design process for preventing brittle failure in strengthening RC slabs with hybrid FRP-HPC retrofit systems based on novel failure limit classifications. The failure limits of the end and interior span of continuous RC slabs in relation to moment- and shear- carrying capacities are discussed. The formulas for determining the design-factored load and the ultimate failure load corresponding to the failure mode are also recommended. Based on ACI 440.2R, the effectiveness of the proposed procedure of strengthened slabs using FRP-HPC retrofit systems is confirmed via the design example. The obtained results can be used to draw the following conclusions.

FRP thickness can significantly affect bearing carrying capacities and failure modes of retrofitted slabs. The case study indicates that the retrofitted slab using 0.53-mm-thick CFRP can increase the ultimate failure load by 2.13 times and fail in ductile failure mode.

The approach can optimize material strength while preventing brittle failure and reducing CFRP quantification by up to 47% without noticeable changes in bearing capacities, resulting in economic and safety benefits.

The proposed design process would encourage this strengthening technique to be applied early in practice due to its simplicity and efficiency.

The study is theoretical. Consequently, further experimental studies are recommended to confirm the suitability of the proposed method, along with evaluating the impacts of overlay regarding thickness and compressive strength, and bond strength of concrete-to-FRP interfaces.

Author Contributions: Conceptualization, J.J.K.; methodology, T.H.H. and J.J.K.; software, H.Q.N. and J.K.P.; validation, T.H.H. and J.K.P.; formal analysis, H.Q.N. and J.K.P.; investigation, T.H.H. and J.J.K.; data curation, J.J.K. and H.Q.N.; writing—original draft preparation, H.Q.N. and J.K.P.; writing—review and editing, H.Q.N., J.K.P., T.H.H. and J.J.K.; visualization, T.H.H.; supervision, J.J.K.; project administration, T.H.H.; funding acquisition, T.H.H. All authors have read and agreed to the published version of the manuscript.

Funding: This research was funded by the Korea Institute of Ocean Science and Technology (KIOST) (project no. PEA0133).

Institutional Review Board Statement: Not applicable.

Informed Consent Statement: Not applicable.

Data Availability Statement: Not applicable.

Conflicts of Interest: The authors declare no conflict of interest.

Nomenclature

A_F, A_s	Area of CFRP and tensile steel
b	Width of an existing slab
c	The distance between the extreme compression fiber and the neutral axis
C_E	Environmental reduction coefficient
$C_{m,Ni}$	Moment coefficients at support section i^{th}
$C_{m,Pi}$	Moment coefficients at mid-span section i^{th}
C_{vi}	Shear coefficients at section i^{th}
d	The distance between the extreme compression fiber and the center of the steel
E_c, E_s, E_{fe}	Elastic modulus of concrete, steel, and CFRP
f'_c, f'_H	Compressive strength concrete of existing slab and overlay
f_{fe}	CFRP effective stress
f_{fu}	Design ultimate strength of CFRP
f^*_{fu}	FRP's ultimate tensile strength, according to the manufacturer
f_s	Tension steel's stress
f_y	Yield stress of tension steel
h	Height of an existing RC slab
I_{cr}	Cracked moment
k	The ratio of the neutral axis depth to tensile steel depth measured from extreme compression fiber
l_{ni}	Length of clear span i^{th}
n	The number of CFRP layers
M_n, V_n	Moment and shear carrying capacity
$M_{n,P}$, $M_{n,N}$	Mid-span and support sections' moment-carrying capacities
M_{ns}, M_{nf}	Moments contributed by tensile steel and CFRP
M_{N1}	Moment-carrying capacities of the N1 section
$M_{D,N2}$	Moment-carrying of the N-2 section
M_u, V_u	Factored moment and shear at sections
t_F, t_H	The thickness of CFRP and HPC overlay
w_f	Ultimate failure load
w_u	Design factored load
w_{uM}, w_{uV}	Design factored load follow moment and shear carrying capacities
ϕ_f, ϕ_v	Flexural and shear strength reduction factors
ψ_f	CFRP strength reduction factor
α_1, β_1	Stress block factors
ε_{bi}	Existing state strain of CFRP installation
ε_{cu}, ε_{fu}	Ultimate strains of concrete and CFRP
ε_{fd}	Debonding strain of CFRP
ε_{fe}, ε_s	Strains of CFRP and tensile steel
γ_c	Concrete unit weight

References

1. Maxineasa, S.G.; Taranu, N. 24-Life cycle analysis of strengthening concrete beams with FRP. In *Eco-Efficient Repair and Rehabilitation of Concrete Infrastructures*; Pacheco-Torgal, F., Melchers, R.E., Shi, X., De Belie, N., van Tittelboom, K., Sáez, A., Eds.; Woodhead Publishing: Cambridge, UK, 2018; pp. 673–721.
2. Ma, Z.; Shen, J.; Wang, C.; Wu, H. Characterization of sustainable mortar containing high-quality recycled manufactured sand crushed from recycled coarse aggregate. *Cem. Concr. Compos.* **2022**, *132*, 104629. [CrossRef]
3. Martínez, S.; de Diego, A.; Castro, V.J.; Echevarría, L.; Barroso, F.J.; Rentero, G.; Soldado, R.P.; Gutiérrez, J.P. Strengthening of Low-Strength Concrete Columns with Fibre Reinforced Polymers. Full-Scale Tests. *Infrastructures* **2020**, *5*, 91. [CrossRef]
4. Hariri-Ardebili, M.A.; Sanchez, L.; Rezakhani, R. Aging of Concrete Structures and Infrastructures: Causes, Consequences, and Cures. *Adv. Mater. Sci. Eng.* **2020**, *2020*, 9370591. [CrossRef]

5. Nguyen, X.T.; Park, J.S. Flexural Behavior of Steel Beams Strengthened with CFRP Under Fire. *Int. J. Steel Struct.* **2022**, *22*, 1769–1785. [CrossRef]
6. Vu, H.D.; Phan, D.N. A framework for predicting the debonding failure modes of RC beams strengthened flexurally with FRP sheets. *Innov. Infrastruct. Solut.* **2022**, *7*, 292. [CrossRef]
7. Haji, M.; Naderpour, H.; Kheyroddin, A. Experimental study on influence of proposed FRP-strengthening techniques on RC circular short columns considering different types of damage index. *Compos. Struct.* **2019**, *209*, 112–128. [CrossRef]
8. Barris, C.; Sala, P.; Gómez, J.; Torres, L. Flexural behaviour of FRP reinforced concrete beams strengthened with NSM CFRP strips. *Compos. Struct.* **2020**, *241*, 112059. [CrossRef]
9. Basaran, B.; Kalkan, I. Development length and bond strength equations for FRP bars embedded in concrete. *Compos. Struct.* **2020**, *251*, 112662. [CrossRef]
10. Mukhtar, F.M.; Arowojolu, O. Recent developments in experimental and computational studies of hygrothermal effects on the bond between FRP and concrete. *J. Reinf. Plast. Compos.* **2020**, *39*, 422–442. [CrossRef]
11. Wang, X.; Yang, Y.; Yang, R.; Liu, P. Experimental Analysis of Bearing Capacity of Basalt Fiber Reinforced Concrete Short Columns under Axial Compression. *Coatings* **2022**, *12*, 654. [CrossRef]
12. Xian, G.; Guo, R.; Li, C.; Wang, Y. Mechanical performance evolution and life prediction of prestressed CFRP plate exposed to hygrothermal and freeze-thaw environments. *Compos. Struct.* **2022**, *293*, 115719. [CrossRef]
13. Li, C.; Guo, R.; Xian, G.; Li, H. Effects of elevated temperature, hydraulic pressure and fatigue loading on the property evolution of a carbon/glass fiber hybrid rod. *Polym. Test.* **2020**, *90*, 106761. [CrossRef]
14. Lu, Z.; Li, W.; Zeng, X.; Pan, Y. Durability of BFRP bars and BFRP reinforced seawater sea-sand concrete beams immersed in water and simulated seawater. *Constr. Build. Mater.* **2023**, *363*, 129845. [CrossRef]
15. Mosallam, A.; Taha, M.R.; Kim, J.; Nasr, A. Strength and ductility of RC slabs strengthened with hybrid high-performance composite retrofit system. *Eng. Struct.* **2012**, *36*, 70–80. [CrossRef]
16. Ho, H.V.; Choi, E.; Park, S.J. Investigating stress distribution of crimped SMA fibers during pullout behavior using experimental testing and a finite element model. *Compos. Struct.* **2021**, *272*, 114254. [CrossRef]
17. *ACI PRC-440.2R-17*; Guide for the Design and Construction of Externally Bonded FRP Systems for Strengthening Concrete Structures. ACI: Farmington Hills, MI, USA, 2017.
18. Nguyen, H.Q.; Nguyen, T.N.M.; Lee, D.H.; Kim, J.J. The Effects of Bond-Slip Laws on the Debonding Failure and Behavior of Flexural Strengthened RC Slabs in Hybrid FRP Retrofit Systems. *Materials* **2022**, *15*, 7453. [CrossRef] [PubMed]
19. Kim, N.; Shin, Y.S.; Choi, E.; Kim, H.S. Relationships between interfacial shear stresses and moment capacities of RC beams strengthened with various types of FRP sheets. *Constr. Build. Mater.* **2015**, *93*, 1170–1179. [CrossRef]
20. Lezgy-Nazargah, M.; Dezhangah, M.; Sepehrinia, M. The Effects of Different FRP/Concrete Bond–Slip Laws on the 3D Nonlinear FE Modeling of Retrofitted RC Beams—A Sensitivity Analysis. *Steel Compos. Struct.* **2018**, *26*, 347–360.
21. Moon, J.; Taha, M.R.; Kim, J. Flexural Strengthening of RC Slabs Using a Hybrid FRP-UHPC System Including Shear Connector. *Adv. Mater. Sci. Eng.* **2017**, *2017*, 4387545. [CrossRef]
22. Gunes, O.; Lau, D.; Tuakta, C.; Büyüköztürk, O. Ductility of FRP–concrete systems: Investigations at different length scales. *Constr. Build. Mater.* **2013**, *49*, 915–925. [CrossRef]
23. Abdallah, M.; Al Mahmoud, F.; Khelil, A.; Mercier, J.; Almassri, B. Assessment of the flexural behavior of continuous RC beams strengthened with NSM-FRP bars, experimental and analytical study. *Compos. Struct.* **2020**, *242*, 112127. [CrossRef]
24. Dai, M.N.; Toong, K.C.; Hee, K.C. Brittle Failure and Bond Development Length of CFRP-Concrete Beams. *J. Compos. Constr.* **2001**, *5*, 12–17.
25. Rabinovitch, O.; Frostig, Y. Experiments and analytical comparison of RC beams strengthened with CFRP composites. *Compos. Part B Eng.* **2003**, *34*, 663–677. [CrossRef]
26. Breveglieri, M.; Barros, J.; Dalfre, G.; Aprile, A. Assessment of the Effectiveness of a NSM-CFRP Flexural Strengthening Technique for Continuous RC Slabs. In Proceedings of the Czech Concrete Society (CBS) 2011, Prague, Czech Republic, 8–10 June 2011.
27. Yasmeen, T.O.; Heyden, S.; Dahlblom, O. Evaluation of Parameters of Bond Action between FRP and Concrete. *J. Compos. Constr.* **2013**, *17*, 626–635.
28. Hawileh, R.A.; Rasheed, H.A.; Abdalla, J.A.; Al-Tamimi, A.K. Behavior of reinforced concrete beams strengthened with externallybonded hybrid fiber reinforced polymer systems. *Mater. Des.* **2014**, *53*, 972–982. [CrossRef]
29. Bocciarelli, M.; Pisani, M.A. Modified force method for the nonlinear analysis of FRP reinforced concrete beams. *Compos. Struct.* **2015**, *131*, 645–653. [CrossRef]
30. Haciyev, V.; Sofiyev, A.; Kuruoglu, N. Free bending vibration analysis of thin bidirectionally exponentially graded orthotropic rectangular plates resting on two-parameter elastic foundations. *Compos. Struct.* **2018**, *184*, 372–377. [CrossRef]
31. Al-Rousan, R.; Issa, H.S.M. Performance of reinforced concrete slabs strengthened with different types and configurations of CFRP. *Compos. Part B Eng.* **2012**, *43*, 510–521. [CrossRef]
32. Duy, N.P.; Dang, V. Limiting Reinforcement Ratios for Hybrid GFRP/Steel Reinforced Concrete Beams. *Int. J. Eng. Technol. Innov.* **2021**, *11*, 1–11.
33. Wei, L.; Ueda, T.; Matsumoto, K.; Zhu, J.-H. Experimental and analytical study on the behavior of RC beams with externally bonded carbon-FRCM composites. *Compos. Struct.* **2021**, *273*, 114291. [CrossRef]

34. Nguyen, H.Q.; Nguyen, T.N.M.; Lee, D.H.; Kim, J.J. A Design Method to Induce Ductile Failure of Flexural Strengthened One-Way RC Slabs. *Materials* **2021**, *14*, 7647. [CrossRef] [PubMed]
35. Kim, J.; Noh, H.-C.; Taha, M.R.; Mosallam, A. Design limits for RC slabs strengthened with hybrid FRP–HPC retrofit system. *Compos. Part B Eng.* **2013**, *51*, 19–27. [CrossRef]
36. Nguyen, H.Q.; Yang, K.; Kim, J.J. An Efficient Method for Optimizing HPC-FRP Retrofit Systems of Flexural Strengthened One-Way Continuous Slabs Based on ACI 440.2R. *Materials* **2022**, *15*, 8430. [CrossRef] [PubMed]
37. *ACI 318M-19*; Building Code Requirements for Structural Concrete and Commentary, Metric. ACI: Farmington Hills, MI, USA, 2019.

materials

Article

The Impact of Different Parameters on the Formwork Pressure Exerted by Self-Compacting Concrete

Yaser Gamil *, Andrzej Cwirzen, Jonny Nilimaa and Mats Emborg

Building Materials, Department of Civil, Environmental and Natural Resources Engineering, Lulea University of Technology, 97187 Luleå, Sweden
* Correspondence: yaser.gamil@ltu.se

Abstract: Despite the advantageous benefits offered by self-compacting concrete, its uses are still limited due to the high pressure exerted on the formwork. Different parameters, such as those related to concrete mix design, the properties of newly poured concrete, and placement method, have an impact on form pressure. The question remains unanswered on the degree of the impact for each parameter. Therefore, this study aims to study the level of impact of these parameters, including slump flow, T500 time, fresh concrete density, air content, static yield stress, concrete setting time, and concrete temperature. To mimic the casting scenario, 2 m columns were cast at various casting rates and a laboratory setup was developed. A pressure system that can wirelessly and continuously record pressure was used to monitor the pressure. Each parameter's impact on the level of pressure was examined separately. Casting rate and slump flow were shown to have a greater influence on pressure. The results also demonstrated that, while higher thixotropy causes form pressure to rapidly decrease, a high casting rate and high slump flow lead to high pressure. This study suggests that more thorough analysis should be conducted of additional factors that may have an impact, such as the placement method, which was not included in this publication.

Keywords: formwork; pressure; self-compacting concrete; parameters

Citation: Gamil, Y.; Cwirzen, A.; Nilimaa, J.; Emborg, M. The Impact of Different Parameters on the Formwork Pressure Exerted by Self-Compacting Concrete. *Materials* **2023**, *16*, 759. https://doi.org/10.3390/ma16020759

Academic Editors: Nikolina Zivaljic, Hrvoje Smoljanović and Ivan Balić

Received: 20 December 2022
Revised: 3 January 2023
Accepted: 9 January 2023
Published: 12 January 2023

1. Introduction

The formwork pressure exerted by self-compacting concrete (SCC) is considerably higher than that of normal concrete due to its high flowability ([1]). This has slowed down the acceptance of SCC in civil engineering projects [2]. The current design models assume full hydrostatic pressure while designing the form ([1,3,4]). International standards such as ACI347R [5] and the European guidelines [6] have also suggested the design to be fully hydrostatic. However, other findings have proven otherwise, showing that the actual form pressure is less than the hydrostatic pressure [7–11]. This raises the question as to how much lower the actual pressure can be when casting with SCC. This also introduces the possibility of using SCC more often and shifting from a reliance on normal concrete.

Previous studies have shown that form pressure is affected by several parameters, and these parameters are generally classified into parameters related to mixture design and material properties, the fresh concrete properties of the mix, and the placement method and speed of casting [12]. However, the degree of impact of these parameters is still unknown.

The parameters that affect the amount of pressure and its distribution over the form's height are relatively interrelated. For example, mixture design affects the flowability of the concrete and has an influence on the pressure and its reduction over time. To understand that phenomenon, [13] conducted a study incorporating factorial design to investigate the impacting parameters of fresh normal concrete on pressure. Their analysis showed that formwork shape and coarse aggregate have a minor impact on pressure, temperature has an inverse relationship with pressure, and form size has a major effect on pressure, with narrow sections producing less pressure. Similarly, [14,15] found that wall friction forces between concrete and formwork affect the amount of pressure, with higher friction leading to higher pressure.

A different study by [16] investigated the effect of water-to-cement ratio and the amount of water-reducing admixture and found that both thixotropy and form pressure are affected by the w/cm ratio, whereby mixes with 0.46 w/cm showed greater initial pressure compared to mixes with 0.4 and 0.36. The reason for this is thought to be the increased water paste content and a reduction in coarse aggregate, which thus lowers the shear strength of fresh concrete. Additionally, [17] studied the effect of viscosity-enhancing admixtures (VEA) on the pressure of SCC and found that the use of VEA results in lower form pressure, while water-reducing admixtures exhibited higher pressure with higher dosages.

The important effect of reinforcing bars has also been demonstrated, as shown in congested reinforcement where pressure was rather lower than it was when few reinforcing bars were used [8,18,19]. The effect of mixture composition has also been shown to have high importance in terms of form pressure [20]. For example, powdered materials in the mix considerably affect the amount of pressure, with higher content increasing pressure and reducing workability [21].

Casting rate and placement methods have a significant effect on pressure, with a higher casting rate leading to higher pressure [1,4,16,22–28]. Similarly, casting characteristics also affect pressure, with concrete temperature accelerating pressure decay and greater casting depth leading to higher pressure [20,29]. In this article, an extended investigation is performed while varying slump flow, T500 time, fresh concrete density, air content in concrete, static yield stress, and the ambient and concrete temperature of concrete. Additionally, their impact on the amount of form pressure and its distribution was also investigated. The use of a state-of-the-art pressure system helped to realize more accurate and real-time pressure monitoring during the casting process, which thus allowed for instant decision making regarding the amount of pressure and casting rate.

2. Methodology

2.1. Experimental Setup

A laboratory setup was developed to simulate the casting process. A 2 m plastic pipe was vertically fixed to the column and given rigid support to avoid any tilt, trembling, or collapse (Figure 1). Pressure sensors were installed and mounted to the pipe by drilling 50 mm holes to enable direct contact with the concrete. Four sensors were installed. The first sensor was placed about 0.05 m from the ground, the second sensor 0.5 m from the ground, the third sensor 1 m from the ground, and the fourth sensor 1.5 m from the ground. The concrete was prepared and mixed in the laboratory, and the number of batches was governed by the casting rate (for example, casting 1 m/h required 2 batches of concrete). The casting was performed from the top and a plastic cone was used to avoid any spillage of concrete.

2.2. The Instrument for Measuring Pressure

With the help of cutting-edge sensors and the Internet of Things, a system developed by PERI was used in the experiments with the idea of turning cast-in-place concrete into a digital process. The sensor system consisted of data transmitters that provided readings from the sensors to the cloud. As seen in Figure 2, the pressure sensors were mounted to the pipe to facilitate direct contact with the concrete. Real-time data regarding pressure during casting were displayed instantly on the computer. This made it simple to check the pressure. If pressure was low, casting pace could be increased; if pressure was high, casting could be slowed down.

Figure 1. Laboratory setup.

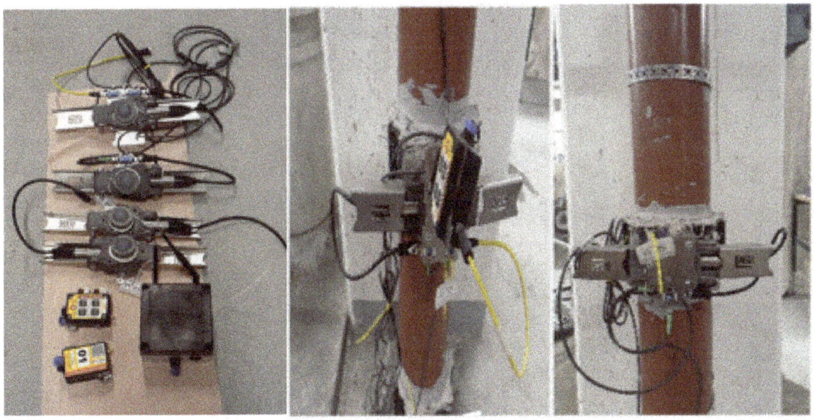

Figure 2. Pressure measuring system (the system in the left is developed by PERI).

2.3. Mix Design and Test Plan

The mixture used for casting was a commercial mixture with fixed proportions, and only the type of cement and superplasticizers were altered depending on the required slump flow. Different variations were then used to study the significance level of different parameters, such as slump flow, T500 time, fresh concrete density, air content, static yield stress, concrete setting time, and concrete temperature.

For all batches, the water-to-cement ratio was kept at 0.59, and the superplasticizer (SP) used was MasterGlenium 592. A small mixer was used in the laboratory to mix the concrete in various batches. New batches of concrete were created for each layer of casting while preserving the same characteristics. To ensure the same flowability, the measurement of the slump was kept track of and maintained for each batch. Table 1 shows the initial mix design used for casting the pipes. The test plan used for varying the input parameters is described in Table 2. The test plan and mix proportions are part of a pre-established plan that was used by the authors in a different study to assess mathematical models for forecasting form pressure when casting with SCC.

Table 1. Mix proportions.

Material	Cement: BAS/ANLFA	Filler	Agg (0–8) mm	Agg (8–16) mm	Superplasticizer (MG592)	Water
Kg/m^3	350	140	978.1	652.08	Varies according to the required slump flow	207

Table 2. Test plan and parameter variations.

<table>
<tr><th>Test No</th><th>Targeted Slump Flow (mm)</th><th>Casting Rate (m/h)</th><th>Cement Type</th><th>Variations</th></tr>
<tr><td>1
2
3
4</td><td>700–750</td><td>0.25
0.5
1
4</td><td rowspan="2">CEM II/A-V 52.5 N (BAS)</td><td>Casting rate</td></tr>
<tr><td>5
6
7
8</td><td>700–750
600–650
500–550
400–450</td><td>0.5</td><td>Slump flow</td></tr>
<tr><td>9
10
11</td><td>700–750</td><td>0.25
0.5
4</td><td rowspan="2">CEM II/A-V 42.5 N (ANLFA)</td><td>Cement type, casting rate</td></tr>
<tr><td>12
13
14</td><td>700–750
600–650
500–550</td><td>0.5</td><td>Cement type, slump flow</td></tr>
</table>

2.4. Cement Types

Two cement types were used. The first cement was Bascement (BAS) CEM II/A-V 52.5 N Portland-fly ash, which is recommended for use in normal concrete work, per the supplier Cementa. This cement has a different setting time. The second cement, known as Anläggningscement FA (ANLFA), is a Portland-fly ash cement type (CEM II/A-V 42.5 N MH/LA/NSR) used in mass concrete which has mild heat requirements in terms of hydration (cementa.se).

2.5. Casting Rate

Casting rate is a significant parameter that affects the amount of lateral pressure when casting with SCC. In the laboratory plan, casting rate was set at 0.25 m/h, 0.5 m/h, 1 m/h, and 4 m/h. Using these diverse rates, the effect of casting rate was then investigated.

2.6. Slump Flow

Different types of slump flow were created by controlling the amount of superplasticizers so as to produce the desired slump flow diameter, with the first interval being 700–750 mm, the second 600–650 mm, the third 500–550 mm, and the fourth interval 400–450 mm.

3. Testing of Fresh Concrete

Before casting, the concrete was tested for different properties, including slump flow, T500 time, air content, fresh density, setting time, and static yield stress. The tests were performed according to the guidelines specified for each test in the European guidelines for SCC [6], except for setting time, which was performed according to the German standard [25], and static yield stress, which was performed according to the portable vane test developed by [1].

The setting bag test was conducted as specified by [25] using a polyethylene plastic bag containing around 8 L of concrete in a bucket. The consistency was checked every 30 min by manually applying a force of around 50 N to the bag surface via thumb press and examining the concrete imprint. Setting time was indicated by the depth of the impression. Setting time is calculated as 1.25 times the amount of time needed to dent something less than 1 mm. For instance, if recording of setting time begins at te and kb = 5 h, the final setting time is te = 1.25 × 5 = 6.25 h.

For each concrete recipe, the portable vane test or torque test was performed instantly to measure static yield stress. This was used to investigate the structural behavior of SCC at rest and involves using four-blade vanes and a torque gauge to measure the torque required to break the structure. The static yield stress (in Pa) of the concrete at rest can be determined from the measured torque value and vane geometry. The method is explained in detail in [1].

4. Results

4.1. Form Pressure Monitoring

An example of pressure monitoring is shown below for one of the tests of SCC with a casting rate of 0.25 m/h. Pressure was monitored over 8 h of casting, as demonstrated in Figure 3.

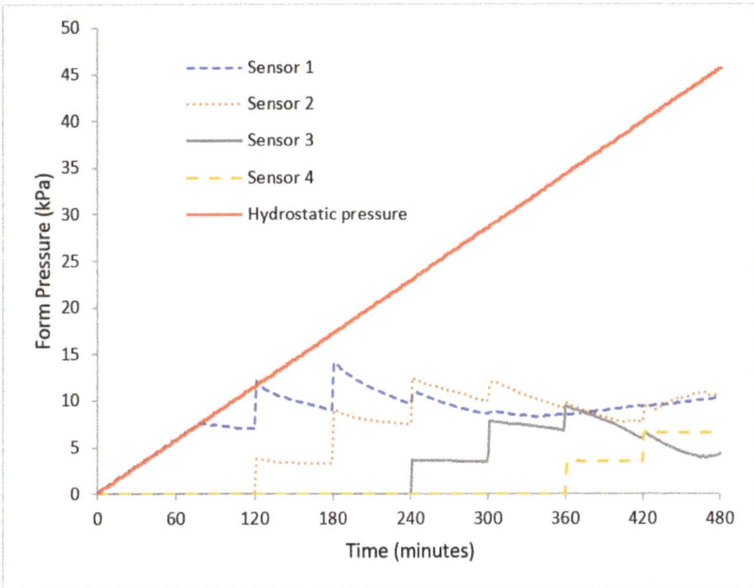

Figure 3. Pressure monitoring over time for 0.25 m/h casting rate.

Figure 3 demonstrates pressure monitoring over an 8 h period of casting on a real-time basis, with the system capable of sending a reading every minute. From the results, it can be seen that the actual pressure exerted by the concrete is far less than the hydrostatic design which the form is supposed to bear, thus suggesting that it is not a cost-effective design and that there is a possibility of casting at a rate faster than 0.25 m/h. Hydrostatic pressure was calculated using the principle formula pressure $P_{hydro} = \rho gh$, where ρ is concrete density in kg/m^3, h is casting height (m), and g is gravitational acceleration (9.8 m/s^2).

Figure 4 shows selected values of the pressure across the height of the form. It is indicated that pressure is affected by casting rate, with a higher casting rate resulting in higher form pressure. This was also proven by [10], who indicated that pressure is highly affected by casting rate. By varying the amount of superplasticizer supplied, a modification in slump flow was also achieved, and the results are shown in Figure 5.

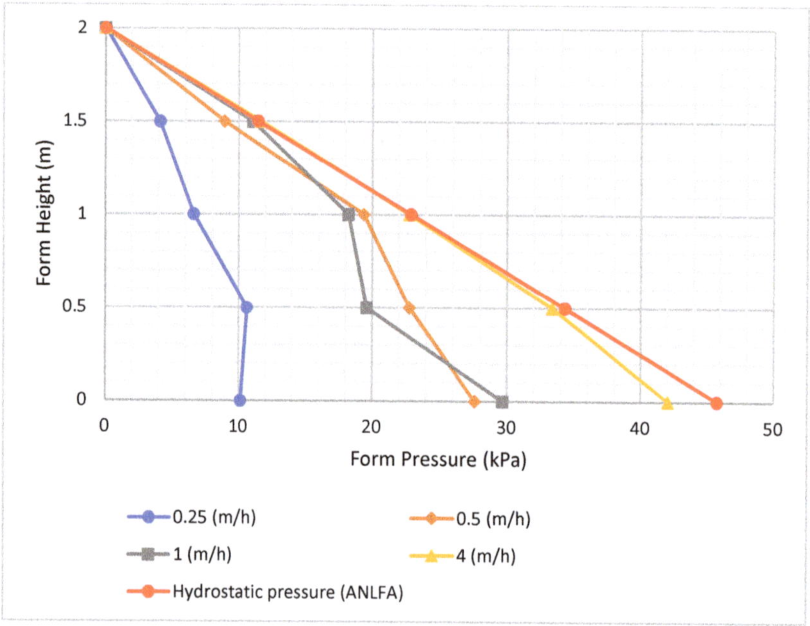

Figure 4. The effect of casting rate on pressure.

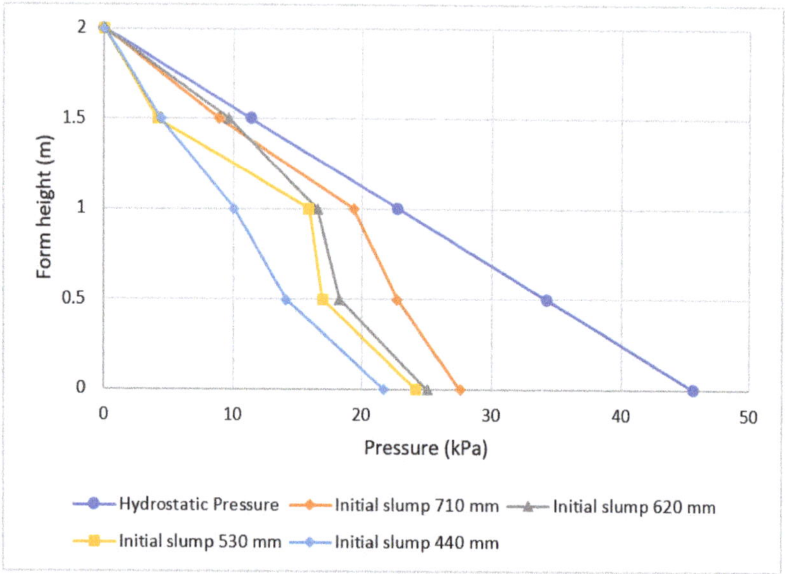

Figure 5. Effect of initial slump flow on pressure.

Figure 5 illustrates how higher form pressure is affected by higher initial slump flow. It is indicated that when the initial slump flow is high, the pressure remains high even if the casting level is near the top of the form. Similar findings were observed by [26]. This is explained by the pressure decay phenomenon, which is influenced by the thixotropy (or structural development) of the concrete over time; concrete that dries out more quickly starts to exert less pressure at the bottom cast layers. Observing how the type of cement affects the form is equally important, as various studies have shown [3,10,17,20].

The impact of employing various types of cement on pressure is demonstrated in Figure 6 through a comparison of BAS and ANLFA cement using the data collected from the bottom sensor (0.25 m/h casting rate). The fresh density of concrete produced with various types of cement varies, measuring 2331 kg/m^3 for BAS concrete and 2382 kg/m^3 for ANLFA concrete. The suppliers of the concrete claim that BAS takes 150 min to set, whereas ANLFA takes 170 min. Figure 6 indicates that cement hardens more quickly, with BAS showing a value of 12.0 kPa and ANLFA a value of 13.14 kPa at the bottom of the form. According to the findings, cement type has a small impact on maximum pressure, though a larger impact is seen when examining pressure decay.

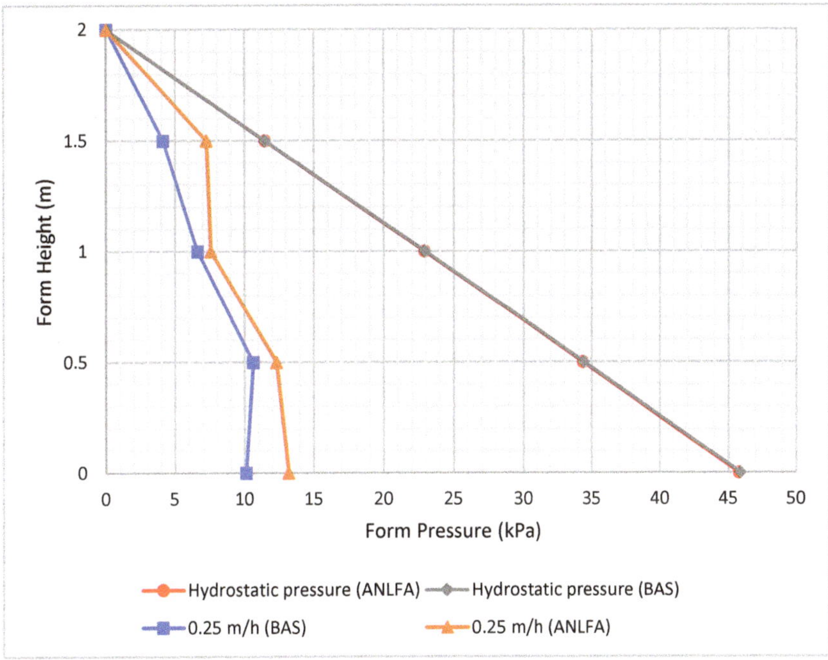

Figure 6. Effect of different types of cement on form pressure.

Figure 7 shows the effect of different concrete densities on pressure with a fixed casting rate for both. It is indicated that higher concrete density results in higher form pressure.

Figure 8 shows the effect of different air content on form pressure while maintaining the same casting rate and cement type, with the only change being the percentage of air in the concrete. It is indicated that higher air content leads to initial low form pressure in early fresh concrete, and the effect is diminished when the concrete starts to harden.

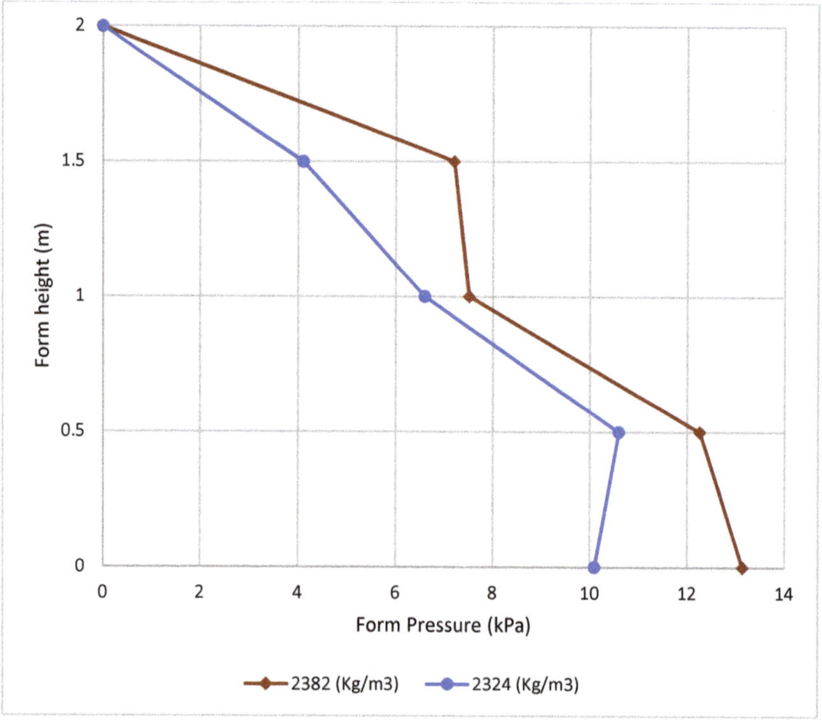

Figure 7. Effect of fresh concrete density on form pressure.

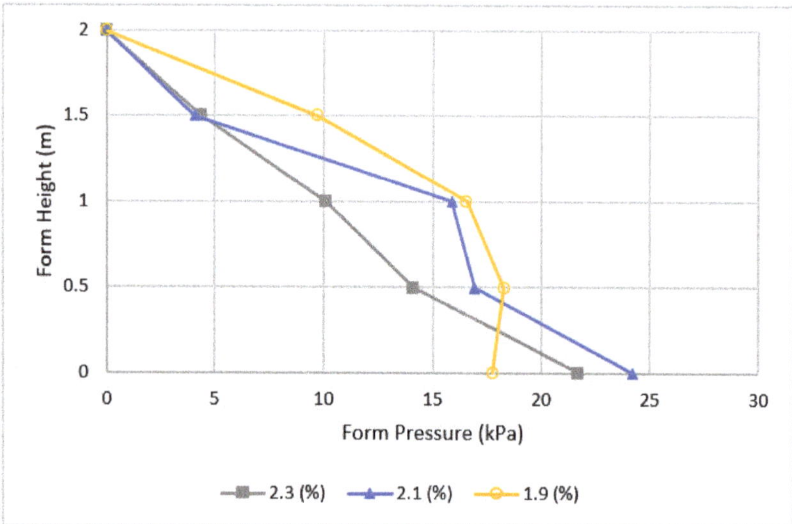

Figure 8. The effect of different air content on pressure.

Figure 9 shows the effect of different ambient temperatures on the amount of form pressure. It is indicated that it is rather difficult to observe the effect of concrete temperature on pressure.

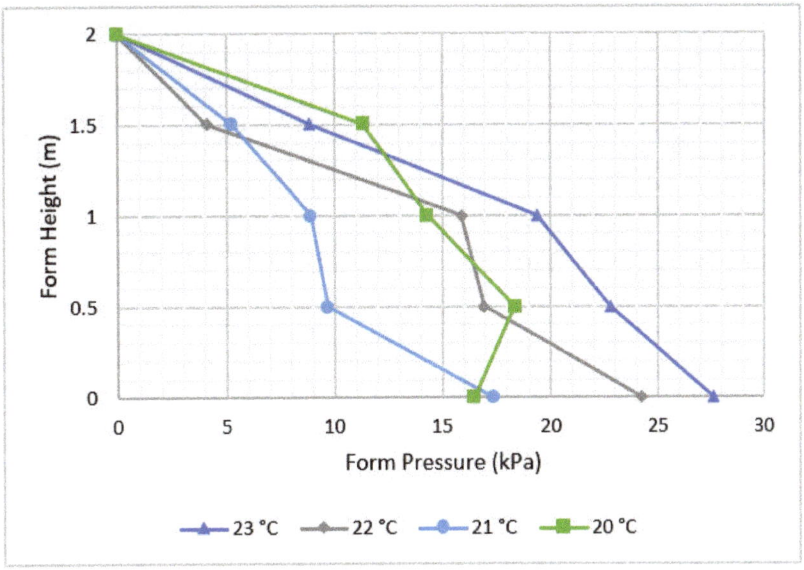

Figure 9. Effect of ambient temperature on form pressure.

Concrete temperature affects the amount of form pressure reduced but not maximum pressure (Figure 10), with concrete at higher temperatures tending to harden faster than at lower temperatures due to the hydration rate of formerly cast concrete.

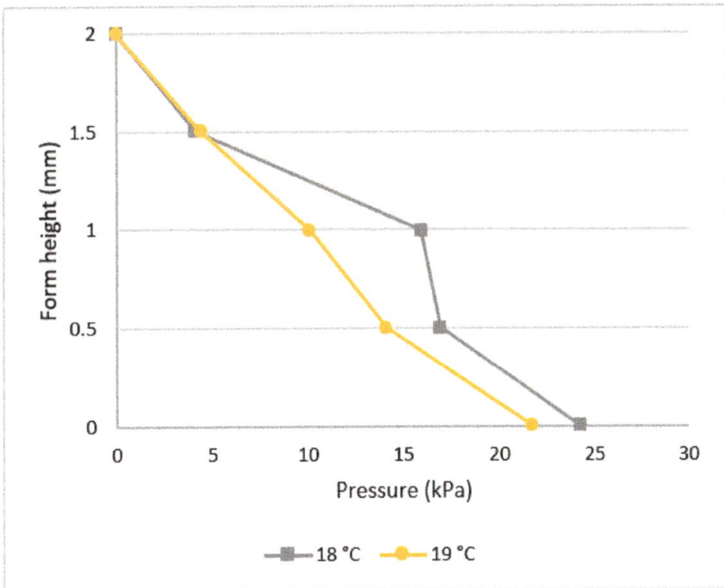

Figure 10. Effect of concrete temperature on form pressure.

Setting time was documented according to DIN18218 [25], which is formally known as the setting bag test. The results in Figure 11 indicate that higher setting time leads to higher pressure, which is due to the rate of hardening (concrete that hardens faster leads to lower form pressure).

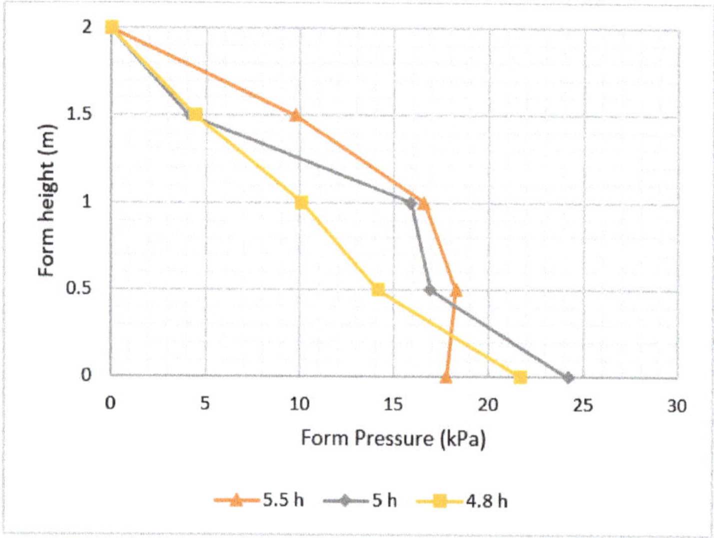

Figure 11. The effect of concrete setting time on pressure.

The effect of static yield stress at 15 min is crucial to understanding the effect of concrete viscosity as well, with high-viscosity concrete leading to lower form pressure. From Figure 12, it is indicated that using concrete with higher stress at 15 min results in lower pressure.

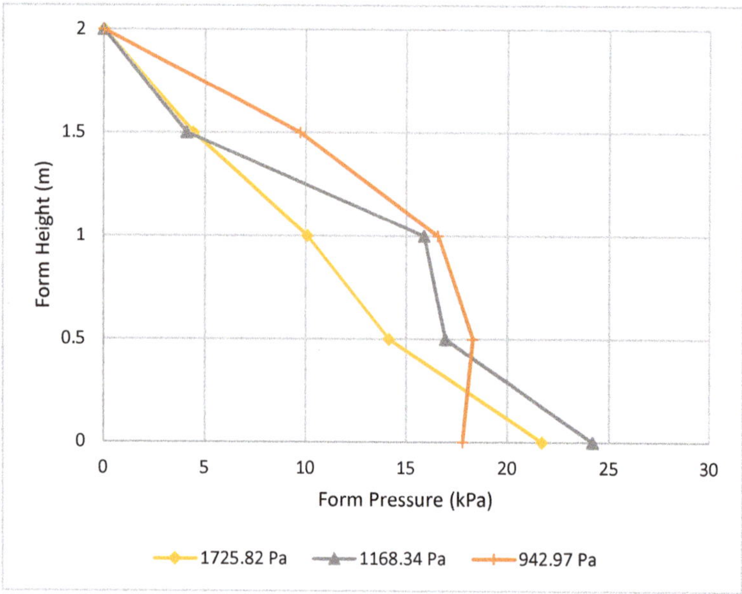

Figure 12. Effect of static yield stress at 15 min on form pressure.

4.2. Correlation between Variables

Correlation denotes the form of relationship between variables, which in this case includes either inputs such as density, height, and casting rate or the output, which is pressure. A negative coefficient denotes that when one variable decreases the other increases, while a positive coefficient denotes that when a variable increases the other associated variable also increases.

The results in Table 3 indicate how important the variables are in terms of their impact on the parameters related to form pressure, not only through their direct importance, but also in a cumulative sense, which was obtained by analyzing all the time series data obtained during the laboratory tests. These data were sourced from random analysis in order to estimate the correlation between the input variables stated in Table 3. Values indicate the degree of relationship between the variables. The correlation falls between −1 and +1. The focus of this study is the relationship between formwork pressure and the input parameters (concrete properties). It is indicated that the parameters of density and air temperature have a different correlation with maximum pressure. Fresh concrete density, as a main parameter in hydrostatic pressure, has been identified in previous studies as a parameter that significantly affects form pressure. This was specifically addressed by [7,12]. Air temperature was also identified by [1] as a parameter that significantly affects form pressure over time.

Table 3. Correlation between input variables and form pressure.

	Density (Kg/m^3)	Air Content (%)	Air Temperature °C	Concrete Temperature °C	Setting Time (h)	Initial Slump Flow (mm)	Initial T500 Time (s)	Static Yield Stress at 15 min	Height at Pmax	Pmax
Density (Kg/m^3)	1.00	−0.16	−0.67	0.01	0.59	−0.29	0.60	0.10	0.22	**−0.37**
Air content (%)	−0.16	1.00	0.19	−0.17	−0.51	−0.35	−0.05	0.28	−0.14	0.03
Air temperature °C	−0.67	0.19	1.00	−0.64	−0.34	0.21	−0.27	−0.18	0.11	**0.32**
Concrete temperature °C	0.01	−0.17	−0.64	1.00	0.07	0.16	−0.25	0.03	−0.25	−0.08
Setting time (h)	0.59	−0.51	−0.34	0.07	1.00	0.58	0.72	−0.71	0.27	−0.04
Initial slump flow (mm)	−0.29	−0.35	0.21	0.16	0.58	1.00	0.22	−0.94	0.15	0.15
Initial T500 time (s)	0.60	−0.05	−0.27	−0.25	0.72	0.22	1.00	−0.51	0.34	−0.08
Static yield stress at 15 min	0.10	0.28	−0.18	0.03	−0.71	−0.94	−0.51	1.00	−0.18	−0.19
Form height (m)	0.22	−0.14	0.11	−0.25	0.27	0.15	0.34	−0.18	1.00	**−0.42**
Pmax	**−0.37**	0.03	**0.32**	−0.08	−0.04	0.15	−0.08	−0.19	**−0.42**	1.00

5. Conclusions

This study explored the degree to which slump flow, T500 time, fresh concrete density, air content, static yield stress, concrete setting time, and concrete temperature have an impact on form pressure when casting with SCC. Extensive laboratory experiments were conducted with a newly developed pressure system to monitor pressure and study the effects of these parameters. Several conclusions are made:

1. Casting rate has a significant effect on the amount of pressure, with faster rates resulting in more pressure.
2. Flowability, as measured by slump flow, also has a significant effect on pressure.
3. Cement type (ANLFA and BAS) had a minimal effect on form pressure during casting, though a larger impact was observed when examining pressure reduction.
4. Higher density results in higher pressure.
5. Higher air content causes early fresh concrete to have lower initial form pressure, and this impact lessens as the concrete begins to solidify.
6. It can be challenging to determine how concrete temperature affects pressure.
7. The temperature of the concrete affects the amount of form pressure that can be reduced but does not affect maximum pressure; higher temperature concrete tends to harden more quickly than lower temperature concrete due to the hydration rate of previously cast concrete.
8. Concrete that sets more slowly leads to higher pressure, whereas concrete that sets more quickly leads to lower form pressure.
9. Concrete with high viscosity has less form pressure, which is supported by evidence that, after 15 min, pressure is lower in concrete with increased stress.

10. The correlation matrix demonstrated that air temperature and density both have a greater correlation with maximum pressure, as shown by measurements recorded throughout the entire testing period.

Author Contributions: Conceptualization, Y.G.; Software, Y.G.; Validation, Y.G. and J.N.; Formal analysis, Y.G. and J.N.; Resources, M.E.; Data curation, Y.G.; Writing—review & editing, A.C.; Supervision, A.C.; Project administration, M.E.; Funding acquisition, M.E. All authors have read and agreed to the published version of the manuscript.

Funding: Swedish Construction Industry Development Found (SBUF) funded this research.

Institutional Review Board Statement: Not applicable.

Informed Consent Statement: Not applicable.

Data Availability Statement: No available data.

Acknowledgments: The authors thank the partners of the study, including NCC AB, Vema Venturi AB, PERI, and SBUF, for their facilities, funding, and technical support, as well as the Lulea University of Technology.

Conflicts of Interest: The authors declare no conflict of interest.

References

1. Khayat, K.H.; Omran, A.F. *State-of-the-Art Review of Form Pressure Exerted by Self-Consolidating Concrete*; Université de Sherbrooke: Sherbrooke, QC, Canada, 2009.
2. Hurd, M.K. Lateral pressures for formwork design. *Concr. Int.* **2007**, *29*, 31–33.
3. Billberg, P. Understanding formwork pressure generated by fresh concrete. In *Understanding the Rheology of Concrete*; Elsevier: Amsterdam, The Netherlands, 2012; pp. 296–330.
4. Teixeira, S.; Puente, I.; Santilli, A. Statistical model for predicting the maximum lateral pressure exerted by self-consolidating concrete on vertical formwork. *J. Build. Eng.* **2017**, *12*, 77–86. [CrossRef]
5. ACI347R. *Guide to Formwork for Concrete Reported by ACI Committee 347*; American Concrete Institute: Farmington Hills, MI, USA, 2005.
6. EFNARC. *The European Guidelines for Self-Compacting Concrete Specification, Production and Use*; BIBM: Bruxelles, Belgium, 2005.
7. Glinicki, M.; Gołaszewski, J.; Cygan, G. Formwork Pressure of a Heavyweight Self-Compacting Concrete Mix. *Materials* **2021**, *14*, 1549. [CrossRef]
8. Němeček, J.; Zacharda, V.; Trávníček, P. Reduction of Formwork Pressures in Self-Compacting Concrete. In Proceedings of the IRF2020: 7th International Conference Integrity-Reliability-Failure, Funchal, Portugal, 6–10 September 2020.
9. Leemann, A.; Hoffmann, C.; Winnefeld, F. Pressure of self-consolidating concrete on formwork. *Concr. Int.* **2006**, *28*, 28–31.
10. Omran, A.F.; Khayat, K.H. Choice of thixotropic index to evaluate formwork pressure characteristics of self-consolidating concrete. *Cem. Concr. Res.* **2014**, *63*, 89–97. [CrossRef]
11. Gowripalan, N.; Shakor, P.; Rocker, P. Pressure exerted on formwork and early age shrinkage of self-technical papers pressure exerted on formwork and early age shrinkage of self-compacting concrete. *Case Stud. Constr. Mater.* **2021**, *15*, e00642. [CrossRef]
12. Gamil, Y.; Nilimaa, J.; Emborg, M.; Cwirzen, A. Lateral Formwork Pressure for Self-Compacting Concrete—A Review of Prediction Models and Monitoring Technologies. *Materials* **2021**, *14*, 4767. [CrossRef]
13. Santilli, A.; Puente, I.; Tanco, M. A factorial design study to determine the significant parameters of fresh concrete lateral pressure and initial rate of pressure decay. *Constr. Build. Mater.* **2011**, *25*, 1946–1955. [CrossRef]
14. Kwon, S.H.; Phung, Q.T.; Park, H.Y.; Kim, J.H.; Shah, S.P. Experimental study on effect of wall friction on formwork pressure of self-consolidating concrete. In Proceedings of the the 6th International RILEM Symposium on Self-Compacting Concrete and 4th North American Conference on the Design and Use of SCC, SCC, Montreal, QC, Canada, 26–29 September 2010.
15. Kwon, S.H.; Phung, Q.T.; Park, H.Y.; Kim, J.H.; Shah, S.P. Effect of wall friction on variation of formwork pressure over time in self-consolidating concrete. *Cem. Concr. Res.* **2011**, *41*, 90–101. [CrossRef]
16. Khayat, K.H.; Assaad, J.J. Effect of w/cm and high-range water-reducing admixture on formwork pressure and thixotropy of self-consolidating concrete. *ACI Mater. J.* **2006**, *103*, 186. [CrossRef]
17. Assaad, J.J.; Khayat, K.H. Effect of viscosity-enhancing admixtures on formwork pressure and thixotropy of self-consolidating concrete. *ACI Mater. J.* **2016**, *103*, 280. [CrossRef]
18. Assaad, J.J.; Matar, P. Regression models to predict SCC pressure exerted on formworks containing vertical and transverse reinforcing bars. *Mater. Struct.* **2018**, *51*, 62. [CrossRef]
19. Matar, P.; Assaad, J.J. Effect of vertical reinforcing bars on formwork pressure of SCC containing recycled aggregates. *J. Build. Eng.* **2017**, *13*, 159–168. [CrossRef]
20. Omran, A.F.; Khayat, K.H.; Elaguab, Y.M. Effect of SCC mixture composition on thixotropy and formwork pressure. *J. Mater. Civ. Eng.* **2012**, *24*, 876–888. [CrossRef]

21. Kim, J.H.; Noemi, N.; Shah, S.P. Effect of powder materials on the rheology and formwork pressure of self-consolidating concrete. *Cem. Concr. Compos.* **2012**, *34*, 746–753. [CrossRef]
22. Assaad, J.J. Correlating thixotropy of self-consolidating concrete to stability, formwork pressure, and multilayer casting. *J. Mater. Civ. Eng.* **2016**, *28*, 04016107. [CrossRef]
23. Assaad, J.J.; Khayat, K.H. Effect of casting rate and concrete temperature on formwork pressure of self-consolidating concrete. *Mater. Struct.* **2016**, *39*, 333–341. [CrossRef]
24. Billberg, P. *Form Pressure Generated by Self-Compacting Concrete: Influence of Thixotropy and Structural Behaviour at Rest*; KTH Royal Institute of Technology: Byggvetenskap, Sweden, 2006.
25. DIN18218. DIN Standard on Formwork Pressures Updated. *Concr. Int.* **2010**, *1*, 27–29. Available online: https://www.concrete.org/publications/internationalconcreteabstractsportal/m/details/id/51663776 (accessed on 20 December 2022).
26. Nilimaa, J. Lateral Form Pressure Induced by SCC. In Proceedings of the the Nordic Concrete Research XXIV NCR Symposium 2022, Stockholm, Sweden, 17–19 August 2022.
27. Perrot, A.; Pierre, A.; Vitaloni, S.; Picandet, V. Prediction of lateral form pressure exerted by concrete at low casting rates. *Mater. Struct.* **2014**, *48*, 2315–2322. [CrossRef]
28. Teixeira, S.; Santilli, A.; Puente, I. Analysis of casting rate for the validation of models developed to predict the maximum lateral pressure exerted by self-compacting concrete on vertical formwork. *J. Build. Eng.* **2016**, *6*, 215–224. [CrossRef]
29. Omran, A.F.; Elaguab, Y.M.; Khayat, K.H. Effect of placement characteristics on SCC lateral pressure variations. *Constr. Build. Mater.* **2014**, *66*, 507–514. [CrossRef]

materials MDPI

Article

Identification of Dynamic Behavior Models of Concrete B22.5

Anatoly M. Bragov [1], Andrey K. Lomunov [1], Mikhail E. Gonov [1], Aleksandr Yu. Konstantinov [1], Leonid A. Igumnov [1] and Victor A. Eremeyev [1,2,*]

[1] Research Institute of Mechanics, National Research Lobachevsky State University of Nizhny Novgorod, 603022 Nizhny Novgorod, Russia
[2] Department of Civil and Environmental Engineering and Architecture (DICAAR), University of Cagliari, Via Marengo 2, 09123 Cagliari, Italy
* Correspondence: eremeyev.victor@gmail.com; Tel.: +39-070-675-5417

Abstract: We discuss experimental and numerical studies of the deformation and destruction of fine-grained concrete B22.5 under dynamic loading. The experiments were carried out using the Kolsky (or split-Hopkinson pressure bar) method, and its modifications in the strain rate range from 400 to 2000 s^{-1}. The rate dependences of ultimate stresses and fracture energy in tension and compression are obtained. Based on experimental data, the identification of the dynamic component of two models from the LS-DYNA computational complex was carried out: *MAT_CONCRETE_DAMAGE and *MAT_CSCM. The results of a comparative analysis of the identified models based on single-element modeling and comparison with experimental data are presented. It is shown that the obtained experimental strain rate dependences of the fracture characteristics can significantly improve the predictive ability of the model compared to the default parameter set. Information about the rate dependence of the fracture energy in *MAT_CSCM model makes it possible to more realistically simulate the behavior of the material beyond the ultimate stress.

Keywords: ultimate stress; concrete; strain rate; Kolsky method; identification; behavior model; fracture energy

Citation: Bragov, A.M.; Lomunov, A.K.; Gonov, M.E.; Konstantinov, A.Y.; Igumnov, L.A.; Eremeyev, V.A. Identification of Dynamic Behavior Models of Concrete B22.5. *Materials* **2023**, *16*, 2259. https://doi.org/10.3390/ma16062259

Academic Editors: Hrvoje Smoljanović, Ivan Balić and Nikolina Zivaljic

Received: 14 February 2023
Revised: 8 March 2023
Accepted: 9 March 2023
Published: 11 March 2023

1. Introduction

For reliable numerical simulation of the behavior of various structures made of concrete and fiber-reinforced concrete under conditions of high-speed impulse loading, the deformation models and failure criteria are needed. These models should adequately describe the behavior of the material in a wide range of strain rates and temperatures under simple and complex loading conditions. These models should also adequately describe the behavior of the material under various types of stress–strain state: uniaxial stress state in compression and tension, uniaxial deformation in compression, shear, and combined state. In addition, it is important to take into account the moisture content of concrete and the temperature at which it is used.

To equip the deformation model and fracture criteria of concrete and fiber-reinforced concrete with the necessary parameters, a complex experimental study of the dynamic deformation of concrete samples under various loading conditions is required. As a result of such experiments, deformation diagrams (strain–stress $\sigma(\varepsilon)$) are obtained. Furthermore, the ultimate strength and deformation characteristics of concrete and fiber-reinforced concrete are determined, taking into account the types of tests (compression, tension, shear, complex stress) at various strain rates and stress growth rates. The energy intensity and crack resistance are also evaluated. The dependences of the specified parameters on the strain rate or the rate of stress growth are obtained as a result of data analysis. Based on the obtained parameters of the mechanical properties of concrete and fiber-reinforced concrete, a particular mathematical deformation model is identified. The parameters of failure criteria are also determined.

Then, the mathematical models and fracture criteria of concrete and fiber-reinforced concrete are verified. That is, the adequacy of the model is assessed under conditions that differ from those in which the mechanical properties were obtained to identify the models.

Thus, in order to obtain adequate deformation models and fracture criteria of concrete and fiber-reinforced concrete, a system of basic experiments should be implemented. Based on these results, a certain model of concrete and fiber-reinforced concrete is identified. A system of verification experiments is needful, based on which the adequacy of the identified model and failure criteria are assessed.

The dynamic properties of brittle materials are important in various fields of military, industrial, and civilian activities. Due to the transient nature of the load, dynamic testing of brittle materials is very different from that of the ductile materials. In addition, dynamic tests are more complex than static ones. Dynamic tests are usually carried out according to the Kolsky method using a split-Hopkinson bar. The Kolsky technique is a fairly reliable method for measuring the dynamic properties of brittle materials at high strain rates. This method was proposed by G. Kolsky in 1949 [1,2] to evaluate the dynamic response of various metals under the influence of high loads or strain rates. Shortly thereafter, various researchers began using various modifications of the Kolsky method to test brittle materials, such as concrete, ceramics, and rocks [3]. Several comprehensive reviews have been carried out regarding the dynamic behavior of brittle materials, such as mortar, ceramics, concrete, and rocks [4–7]. Dynamic experimental methods [8–12] were considered to obtain the properties of brittle materials under various types of stress–strain states. Significant progress has been made regarding the quantification of the various dynamic properties of rock-like materials thanks to advances in split=Hopkinson pressure bar (SHPB) experimental techniques.

In the work, the Kolsky method and its modifications are used to obtain experimental data on the natural laws of high-speed deformation and fracture of fine-grained concrete samples, and several models from the library of the LS-DYNA calculation complex are identified based on this data.

2. Materials and Methods

To test samples under dynamic compression and tension, we used installations that implement the classical split-Hopkinson pressure bar (SHPB) method (Figure 1a) and its modification (Figure 1b) [13]. The histories of changes in strains, strain rates, and stresses in the samples were calculated using the formulas proposed by Kolsky [1,2,13] using the strain pulses recorded in elastic measuring bars. The loading condition (strain rate) was varied by changing the impactor speed. The speed of the impactor depends on the pressure in the chamber of the air gun. It was measured using a light speed meter set.

The experimental setup for direct tension (Figure 1b) [13] differs from the traditional SHPB setup for compression in that the tubular impactor is accelerated in the barrel by a gas gun in the direction opposite to the sample. The tubular impactor hits the anvil, which is fixed at the end of the incident measuring bar, thereby exciting an elastic tension wave in it. The sample is glued either directly to the measuring bars or to special threaded nozzles that are screwed into threaded sockets at the ends of the measuring bars. The processing of experimental data in the split-Hopkinson tension bar (SHTB) setup is carried out using the main dependences of the SHPB method.

Samples for direct tension were glued with epoxy glue with the addition of tungsten powder to increase the strength of the adhesive seam with the replaceable threaded nozzles (Figure 2). Threaded nozzles were screwed into the threaded holes of the measuring bars. The test results showed that the strength of the adhesive joint exceeds the tensile strength of the sample (Figure 2).

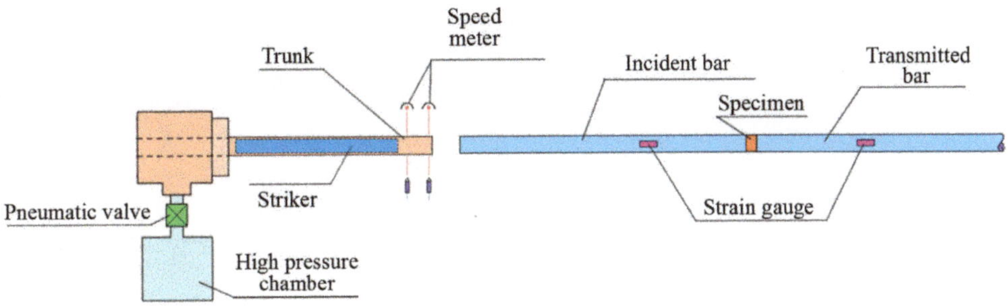

(**a**)

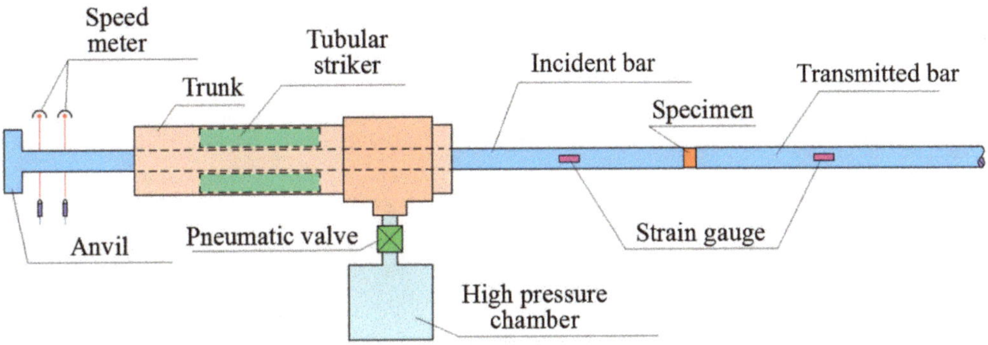

(**b**)

Figure 1. Schemes of test setups: (**a**) for compression, (**b**) for tension.

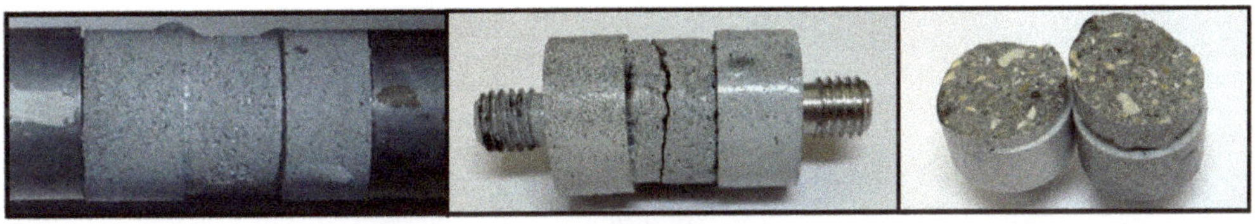

Figure 2. Fixing the sample in a tensile experiment.

For experimental studies according to the SHPB and SHTB methods, both in compression and in tension, cylindrical specimens of a circular cross section are required. The test specimen is placed between the ends of two long coaxial measuring bars. To transfer the shock pulse from the incident bar to the sample and then to the transmitted bar, it is necessary to ensure a tight fit of the test sample to the end of the transmitted bar. In this regard, the Kolsky method imposes a number of high requirements on the samples, such as uniformity of the material and minimum roughness of the end surfaces of the sample. The parallelism of the ends of the sample and their perpendicularity to the axis of the cylinder must necessarily be ensured so that eccentricity does not occur, causing bending of the sample. The optimum ratio of sample diameter to sample length should be between 1 and 2 to reduce the effects of inertia and friction. To reduce the scale effect when using measuring bars with a diameter of 20 to 60 mm, it is desirable to test fine-grained concrete with filler fractions from 1 to 3 mm.

Tests were carried out on samples of fine-grained concrete of class B22.5 obtained via semi-dry vibrocompression. As part of fine-grained concrete, cement grade 500D (consumption 430 kg per 1 m^3), coarse sand (module 3 mm, consumption 1435 kg per 1 m^3), fine sand (module 1.6 mm, consumption 355 kg per 1 m^3), plasticizing Murasan BWA

16 additive (consumption 1.6 kg per 1 m^3), and water (consumption 120 kg per 1 m^3) are used.

Concrete workpieces were cut into plates using a Cedima CTS-57-G stone-cutting machine, and specimens of the required size were drilled from the plates using an NS-12M table drilling machine and diamond crowns. Concrete samples for dynamic tests were made with a diameter of 20 mm and a length of 10 mm.

Concrete workpieces for the manufacture of specimens were obtained in the factory. Quasi-static strength, which corresponds to the concrete class B22.5, was provided by the manufacturer.

3. Results of an Experimental Study

The test parameters for dynamic compression and tension, as well as some parameters determined from their results, are presented in Tables 1 and 2.

Table 1. Parameters of experiments for uniaxial compression.

Mode	Experiment Code	Stricker Velocity, m/s	Max Stress, MPa	Ultimate Strain, %	Strain Rate, 1/s	Lifetime, ms	DIF	Specific Fracture Energy, MPa
1	c637-19	14.13	78.14	0.89	665.74	25.00	3.47	2.70
	c637-20	13.67	80.49	0.80	666.23	17.50	3.58	2.51
	c637-21	13.44	65.24	0.83	676.59	21.25	2.90	2.30
	c637-22	13.62	79.86	0.79	676.59	21.25	3.55	2.38
	c637-23	14.2	71.28	0.83	748.59	20.00	3.17	2.47
2	c637-01	18.02	84.55	0.93	1019.51	16.25	3.76	3.61
	c637-02	18.33	83.21	0.98	1032.09	17.50	3.70	3.51
	c637-03	18.62	93.81	1.08	974.41	17.50	4.17	3.79
	c637-04	18.68	83.54	1.10	1084.05	17.50	3.71	3.70
	c637-05	18.09	100.56	1.12	1003.36	17.50	4.47	4.02
3	c637-06	25.86	98.75	1.46	1717.47	17.50	4.39	5.06
	c637-07	25.77	100.64	1.29	1560.29	17.50	4.47	5.15
	c637-08	26.45	105.76	1.50	1698.34	17.50	4.70	5.62
	c637-09	26.2	99.50	1.32	1764.55	18.75	4.42	5.16
	c637-10	25.75	105.75	1.47	1704.93	18.75	4.70	5.57
4	c637-11	31.05	108.61	1.65	1870.60	15.00	4.83	6.03
	c637-12	30.72	108.94	1.44	2019.84	13.75	4.84	5.51
	c637-13	31.11	103.71	1.46	1900.78	13.75	4.61	5.97
	c637-14	30.73	108.00	1.80	1931.89	16.25	4.80	5.47
	c637-15	31.45	108.20	1.59	1944.06	13.75	4.81	6.29
5	c637-24	18.4	66.96	1.21	403.13	50.00	2.98	2.34
	c637-25	19.09	60.98	1.81	604.31	52.50	2.71	2.97
	c637-26	18.46	59.65	1.72	572.70	52.50	2.65	2.83
	c637-27	18.55	64.62	1.53	501.15	47.50	2.87	2.94
6	c637-28	18.6	64.45	1.51	487.37	51.25	2.86	2.77
	c637-30	35.25	71.92	1.51	860.89	30.00	3.20	6.51
	c637-31	35.65	72.07	1.30	800.43	28.75	3.20	6.43
	c637-32	35.76	72.76	1.24	714.93	26.25	3.23	6.35
	c637-33	35.63	69.39	1.62	937.20	32.50	3.08	6.64
	c637-34	35.63	64.27	1.32	776.42	28.75	2.86	5.78
7	c637-46	60.98	90.10	1.68	1109.74	28.75	4.00	8.07
	c637-47	60.27	85.94	1.33	961.21	21.25	3.82	7.97
	c637-48	60.73	79.81	1.63	1258.44	25.00	3.55	8.46
	c637-49	59.77	77.84	1.63	1215.06	27.50	3.46	7.42
	c637-50	61.15	72.97	1.69	1245.61	26.25	3.24	7.19

Table 2. Parameters of direct tensile experiments.

Mode	Experiment Code	Stricker Velocity, m/s	Max Stress, MPa	Ultimate Strain, %	Strain Rate, 1/s	Lifetime, ms	DIF	Specific Fracture Energy, Mpa
1	t637-01	7.11	13.43	1.31	539.07	43	5.97	0.054
	t637-02	7.87	13.79	1.1	567.69	34.5	6.13	0.108
	t637-03	7.64	11.2	0.89	525.76	35	4.98	0.088
	t637-04	6.38	13.04	0.831	380.64	34	5.80	0.052
	t637-05	5.94	13.01	1	373.56	39	5.78	0.058
2	t637-06	10.85	8.94	0.92	723.52	25.5	3.97	0.093
	t637-07	11.23	8.24	1.11	844.88	26.5	3.66	0.100
	t637-08	11.39	10.75	0.95	750.21	21	4.78	0.080
	t637-09	10.54	6.44	0.66	545	24	2.86	0.110
	t637-10	10.95	8.49	0.93	727.69	29.5	3.78	0.079

Under conditions of dynamic compression, the concrete was tested at seven different speed modes. Five repeated tests were performed for each mode. The strain rates were in the range from 400 to 2000 1/s. To ensure a smooth increase in the load, three modes out of seven were carried out using copper pulse shapers. Figure 3 shows the results of checking the fulfillment of the force equilibrium conditions in the sample. It can be seen in the case of using a pulse shaper (Figure 3b), the assumption of the equality of forces at the ends is fulfilled more qualitatively.

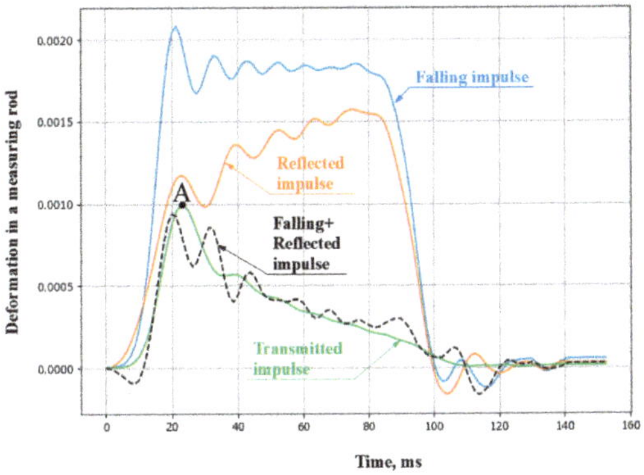

(a)

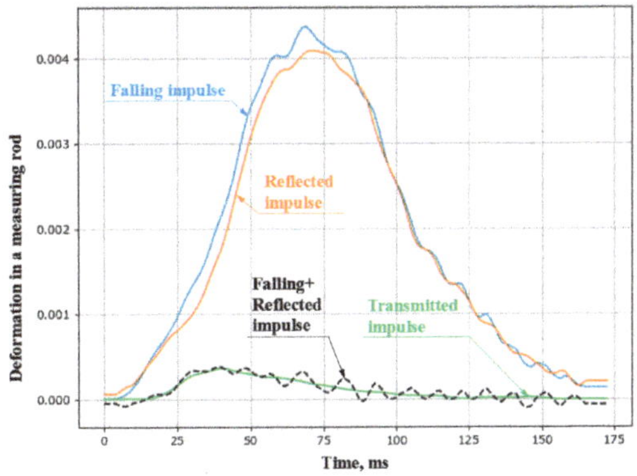

(b)

Figure 3. Pulses in measuring bars: (**a**) without an impulse shaper at a strain rate of 1100 1/s and (**b**) with an impulse shaper at a strain rate of 1250 1/s.

The strain pulses recorded in the measuring bars were synchronized in time. According to the Kolsky formulas, for each experiment, the stress σ(t), strain ε(t), and strain rate $\dot{\varepsilon}$(t) as a function of time were plotted. The obtained dependences were synchronized in time for each strain rate condition. The time parameter having been excluded, a dynamic deformation diagram $\sigma(\varepsilon)$ was constructed for the known strain rate history $\dot{\varepsilon}(\varepsilon)$. Furthermore, for loading modes of the same type, averaged diagrams $\sigma(\varepsilon)$ and $\dot{\varepsilon}(\varepsilon)$ as well as the dependences of stresses on time σ(t) and strain rates on time $\dot{\varepsilon}$(t) were plotted. For each of the resulting diagrams, characteristic points with the maximum achieved stresses (σ_{max}) were identified. Those points correspond to moment after which the fracture of the samples occurred. For these points, the corresponding values of ultimate strains (ε_{max}) and time to failure (τ_{max}) are determined. The strain rate ($\dot{\varepsilon}$) corresponding to the loading regime was taken to be the maximum value in the interval before the onset of fracture of the specimens.

The ultimate (maximum) stress reached at point A (Figure 3a) is the strength of the material. It should be noted that the process of loading from zero stress values to their maximum value proceeds at a variable strain rate both in the case of traditional loading without a pulse shaper in the transmitting bar and when a shaper is used. As shown by our studies with various materials (concrete, fiber-reinforced concrete, rocks), this effect occurs for all tested brittle materials. In this regard, when analyzing the influence of the strain rate, in some cases it is preferable to choose the value of the stress growth rate as a characteristic of the process of high-speed loading. As experience shows, it remains almost constant in active loading.

In the present study, the strain rate characteristic of ultimate stress was determined from the moment of the maximum stress achievement (the moment of failure) based on the history of stress and strain rate changes over time (Figure 4).

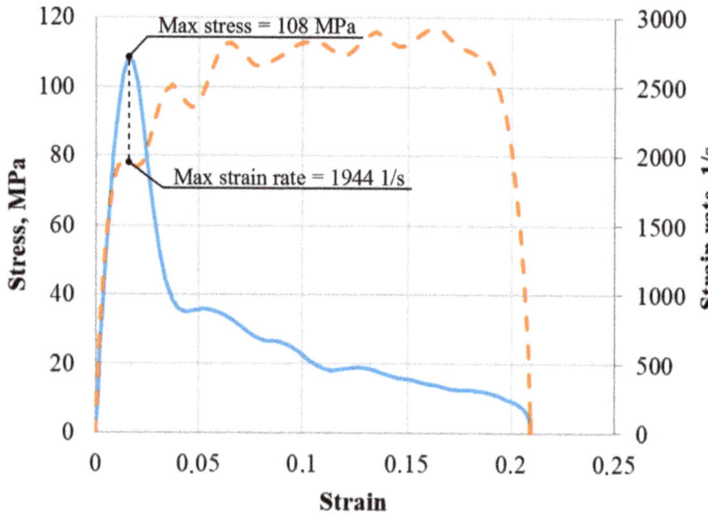

Figure 4. Strain rate characteristic of ultimate stress determination.

Figure 5 shows average strain diagrams over modes with a chronology of strain rate changes. Figure 5a shows the strain dependencies of the stress (solid lines and left vertical axis) and strain rate (dotted lines and right vertical axis), and the Figure 5b shows the corresponding time dependencies. The color represents the loading mode. In the above diagrams in the stress–strain axes $\sigma(\varepsilon)$, the initial loading section is close to linear. With further deformation, when the ultimate stress values are reached, the concrete is intensively destroyed and is accompanied by a decrease in stresses and an increase in deformations. It can be seen that in the cases with and without a copper pulse shaper, the influence of the strain rate on the initial section of the diagrams is not observed under all loading modes. The magnitude of the maximum stresses is determined by the amplitudes of the incident pulses, that is, concrete samples during loading can withstand stresses that are significantly greater than its structural strength. This is due to the presence of two opposing processes: the rate of growth of longitudinal compressive loads, determined by the amplitudes of the loading waves, and the rates of formation and merging of microcracks during transverse expansion of the samples.

The dependences obtained demonstrate that with an increase in the strain rate, the maximum stresses increase. The ultimate strains corresponding to them also grow according to a linear law. The time before the onset of destruction decreases according to a power law. It can be seen that the use of pulse shapers reduced the range of strain rates and resulting stresses and increased the level of ultimate strains, and the process of loading the sample was extended in time, as was the destruction of the sample itself.

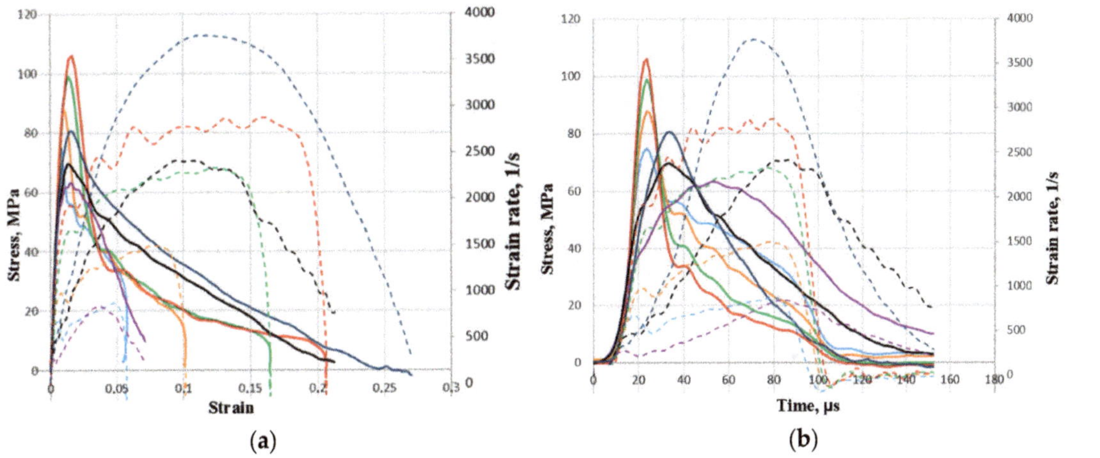

Figure 5. Average diagrams of dynamic deformation under compression: (**a**)—stress and strain rate dependencies on strain, (**b**)—time dependencies of stress and strain rate.

Direct tensile tests were carried out with two speed modes. In the first mode (Figure 6), there were five experimental shots; the average velocity of the impactor was 7 m/s, the average strain rate was about 500 reverse seconds, and the average maximum stress was 12 MPa. In Figure 6, the straight line shows the dependence on stress; the dashed line shows the dependence on the strain rate. In the second mode, there were also five experimental shots; the average velocity of the impactor was 11 m/s, the average strain rate was about 700 inverse seconds, and the average maximum stress was 8 MPa.

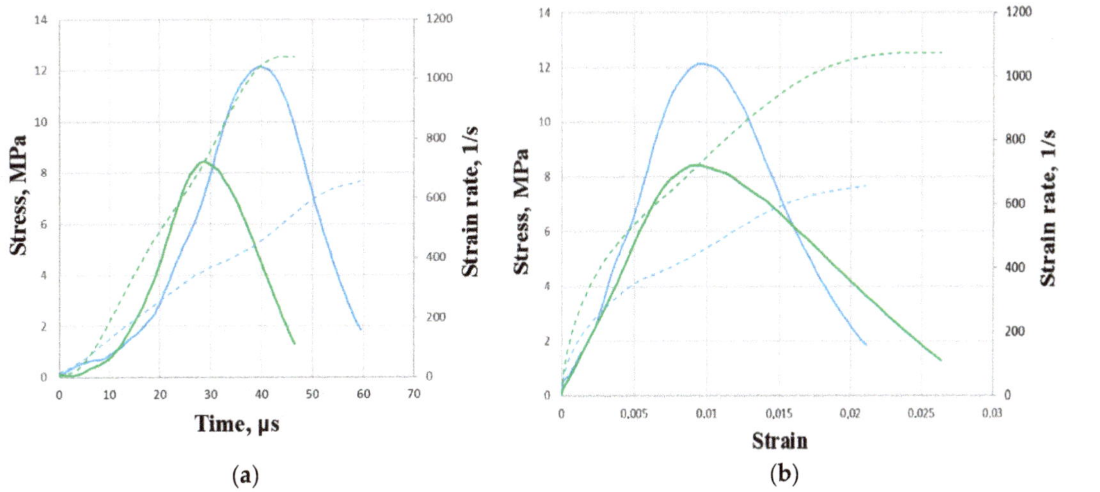

Figure 6. Averaged diagrams of the dependence of stress on strain and stress on time of modes No. 1 and No. 2 (**a**)—time dependencies of stress and strain rate, (**b**)—stress and strain rate dependencies on strain.

The dynamic strength of concrete is characterized by the dynamic increase factor (DIF), which is obtained from the ratio of the maximum dynamic stress achieved in the experiment to the static compressive strength of concrete 22.5 MPa. The DIF value in the obtained strain rate range from ~400 to 2000 1/s varies from 1 to 5 (Figure 7).

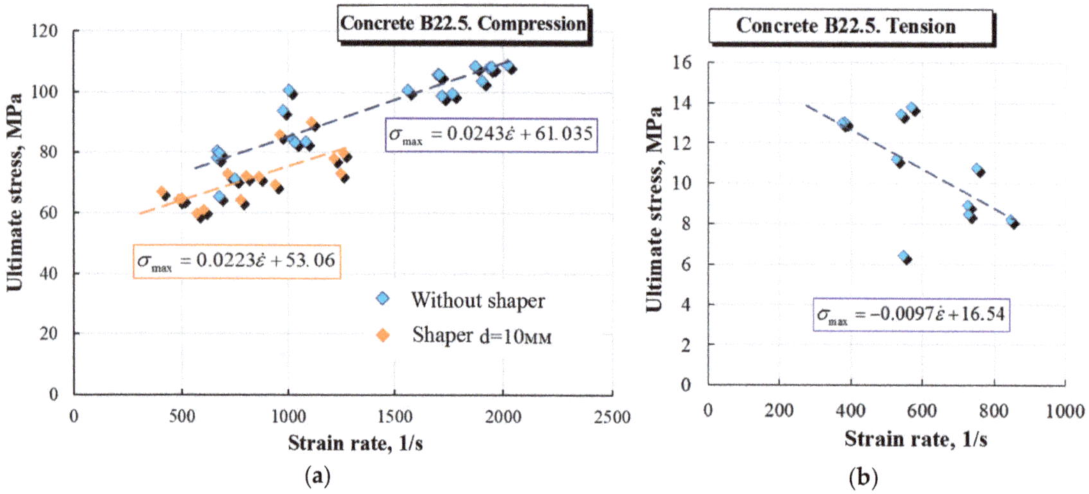

Figure 7. Rate dependences of ultimate stresses in compression (**a**) and tension (**b**).

4. Identification of Computational Models

Explicit and implicit finite element analysis (FEA) offers a fast, reliable, and cost-effective solution for investigating the behavior of concrete and reinforced concrete structures under various load and strain rate conditions. Mathematical modeling is more profitable in contrast to time-consuming and expensive full-scale tests. However, when using the FEA approach, accurate material models are needed to describe the actual behavior of materials under various loading conditions. In case of impulse high-intensity loads, such as explosion and impact, FEA modeling of building structures evokes a serious problem.

The development of mathematical models of concrete behavior has been carried out for a long time [14–18], but due to the complex nonlinear behavior, a simple but accurate behavior model has not yet been formulated. Several complex models of concrete materials have been successfully used for dynamic analysis of concrete structures subjected to intense impact and explosive loads using the programs LS-DYNA and AUTODYN [19,20]. The number of required input parameters in these models ranges from 32 to 78.

At present, the materials library of the LS-DYNA software package contains about 250 models of materials of various physical natures [21], 31 of which are proposed to be used to describe the behavior of geomaterials, concretes, or rocks. In this work, the two most suitable models were selected from this list to describe the behavior of concrete: MAT_CONCRETE_DAMAGE_REL3 (No. 72) and MAT_CSCM (No. 159).

Model No. 72 (concrete damage) and its improved version MAT_CONCRETE_DAMAGE_REL3 [21,22] are used to simulate the behavior of concrete considering the fracture and the influence of the strain rate. This is a tri-invariant model that uses three fracture surfaces. Initially, it was based on the pseudotensor model. The most significant improvement for the user is the ability to generate model parameters based on the uniaxial compressive (uniaxial stress) strength of the concrete. The strength scaling factor is given by the curve for the entire range of strain rates. Tensile strain rates are defined by negative values. Numerical simulation results can be improved using custom input parameters that control the degradation of properties in compression and tension, the effect of strain rate, and the kinetics of damage accumulation.

Model No. 159 (CSCM) [23,24] uses a smooth "cap"-type flow surface. The strain rate is taken into account using the viscoplasticity model. The flow surface is formulated in terms of three stress tensor invariants: j'_1 is the first stress tensor invariant, j'_2 is the second stress tensor invariant, and j'_3 is the third stress tensor invariant. These invariants are defined in terms of the deviator components of the stress tensor S_{ij} and pressure P. At each calculation step, the viscoplasticity algorithm interpolates between the elastic trial stress ${\sigma_{ij}}^T$ and the inviscid stress (without considering the strain rate) ${\sigma_{ij}}^P$ to calculate the viscoplastic stress ${\sigma_{ij}}^{VP}$. The model makes it possible to take into account the influence of the strain rate on the strength characteristics of the material under tension, uniaxial compression, and shear. In addition, the user can take into account the influence of the strain rate on the fracture energy. To prevent an endless rise in DIF with strain rate, it is necessary to use an overstress intercept. For this purpose, Model 159 has two parameters: maximum overstress allowed in compression (OVERC) and maximum overstress allowed in tension (OVERT). In the present study, the maximum stresses were limited to the maximum stresses obtained in the experiments.

In modeling the process of dynamic deformation and destruction, the strain rate significantly affects the strength characteristics along with a significant difference of concrete properties in tension and compression. These effects were studied experimentally and then used to equip mathematical models with the necessary parameters.

The experimentally obtained strain rate dependences of strength are transferred to the model as a dependence of DIF on the strain rate. DIF was approximated by a power function:

$$\mathrm{DIF}(\dot{\varepsilon}) = 1 + k \cdot \dot{\varepsilon}^n$$

The parameters of approximations of DIF velocity are presented in Table 3. A comparison of data with approximating functions is shown in Figure 8a.

Table 3. Approximation parameters for the speed dependences of the strength of concrete class B22.5.

	k	n
Compression	0.125	0.446
Tension	0.65	0.28

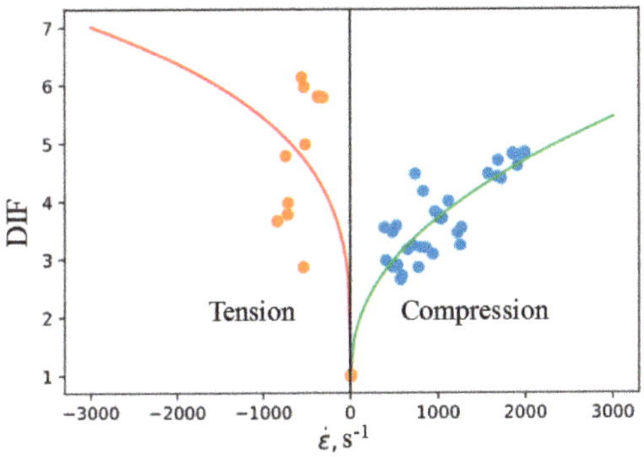

(a)

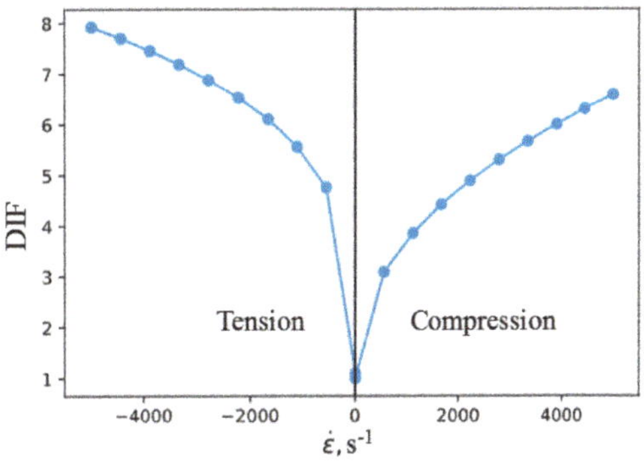

(b)

Figure 8. Rate dependences of ultimate stresses in compression and tension: (**a**) is an approximation of experimental data; (**b**) is a tabular dependence.

In the MAT_72 model, the speed dependence DIF is given by a table function. Negative strain rates correspond to tension. The value of DIF at a strain rate equal to 0 should be equal to 1. To obtain such a function, 20 points were uniformly selected in the range of strain rates from −5000 1/s to 5000 1/s. The DIF values at the corresponding points were calculated using the approximating dependences of the experimental data. The obtained data are presented in Figure 8b.

There are two modes of using Model 159: automatic generation of model parameters for concrete specified by static compressive strength (*fc*) and manual input of model parameters. In the first case, it is impossible to correct part of the model parameters in order to refine the behavior of a particular material. In the second case, it is necessary to set the values of 37 material constants. The work [23] describes the algorithms for calculating the parameters that are used by default. These algorithms are based on the interpolation of the values of the corresponding material parameters according to the data obtained for concretes with different static compressive strength. The parameters for concrete B22.5, independent of the strain rate and calculated in accordance with the dependencies embedded in the model, are presented in Table 4.

Table 4. B22.5 concrete model parameters.

Modulus of elasticity	E [GPa]	23.947	**Parameters defining the triaxial expansion surface**	α_2 [-]	0.66
	K [GPa]	11.4		λ_2 [-]	0.16
	G [GPa]	10.4		β_2 [MPa^{-1}]	0.077
	ν [-]	0.15		θ_2 [MPa^{-1}]	0.0016
Static tensile strength	f_t [MPa]	1.79			
Parameters defining the surface of triaxial compression	α [MPa]	13.3	**Parameters of the "cap" surface**	R [-]	5
	λ [MPa]	10.5		X_0 [MPa]	87.8
	β [MPa^{-1}]	0.01929	**Porosity of concrete**	W [-]	0.05
	θ [-]	0.27	**Parameters that determine hardening on the "pressure-volume deformation" curve**	D_1 [MPa]	2.5×10^{-4}
Parameters defining the surface of torsion	α_1 [-]	0.747		D_2 [MPa^2]	3.49×10^{-7}
	λ_1 [-]	0.17			
	β_1 [MPa^{-1}]	0.0766			
	θ_1 [MPa^{-1}]	0.0013			

To take into account the influence of the strain rate on the ultimate stresses of concrete, the model provides the following six parameters:

- $\eta 0c$ and ηc—for compression;
- $\eta 0t$ and ηt—for tensile;
- $\eta 0s$ and ηs—for shear.

The dynamic limit stress is calculated using the formulas:

$$f_c^{dyn} = f_c + E \cdot \dot{\varepsilon} \cdot \frac{\eta 0c}{\dot{\varepsilon}^{\eta c}}$$

$$f_t^{dyn} = f_t + E \cdot \dot{\varepsilon} \cdot \frac{\eta 0t}{\dot{\varepsilon}^{\eta t}}$$

$$f_c^{dyn} = f_s + E \cdot \dot{\varepsilon} \cdot \frac{\eta 0s}{\dot{\varepsilon}^{\eta s}}$$

where the effective strain rate $\dot{\varepsilon}$ is determined by the expression:

$$\dot{\varepsilon} = \sqrt{\frac{2}{3}\left[\left(\dot{\varepsilon}_x - \dot{\varepsilon}_y\right)^2 + \left(\dot{\varepsilon}_x - \dot{\varepsilon}_z\right)^2 + \left(\dot{\varepsilon}_z - \dot{\varepsilon}_y\right)^2 + \dot{\varepsilon}_x{}^2 + \dot{\varepsilon}_x{}^2 + \dot{\varepsilon}_z{}^2\right]}$$

To determine these parameters, the previously obtained experimental velocity dependences of the ultimate stresses in compression and tension were approximated as follows, see Figure 9. Parameter values are given in Table 5.

Table 5. Parameters that determine the effect of strain rate on ultimate stresses.

$\eta 0c$ [-]	ηc [-]	$\eta 0t$ [-]	ηt [-]
2.6×10^{-3}	0.55	4.4×10^{-4}	0.72

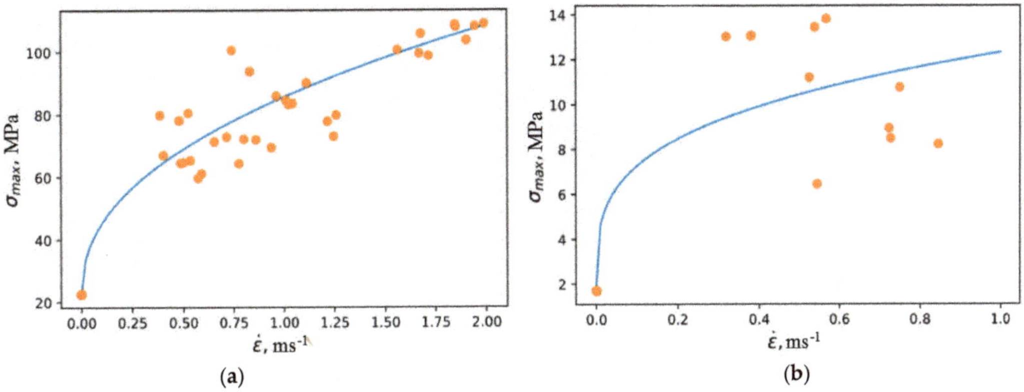

Figure 9. Approximation of the rate dependences of the ultimate stress in compression (**a**) and tension (**b**).

Model No. 159 includes the possibility of taking into account the influence of the strain rate on the fracture energy of the material (separately in compression, tension, and shear). These dependencies are determined by the formulas:

$$G_{fc}^{dyn} = G_{fc} \cdot \left(1 + \frac{E \cdot \dot{\varepsilon}}{f_c} \cdot \frac{\eta 0c}{\dot{\varepsilon}^{\eta c}}\right)^{repow}$$

$$G_{ft}^{dyn} = G_{ft} \cdot \left(1 + \frac{E \cdot \dot{\varepsilon}}{f_t} \cdot \frac{\eta 0t}{\dot{\varepsilon}^{\eta t}}\right)^{repow}$$

To determine the values of the parameters G_{fc}, G_{ft}, and *repow* for each test, the fracture energy was determined. This energy corresponds to the area under the strain diagram of the sample beyond the point of maximum stress that leads to fractures.

The results of the approximation of the rate dependences of the fracture energy in compression and tension are shown in Figure 10. The parameters are given in Table 6.

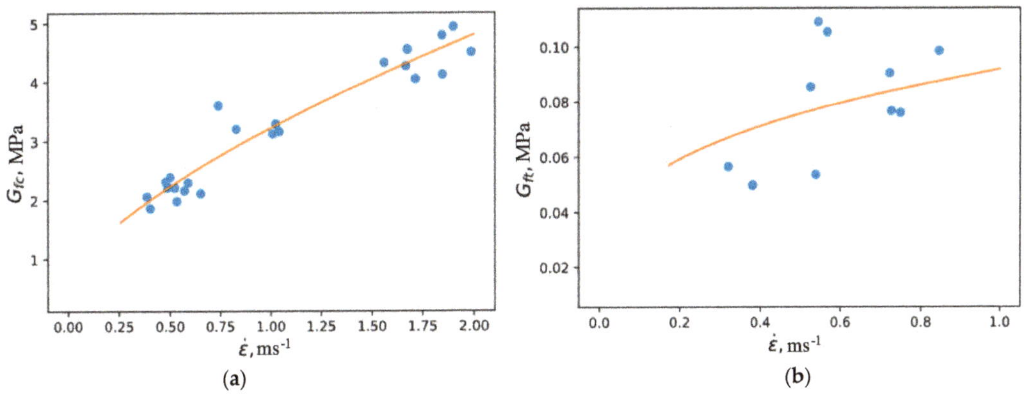

Figure 10. Approximation of the rate dependence of the fracture energy in compression (**a**) and tension (**b**).

Table 6. Parameters of rate dependences of the fracture energy.

G_{fc} [MPa]	G_{ft} [MPa]	*Repow* [-]
0.341	$3.5 \cdot 10^{-3}$	1.681

Thus, a complete set of parameters has been defined to describe the dynamic deformation and fracture of B22.5 concrete within the framework of the MAT_159 model.

Stress–strain diagrams obtained during compression and tension of an elementary volume at a constant strain rate are compared in different models. In simulation, one hexahedral element was loaded under one-dimensional stress conditions. Symmetry conditions were set on three faces of the element. Two faces were free and normal displacement was predefined of one face to obtain a constant specified strain rate. The results are shown in Figure 11. The dotted line shows the results of the real-world tests; the mat_52 legend corresponds to the calculation with the MAT_72 material, and the mat_159 legend corresponds to the calculation with the MAT_159 model, in which the main part of the parameters was determined from the dependencies embedded in the model and the parameters characterizing the strain rate effects were determined from the experimental data. The mat_159_concrete curve was obtained using a model in which all parameters, including those responsible for strain rate effects, were calculated according to the algorithms embedded in the model for a given concrete strength class. It should be noted that for tension, the slope of the initial section of the diagrams in the experiment and calculation are very different, which is apparently due to the error in measuring deformations using measuring bars. The strains realized under tension are an order of magnitude smaller than the strains under compression.

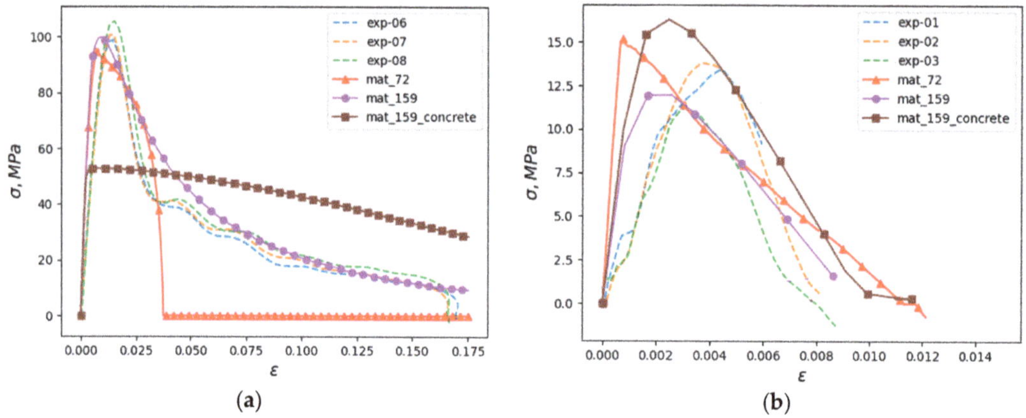

Figure 11. Comparison of stresses in an elementary volume: experiment and prediction using different models (**a**) compression, (**b**) tension.

It can be noted that the fracture branch of the curves is qualitatively best described by Model 159, in which some parameters were set manually. Apparently, this is due to the fact that in this case, the strain rate dependence of the fracture energy of the material was explicitly specified. The model automatically generated from the static strength of the material (line with square markers) predicts a significantly lower maximum stress and correlates poorly with the experimental data beyond the fracture initiation point.

5. Conclusions

As a result of the experimental study, the strain rate dependences of the ultimate stresses of concrete—as well as the fracture energy—under compression and tension were obtained. The ultimate stresses during compression and tension are affected by the strain rate. The dynamic strength values differ from static values by six times in the studied range. Based on the series of experiments on concrete samples, the identification was carried out, and the parameters for the MAT_CONCRETE_DAMAGE_REL3 (No. 72) and MAT_CSCM (No. 159) models were determined taking into account the effect of strain rate on strength characteristics as well as a significant difference in the properties of concrete in tension and compression. As a result of testing the models by the single-element modeling method, it was shown that the verified models qualitatively and quantitatively, with an accuracy acceptable for practical needs, make it possible to predict the results of real-world tests.

Author Contributions: Conceptualization, A.M.B. and L.A.I.; methodology, A.M.B.; software, M.E.G. and A.Y.K.; validation, A.M.B., A.Y.K. and A.K.L.; formal analysis, A.Y.K. and A.K.L.; investigation, A.M.B., A.Y.K., A.K.L. and V.A.E.; resources, A.M.B.; data curation, A.M.B.; writing—original draft preparation, A.M.B., M.E.G. and A.Y.K.; writing—review and editing, V.A.E. and L.A.I.; visualization, M.E.G.; supervision, A.M.B. and L.A.I.; project administration, A.M.B. and L.A.I.; funding acquisition, L.A.I. All authors have read and agreed to the published version of the manuscript.

Funding: This research was funded by the Russian Science Foundation, grant number 22-19-00138.

Data Availability Statement: The data presented in this study are available on request from the corresponding author.

Conflicts of Interest: The authors declare no conflict of interest.

References

1. Kolsky, H. An investigation of the mechanical properties of materials at very high rates of loading. *Proc. Phys. Soc. B* **1949**, *62*, 676. [CrossRef]
2. Kolsky, H. *Stress Waves in Solids*; Clarendon Press: Oxford, UK, 1963.
3. Dai, F.; Xia, K.; Tang, L. Rate dependence of the flexural tensile strength of Laurentian granite. *Int. J. Rock Mech. Min. Sci.* **2010**, *47*, 469–475. [CrossRef]
4. Toutlemonde, F.; Gary, G. Dynamic behavior of concrete. Experimental aspects. In *Dynamic Behavior of Concrete and Seismic Engineering*; Mazars, J., Millard, A., Eds.; ISTE Ltd.: London, UK, 2009; pp. 1–54.
5. Walley, S.M. Historical review of high strain rate and shock properties of ceramics relevant to their application in armour. *Adv. Appl. Ceram.* **2010**, *109*, 446–466. [CrossRef]
6. Zhao, J. An overview of some recent progress in rock dynamics research. In *Advances in Rock Dynamics and Applications*; Zhou, Y.X., Zhao, J., Eds.; CRC Press: Boca Raton, FL, USA, 2011; pp. 5–33.
7. Zhao, J.; Zhou, Y.X.; Hefny, A.M.; Cai, J.G.; Chen, S.G.; Li, H.B.; Liu, J.F.; Jain, M.; Foo, S.T.; Seah, C.C. Rock dynamics research related to cavern development for ammunition storage. *Tunn. Undergr. Space Technol.* **1999**, *14*, 513–526. [CrossRef]
8. Field, J.E.; Walley, S.M.; Proud, W.G.; Goldrein, H.T.; Siviour, C.R. Review of experimental techniques for high rate deformation and shock studies. *Int. J. Impact Eng.* **2004**, *30*, 725–775. [CrossRef]
9. Bragov, A.M.; Gonov, M.E.; Lamzin, D.A.; Lomunov, A.K.; Modin, I.A. Response of fine-grained fiber-reinforced concretes under dynamic compression. *Mater. Phys. Mech.* **2021**, *47*, 962–967.
10. Gonov, M.E. Comprehensive study of the properties of fine-grained concrete under dynamic tension and shear. *Probl. Strength Plast.* **2020**, *82*, 442–457. [CrossRef]
11. Gonov, M.E. Mechanical properties of fiber-reinforced concrete under dynamic compression. *Probl. Strength Plast.* **2022**, *84*, 130–148. [CrossRef]
12. Ramesh, K.T. High rates and impact experiments. In *Springer Handbook of Experimental Solid Mechanics*; Sharpe, W.N., Ed.; Springer: New York, NY, USA, 2004; pp. 929–960.
13. Bragov, A.M.; Igumnov, L.A.; Konstantinov, A.Y.; Lomunov, A.K. *High-Speed Deformation of Materials of Various Physical Nature*; N. Novgorod University: N. Novgorod, Russia, 2020; 299p. (In Russian)
14. Buyukozturk, O.; Shareef, S.S. Constitutive modeling of concrete in finite element analysis. *Comput. Struct.* **1985**, *21*, 581–610. [CrossRef]
15. Han, D.; Chen, W. Constitutive modeling in analysis of concrete structures. *J. Eng. Mech.* **1987**, *113*, 577–593. [CrossRef]
16. Hu, H.; Schnobrich, W. Constitutive modeling of concrete by using nonassociated plasticity. *J. Mater. Civ. Eng.* **1989**, *1*, 199–216. [CrossRef]
17. Malvar, L.J.; Crawford, J.E.; Wesevich, J.W.; Simons, D. A plasticity concrete material model for DYNA3D. *Int. J. Impact Eng.* **1997**, *19*, 847–873. [CrossRef]
18. Liu, Y.; Huang, F.; Ma, A. Numerical simulations of oblique penetration into reinforced concrete targets. *Comput. Math. Appl.* **2011**, *61*, 2168–2171. [CrossRef]
19. Tu, Z.; Lu, Y. Evaluation of typical concrete material models used in hydrocodes for high dynamic response simulations. *Int. J. Impact Eng.* **2009**, *36*, 132–146. [CrossRef]
20. Yonten, K.; Manzari, M.T.; Marzougui, D.; Eskandarian, A. An assessment of constitutive models of concrete in the crashworthiness simulation of roadside safety structures. *Int. J. Crashworthiness* **2005**, *10*, 5–19. [CrossRef]
21. LS-DYNA. Keyword User's Manual, Volume 2, LS-DYNA R11, 10/18/18 (r:10580). Available online: https://www.dynasupport.com/manuals/ls-dyna-manuals/ls-dyna_manual_volume_ii_r11.pdf (accessed on 20 December 2021).
22. Malvar, L.J.; Crawford, J.E.; Morrill, K.B. *Concrete Material Model Release III—Automated Generation of Material Model Input*; K&C Technical Report TR-99-24-B1, 18 August 2000 (Limited Distribution); Defense Nuclear Agency: Washington, DC, USA, 2000.

23. Murray, Y.D. *User's Manual for LS-DYNA Concrete Material Model 159*; Report No. FHWA-HRT-05-062; United States Department of Transportation: Washington, DC, USA, 2007.
24. Murray, Y.D.; Abu-Odeh, A.; Bligh, R. *Evaluation of Concrete Material Model 159*; Report No. FHWA-HRT-05-063; Texas Transportation Institute: College Station, TX, USA, 2007. Available online: https://www.fhwa.dot.gov/publications/research/infrastructure/structures/05063/05063.pdf (accessed on 27 December 2021).

materials

MDPI

Article

Modification Effect of $Ca(OH)_2$ on the Carbonation Resistance of Fly Ash-Metakaolin-Based Geopolymer

Yigang Lv [1,2], Jie Qiao [2,3], Weiwei Han [4,5,*], Bei Pan [2], Xiafei Jin [2,*] and Hui Peng [2]

1 Key Laboratory of Advanced Engineering Materials, Structure Behavior and Functional Control of University of Hunan Province, Changsha University of Science & Technology, Changsha 410114, China
2 School of Civil Engineering, Changsha University of Science & Technology, Changsha 410114, China
3 Zhejiang Communications Construction Group Co., Ltd., Hangzhou 310051, China
4 School of Traffic & Transportation Engineering, Changsha University of Science & Technology, Changsha 410114, China
5 National Engineering Research Center of Highway Maintenance Technology, Changsha University of Science & Technology, Changsha 410114, China
* Correspondence: vv@csust.edu.cn (W.H.); jxf002859@csust.edu.cn (X.J.)

Abstract: Compared with Portland cement, geopolymers have poor carbonization resistance, which will greatly limit the application their application. To improve the carbonization resistance of geopolymers, firstly, the carbonization behavior of the fly ash-metakaolin-based geopolymer was studied through accelerated carbonization tests. Secondly, different amounts of $Ca(OH)_2$ were introduced into the composite system, and the modification effect of the carbonization resistance of the modified geopolymer was studied. Finally, the modification effect of $Ca(OH)_2$ on the fly ash-metakaolin-based geopolymers was analyzed, and the modification mechanism was explored. It was found that adding $Ca(OH)_2$ to the fly ash-metakaolin-based geopolymer could significantly improve its initial compressive strength, but its strength after carbonization remained basically unchanged; meanwhile, the compressive strength of the terpolymer after carbonization clearly decreased after adding $Ca(OH)_2$. Compared with ordinary Portland cement, the carbonization rate of fly ash-metakaolin-based geopolymer is faster, and the addition of $Ca(OH)_2$ can inhibit the development of its carbonization depth. With increased carbonization age, the alkalinity of the geopolymer decreased, and the addition of $Ca(OH)_2$ inhibited the decrease in the alkalinity of the geopolymer. The addition of $Ca(OH)_2$ improved the microstructure of the geopolymers, the pore structure became denser, and the pore size became smaller size after carbonization. The hydration products of fly ash-metakaolin-based geopolymer are mainly amorphous silicaluminate gel and C–S–H gel, and $Ca(OH)_2$ forms in the hydration products of terpolymer with the incorporation of $Ca(OH)_2$, which is conducive to improving the carbonization resistance. In summary, $Ca(OH)_2$ can play a good role in modifying the carbonization resistance of fly ash-metakaolin-based geopolymers.

Keywords: fly ash-metakaolin; geopolymer; $Ca(OH)_2$; carbonization resistance; modification study

Citation: Lv, Y.; Qiao, J.; Han, W.; Pan, B.; Jin, X.; Peng, H. Modification Effect of $Ca(OH)_2$ on the Carbonation Resistance of Fly Ash-Metakaolin-Based Geopolymer. *Materials* **2023**, *16*, 2305. https://doi.org/10.3390/ma16062305

Academic Editors: Hrvoje Smoljanović, Ivan Balić and Nikolina Zivaljic

Received: 14 February 2023
Revised: 9 March 2023
Accepted: 12 March 2023
Published: 13 March 2023

1. Introduction

Geopolymers are inorganic gel materials with a spatial network structure dominated by ionic and covalent bonds, which dissolve and polymerize natural minerals or industrial wastes that are rich in active silicon and aluminum by alkaline solution. The carbon emission of geopolymers in the preparation process is much less than that of traditional Portland cement and it is a green cementitious material with the potential to replace Portland cement [1–3]. There are a variety of raw materials for the preparation of geopolymers, among which fly ash is the most common. The geopolymers prepared with fly ash have good stability, but their reaction and curing process is slow, and they usually need to be cured in a high-temperature environment above 60 °C. At the same time, geopolymers prepared with fly ash have poor permeability resistance [4–6]. Metakaolin has high pozzolanic activity and can also be used to prepare geopolymers. It can control the curing time well at room temperature, and the resulting geopolymer prepared with it is more stable in composition and performance. [7,8]. In addition, it was found that the hydration reaction when the geopolymer is prepared by mixing raw materials can produce a variety of gels. The coexistence of various gels improves the internal pore structure, and the prepared geopolymer can have a more dense and stable internal structure, resulting in better performance [9–12].

At present, many scholars have carried out a large number of studies on the properties of geopolymers [13–15]. It has been found [3,16–18] that geopolymers are made of silico-aluminum inorganic raw materials through mineral condensation, which is an inorganic polycondensation three-dimensional oxide network structure composed of tetrahedral units of $[AlO_4]$ and $[SiO_4]$, the chemical formula is $Mn\{(SiO_2)zAlO_2\}n \cdot wH_2O$. Compared with ordinary Portland cement, geopolymers have excellent properties such as acid resistance, alkali resistance, impermeability, frost resistance and high-temperature resistance [19–22]. However, since the hydration products of geopolymers do not contain $Ca(OH)_2$ as a buffer, their carbonation resistance is much lower than that of ordinary silicate concrete [23,24]. The carbonization resistance of geopolymer is a key factor affecting its durability and is an urgent problem to be solved. Many scholars have carried out some research on carbonization behavior, and the modification of carbonization resistance in geopolymers [25,26]. Lv et al. [27] summarized the current research status of the carbonization properties of geopolymer cementing materials from aspects of the carbonization mechanism, carbonization rate, carbonization influence, and improvement in carbonization resistance, and outlined the achievements and problems in this research field. Gao K et al. [28] added nano SiO_2 into the activator for modification when preparing the geopolymer based on metakaolin. It was found that the compressive strength and structure of the geopolymer improved after the addition of nano SiO_2, and the permeability resistance of the geopolymer improved; thus, the durability of the geopolymer, such as its carbonation resistance, was improved. Park et al. [29] modified alkali-excited slag cementing material by adding active MgO. It was found that the carbonization depth of alkali-excited slag cementing material decreased significantly with an increase in MgO content, and its carbonization resistance was enhanced. He et al. [30] added $Ca(OH)_2$ into alkali-excited slag cementing material and studied its carbonization resistance, finding that CO_2 could be absorbed and consumed by $Ca(OH)_2$, thus improving the carbonization resistance of alkali-excited slag cementing material. At the same time, Glukhovsky [31] believed that the reaction of geopolymers in systems with high calcium content was mainly a process of depolymerization and reaggregation, and C–(A)–S–H gel similar to cement hydration products would be generated in the hydration products.

In summary, nano-SiO_2, $Ca(OH)_2$, MgO, and other admixtures can be used to modify the carbonization resistance of geopolymers. However, at present, most studies on the modification of admixtures on carbonization resistance in geopolymers are based on high-calcium geopolymers prepared from slag and other raw materials, while few studies are focused on geopolymers prepared from low-calcium raw materials such as metakaolin and fly ash [24,27]. However, no $Ca(OH)_2$ is generated in the hydration products of the geopolymer system, whether or not there is a calcium source [32]. Based on this, in order to explore the modification method of carbonization resistance of geopolymers with good modification effect, this study prepared fly ash-metakaolin-based geopolymer with low calcium content fly ash and metakaolin; As is shown in Figure 1, $Ca(OH)_2$ was used as an admixture to increase the content of calcium in the system. Accelerated carbonization and carbonization behavior tests of the geopolymer, before and after modification, were carried out. The modification effect of the carbonation resistance of the composite system was analyzed, and the carbonization resistance of the composite system was compared with that of ordinary Portland cement, to explore the modification mechanism of $Ca(OH)_2$ admixture on the carbonization resistance of the system.

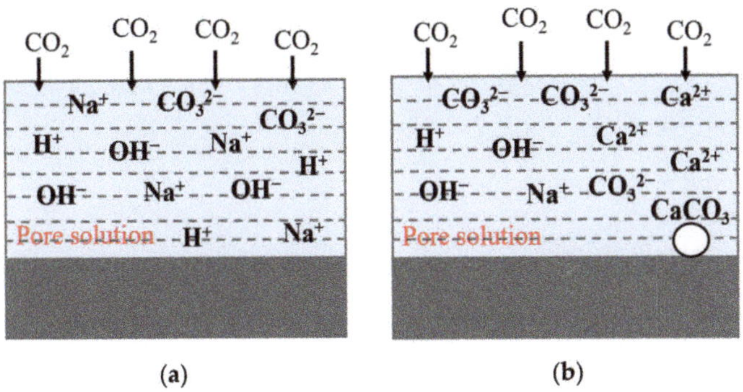

Figure 1. Carbonization process of geopolymers. (**a**) Geopolymers with low calcium content; (**b**) Geopolymers with high calcium content.

2. Experiment Design

2.1. Test Materials and Parameters

In this experiment, fly ash and metakaolin were used to prepare composite cementing material geopolymer. Among them, metakaolin a KAOOPOZZ series of highly active metakaolin produced by China Inner Mongolia Chaobrand Building Materials Technology Co., Ltd., fly ash from China Henan Gongyi City Yuanheng water purification materials factory. Industrial alkaline sodium silicate solution (SiO_2/Na_2O molar ratio of 3.28, solid content of 34.89%) was used as the raw material of the activator. Sodium hydroxide solid (purity of 99.5%) and deionized water were added to the raw material of the activator to make modified sodium silicate as the activator. Among them, the industrial alkaline sodium silicate solution comes from Qiulitian Chemical Co., Ltd., Xingtai City, Hebei Province, China, solid sodium hydroxide is an industrial-grade flake sodium hydroxide produced by Zhengzhou Qingyuan Chemical Products Co., Ltd., Henan Province, China. Before the test, the water–binder ratio was controlled to 0.65, and the modulating modulator and concentration were 1.2 and 36%, respectively. $Ca(OH)_2$ used in this experiment was produced by Xilong Science Co., Ltd., Shantou, China, and its main parameters were as follows: white powdery solid with a mass fraction of $Ca(OH)_2$ greater than 95%.

An X-ray fluorescence spectrometer (XRF) was used to determine the chemical components of metakaolin (MK) and fly ash (FA). The specific mass fraction of chemical composition is shown in Table 1.

Table 1. Mass fraction of chemical composition of metakaolin and fly ash.

Component (%)	CaO	SiO_2	Al_2O_3	Fe_2O_3	MgO	Na_2O	SO_3	K_2O	TiO_2	Others
MK	-	54.5	43	1	0.8	0.1	-	0.4	-	0.2
FA	3.05	59.91	31.24	3.80	0.54	-	0.61	2.06	1.34	0.45

2.2. Test Mix Design and Specimen Preparation

With the total mass of fixed raw materials unchanged, fly ash mass ratios of 0% (F0), 20% (F2), and 40% (F4) were selected as test variables to prepare a batch of 50 × 50 × 50 mm fly ash-metakaolin-based geopolymer net slurry samples. The modification effect of $Ca(OH)_2$ on the carbonation resistance of the polymer was studied by adding $Ca(OH)_2$, with mass fractions of raw material of 5% and 10%. A group of cement slurries with a water-binder ratio of 0.65 were prepared as the control group, and the specific mix ratio design is shown in Table 2.

Table 2. Mixture proportions of fly ash-metakaolin-based geopolymer.

Serial Number	Incorporation Amount of $Ca(OH)_2$ (%)	FA (%)	MK (%)	The Water–Binder Ratio	Mode of Maintenance
F0 Ca5F0 Ca10F0	- 5 10	0	100	0.65	Seal curing
F2 Ca5F2 Ca10F2	- 5 10	20	80	0.65	Seal curing
F4 Ca5F4 Ca10F4	- 5 10	40	60	0.65	Seal curing
PC0.65	-	-	-	0.65	Seal curing

In case of damage to samples during the test, 15 specimens were prepared for each mix ratio for the test. The quantitative geopolymer raw materials were weighed and added into the blender together with the prepared and aged activator for 24 h. First, the materials were stirred in low-speed mode for 2 min, and then stirred in high-speed mode for 3 min, in order to ensure that the mixture was uniform. The mixed slurry was poured into the mold, and then cured in a standard curing box for 24 h. To ensure that no shrinkage cracking occurs before sample carbonization, the temperature T = 20 °C and relative humidity RH = 95% were maintained in the curing box. After 24 h from mold removal, the mixtures continued curing for 28 days; then, accelerated carbonization tests and pre-carbonization compressive strength tests were conducted.

2.3. Accelerated Carbonization Design

An accelerated carbonization test was conducted according to the GB/T 50082-2009 standard [33]. Samples were removed 2 d before the carbonization test, and one surface was selected for exposure carbonization, while the remaining surfaces were sealed with heated paraffin wax. The concentration of CO_2 in the carbonization box was maintained at (20 ± 3)%, the relative humidity was controlled at (70 ± 5)%, and the ambient temperature was controlled within a range of (20 ± 2) °C. The samples were exposed to carbonization for 0 d, 1 d, 3 d, 7 d, 14 d, 21 d, and 28 d. The carbonization depth and material alkalinity were measured for specimens at different carbonation ages, and the compressive strength was tested for specimens at 28 d carbonation ages.

2.4. Compressive Strength Test

A universal pressure testing machine was used to test the compressive strength of the uncarbonized and 28 d carbonated clean pulp samples, respectively, to compare the changes in the compressive strength of specimens before and after carbonization. The number of compressive strength specimens in each group was three, and the arithmetic mean value was taken as the compressive strength value of the group of samples. If the difference between one of the maximum or minimum values of the three values and the median value exceeds 15% of the median value, the median value is taken as the compressive strength value of the group of specimens. If the difference between two measured values and the median value is more than 15% of the median value, the test results of this group of specimens are invalid, and another three specimens are selected for test until they meet the requirements.

2.5. Carbonization Depth Test

The carbonation depths of samples with different carbonation times were measured with phenolphthalein alcohol solution. The cube samples with different carbonation times were taken out, and the samples were separated into two halves from the center line of the cube by a dry sawing method. The fresh section was sprayed with 1% phenolphthalein alcohol solution, the discoloration range of the section was observed, and the distance from the carbonation boundary of multiple points to the surface was measured. The measurement is accurate to 0.1 mm, and the average value was taken as the carbonation depth.

2.6. Material Alkalinity Test

After carbonation depth measurement, the other half of the cube specimen was taken without spraying phenolphthalein, and the carbonized surfaces of each group of samples were sliced from 0 to 10 mm inside. After crushing, the samples were dried to a constant weight of 50 °C, and then ground. Powder (powder particles less than 75 μm) passed through a 200 target quasi sieve was sealed and stored. The powder and deionized water were mixed and soaked in a 1:3 ratio of powder to water. After standing, the pH of the supernatant liquid was measured by a solid–liquid extraction method [34,35], in order to characterize the change in material alkalinity. The result is accurate to 0.01. The pH formula is as follows:

$$pH = 14 + lg\frac{a(OH^{-})}{5} \quad (1)$$

where a is the amount of substance of OH^{-}, and the unit is mol.

2.7. Microstructure Test

In order to study the effect of adding $Ca(OH)_2$ to the microstructure of the fly ash-metakaolin-based geopolymers, F2, Ca10F2, and PC0.65 were taken as examples, and the pore structure and micromorphology of the geopolymers were analyzed via mercury injection and scanning electron microscopy, before carbonization and at 28 days of carbonization.

MIP: The samples before carbonization, and 28 d after carbonization, were crushed from the carbonized surface to the inner (0~10) mm section. The pieces with a regular shape and no larger than 1 cm^3 were soaked in anhydrous ethanol for 48 h and then dried at 50 °C to a constant weight. A Kanta PoreMaster 33 series automatic pore size analyzer was used to test the pore size distribution and porosity.

SEM: The pre-carbonized and 28 d carbonized samples were crushed from the carbonized surface to the inner (0~10) mm section. The pieces with a flat surface and no more than 1 cm^3 in size were soaked in anhydrous ethanol for 48 h and then dried at 50 °C to constant weight. A Zeiss EVO MA 25 series high-resolution scanning electron microscope was used to spray gold on the smooth section of the fragments and observe the micromorphology of the samples.

2.8. Phase Composition Test

In order to study the influence of $Ca(OH)_2$ on the phase composition and change of the fly ash-metakaolin-based geopolymers before and after carbonization, F2, Ca10F2 and PC0.65 were taken as examples. X-ray diffraction (XRD) and Fourier transform infrared spectroscopy (FT-IR) was used to analyze the phase composition of the material before carbonization and at 28 days of carbonization.

XRD: An Ultima IV series X-ray diffractometer was used to test the powder from the sample carbonized surface to the inside (0~10) mm, before carbonization and at 28 d carbonization. The XRD scanning range was 5~70°, and the scanning speed was 4°/min.

FT-IR: A Fourier transform infrared spectrum analyzer Shimazu IRTracer-100 series in Japan was used to take powder from the sample carbonization surface to the inner (0~10) mm, before carbonization and at 28 d carbonation age, after passing a 200 target sieve mixed with KBr tablet for testing. The FT-IR scanning range was 400~4000 cm^{-1}. The resolution was 4 cm^{-1}.

3. Results and Discussion

3.1. Compressive Strength

Figure 2 shows the change in compressive strength of fly ash-metakaolin-based geopolymer and cement before carbonization, and at 28 d carbonization age. As shown in Figure 2, when the dosage of $Ca(OH)_2$ was 5%, the initial compressive strengths of the Ca5F0, Ca5F2, and Ca5F4 groups increased by 3.30%, 16.90%, and 10.60%, respectively. When the dosage of $Ca(OH)_2$ was 10%, the initial compressive strengths of Ca10F0, Ca10F2, and Ca10F4 increased by 7.70%, 28.70%, and 38.80%, respectively. The results showed that $Ca(OH)_2$ could effectively improve the initial compressive strength of fly ash-metakaolin-based geopolymer. With an increase in $Ca(OH)_2$ content, the growth rate of initial compressive strength of fly ash-metakaolin-based geopolymer became greater, and the higher the mass proportion of fly ash in the composite system, the more obvious the growth effect, because the increase in calcium content could promote the hydration reaction of fly ash to some extent [31].

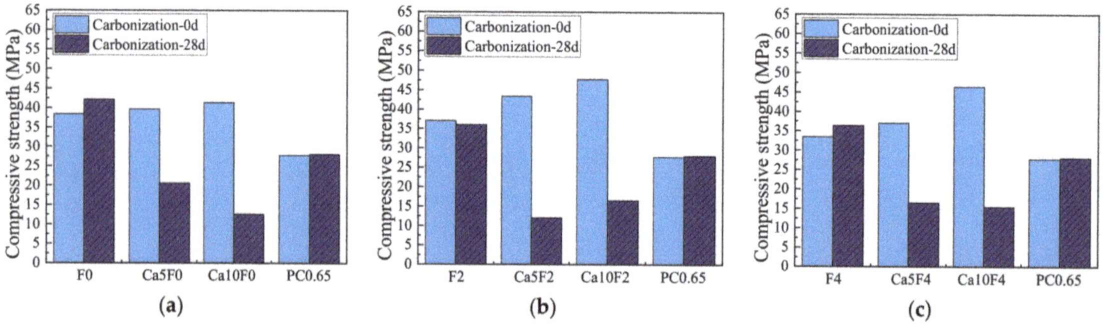

Figure 2. Compressive strength of cement and geopolymers before carbonization, and at 28 days of carbonization. (**a**) F0; (**b**) F2; (**c**) F4.

By comparing the strength before and after carbonization, it can be found that the change in compressive strength of the geopolymers without $Ca(OH)_2$ before and after carbonization was between −2.83% and 9.77% and that the change rate was small. The change in compressive strength of cement paste before and after carbonization is 1.01%, which is basically unchanged. Li [32] found through research that C-S-H and C-A-S-H gels with low Ca/Si ratios would be generated in the geopolymers containing calcium sources. The gels with low Ca/Si ratio were prone to decalcification in the carbonization process, which would lead to the instability and destruction of the gel structure and adversely affect the compressive strength. Therefore, the compressive strength of each group of geopolymer mixed with $Ca(OH)_2$ decreased significantly after carbonization, with a decreased range of more than 48%, and the Ca10F2 group even reached 72.25%. The results showed that the addition of $Ca(OH)_2$ resulted in a significant decrease in the compressive strength of the fly ash-metakaolin-based geopolymers after carbonization.

3.2. Carbonization Depth

Figure 3 shows the evolution law of carbonization depth of cement and fly ash-metakaolin-based geopolymers after carbonization. As can be seen from Figure 3, with an increase in carbonization age, the carbonization depth of the fly ash-metakaolin-based geopolymers gradually increased, indicating that the erosion degree of the geopolymers by CO_2 continuously intensified. The carbonization depths of the F0, F2, and F4 groups reached 25.5 mm, 33.4 mm, and 15.4 mm at 28 days of carbonization, respectively, while the carbonization depth of the PC0.65 cement control group was 17.1 mm after 28 days of accelerated carbonization. When 5% $Ca(OH)_2$ was added into the geopolymer, the carbonation depths of the Ca5F0, Ca5F2, and Ca5F4 groups at the carbonation age of

28 days decreased to 17.7 mm, 13.8 mm, and 12.5 mm, respectively. When the content of $Ca(OH)_2$ was 10%, the carbonation depths of the Ca10F0, Ca10F2, and Ca10F4 groups at the 28 d carbonation stage were 9.5 mm, 12.2 mm, and 11.7 mm, respectively. This indicates that $Ca(OH)_2$ inhibited the development of the carbonization depth of fly ash-metakaolin-based geopolymers under the same carbonization age, and with an increase in $Ca(OH)_2$ content in the system, the modification effect on the carbonization resistance of geopolymer became better.

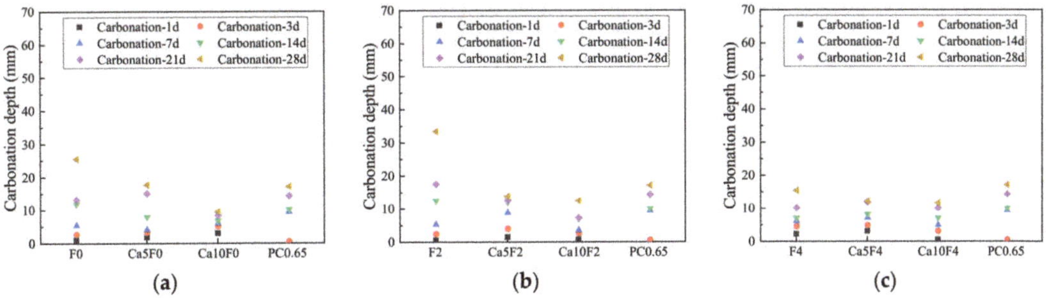

Figure 3. Evolution law of carbonization depth between geopolymers and cement. (**a**) F0; (**b**) F2; (**c**) F4.

According to the literature [36–38], based on Fick's first law, carbonation depth and carbonation age are related as follows:

$$y = D_e\sqrt{t} \tag{2}$$

where y is carbonization depth (mm); t is carbonation age (d); and D_e is the carbonization coefficient.

Since there is a linear relationship between the square root of carbonation depth and carbonation age, the carbonation coefficient D_e reflects the magnitude of the material carbonation rate. The carbonation depth and carbonation age of each group were fitted, and the fitting curve is shown in Figure 4. The carbonization coefficients D_e of F0, F2, F4, and PC0.65 were 3.51, 4.36, 2.44, and 2.97, respectively. When the content of $Ca(OH)_2$ was 5%, the carbonization coefficients D_e of Ca5F0, Ca5F2, and Ca5F4 were 2.85, 2.81, and 1.89, respectively. When the content of $Ca(OH)_2$ was 10%, the carbonation coefficients D_e of Ca10F0, Ca10F2, and Ca10F4 were 1.92, 2.43, and 2.08, respectively. The results show that the addition of $Ca(OH)_2$ in the fly ash-metakaolin-based geopolymers significantly inhibited the carbonization rate of the geopolymers. With an increase in $Ca(OH)_2$ content, the carbonization rate of the composite system decreased gradually, and the modification effect of carbonization resistance became more obvious.

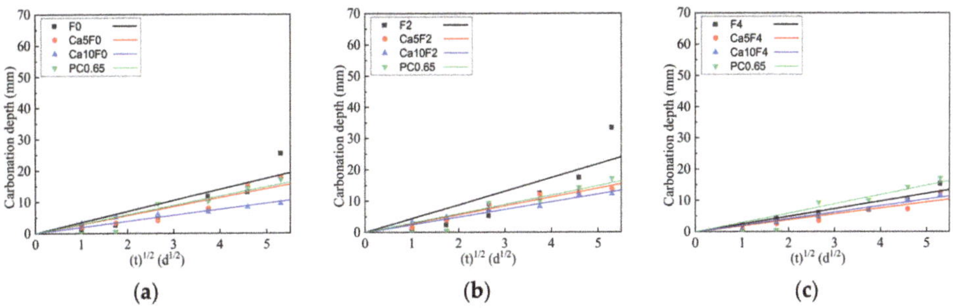

Figure 4. The fitting curve of carbonation depth of cement and geopolymers. (**a**) F0; (**b**) F2; (**c**) F4.

3.3. Material Alkalinity

The change in material alkalinity of fly ash-metakaolin-based geopolymers and cement with carbonization time is shown in Figure 5. As carbonization proceeded, the measured pH decreased gradually. In the first seven days of carbonization aging, the material alkalinity of the F0, F2, and F4 groups decreased rapidly, because a large amount of CO_2 was dissolved in the pore solution to generate H_2CO_3, which then neutralized the OH^- in the pore solution and made the pore solution decline in alkalinity continuously. After seven days of carbonization, the pH decline slowed down and gradually became stable; the pH of some groups even showed a slight rising trend. This was because the ions in the pore solution gradually reached the ionization equilibrium, and part of the alkali metal ions that were solidified in the gel was re-released into the pore solution, affecting the ionization equilibrium of the material, thus affecting its alkalinity. Lv et al. [27] believed that the geopolymer was prepared with a strong alkaline activator, and the pore solution also had a high pH value after the hydration reaction. However, due to the lack of $Ca(OH)_2$ as a base reserve, the pH value of the pore solution would rapidly decline with the diffusion and dissolution of CO_2. This was also verified in this experiment, the material alkalinity of the fly ash-metakaolin-based geopolymers decreased more significantly in this experiment. The material alkalinity of the F2 group at the carbonization age of 28 d even dropped to 9.793, reaching a state where the passivation film on the steel bar surface was at risk of failure.

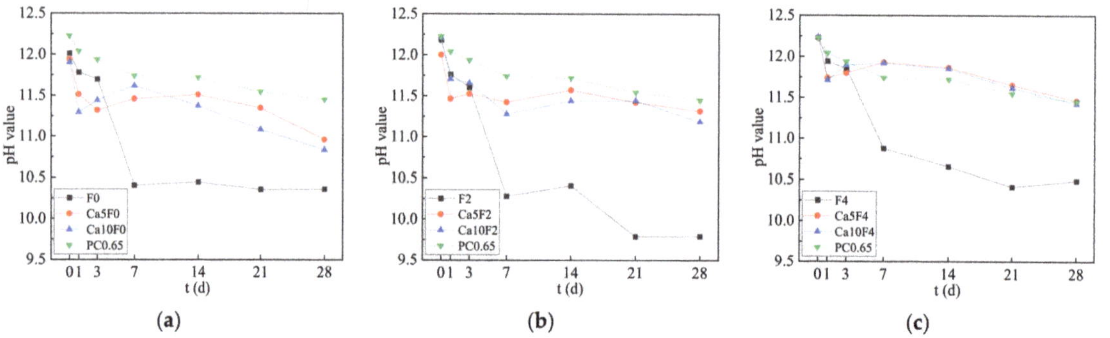

Figure 5. The changes in material alkalinity of geopolymer and cement with carbonation age. (**a**) F0; (**b**) F2; (**c**) F4.

It can be seen from Figure 5 that the addition of $Ca(OH)_2$ played the role of an alkali reserve in the composite system, and could better restrain the decline in alkalinity of the fly ash-metakaolin-based geopolymers. When 5% $Ca(OH)_2$ was added into the system, the material alkalinities of the Ca5F0, Ca5F2, and Ca5F4 groups at 28 d carbonization age were 10.958, 11.318, and 11.461, respectively. When the content of $Ca(OH)_2$ was 10%, the material alkalinities of the Ca10F0, Ca10F2, and Ca10F4 groups at the 28 d carbonization stage were 10.835, 11.188, and 11.423, respectively. Since the ions in the pore solution of the geopolymer basically reached ionization equilibrium after 28 days of carbonization, the pH values of the final pore solutions of the F0, F2 and F4 groups with 5% or 10% $Ca(OH)_2$ content were not significantly different.

3.4. Microstructure Analysis

3.4.1. Mercury Intrusion Method

Figure 6 shows the pore structure parameters of the fly ash-metakaolin-based geopolymers and cement before and after carbonization. In Figure 6a, the solid line represents the pore size distribution of each group before carbonization, while the dashed line represents the pore size distribution of each group at the 28 d carbonization age. As can be seen from Figure 6a, the pore size of both geopolymers and cement was mostly distributed between

10 and 100 nm. Compared to the F2 group, the peak value of the pore size distribution of the Ca10F2 group with $Ca(OH)_2$ shifted to the left before and after carbonization, indicating that the addition of $Ca(OH)_2$ improved the microstructure of the geopolymer, and the pore structure became denser. As can be seen from Figure 6b, the porosity of the F2 group before carbonization was 22.14%, and that of PC0.65 before carbonization was 21.40%. The addition of $Ca(OH)_2$ reduced the porosity of the geopolymer before carbonization, and the porosity of the Ca10F2 group before carbonization was 19.24%. For fly ash-metakaolin-based geopolymers, the main hydration product was N–A–S–H gel. The incorporation of $Ca(OH)_2$ increased the calcium content in the composite system, and C–(A)–S–H gel was generated in the system while the hydration reaction rate improved. The mutual filling of various gels played a positive role in the development of the microstructure, making the microstructure denser.

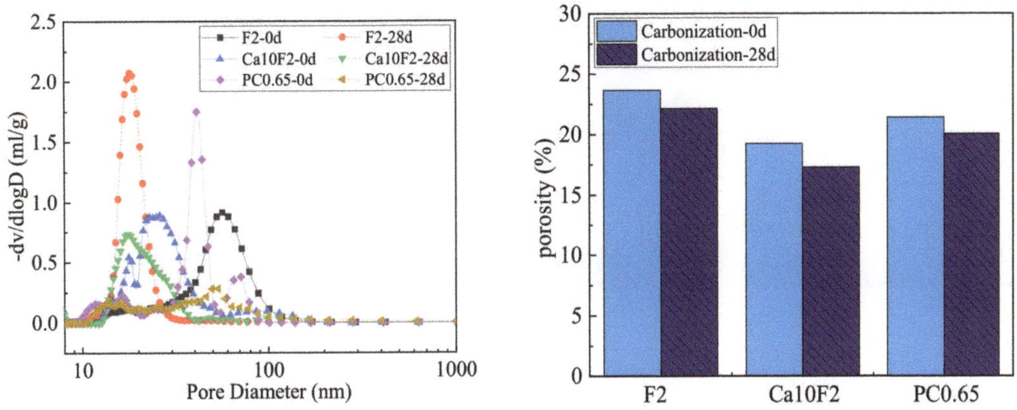

(a) (b)

Figure 6. Effect of carbonization on geopolymers and cement pore structure. (**a**) Aperture distribution variation; (**b**) Porosity variation.

After carbonization, the pores of both geopolymers and cement develop into a smaller pore size. $CaCO_3$ was generated after $Ca(OH)_2$ carbonization in the cement PC0.65 group, which filled some large pores and reduced pore connectivity. However, for the pre-modified geopolymer F2 group, there was a lack of Ca^{2+} in the system, and the densification of the pore structure was mainly due to the precipitation of sodium salt crystals in the gel to fill the pores. As can be seen from Figure 6b, the porosity of the F2 and PC0.65 groups at the 28 d carbonization age both decreased, which is consistent with the change in pore size distribution. Existing studies have shown that for the geopolymers containing calcium source, N–A–S–H gel hardly changes in the carbonization process, and the calcium content in pore solution is low, so the decalcification of C–S–H gel is the main carbonization reaction in the sample [27,32]. For the Ca10F2 group, $Ca(OH)_2$ was introduced into the system as a calcium supplement after the addition of $Ca(OH)_2$, and the $CaCO_3$ generated in the carbonization process played a role in refining pores. However, compared with cement, it was still a system with low calcium content. When the Ca^{2+} concentration in the pore solution dropped to a certain extent, the C–S–H gel generated in the system after the introduction of the calcium source showed a decalcification phenomenon, and the gel structure became unstable, which led to a significant reduction in the compressive strength of the carbonized geopolymer mixed with $Ca(OH)_2$ on a macro level [32].

3.4.2. Scanning Electron Microscopy

Figure 7 shows the comparison of the micro-morphology of fly ash-metakaolin-based geopolymers and cement, before carbonization and at 28 d carbonization age. From

Figure 7a, it can be found that there was unreacted metakaolin with plate structure on the surface of the F2 group, as well as some hollow cavities and unreacted fly ash particles. According to the comparison shown in Figure 7a,b, there was no obvious change in the microstructure of the F2 group before and after carbonization. There were some particles of sodium salt crystal precipitated by gel on the surface of the F2 group after carbonization, but the sodium salt crystals had a limited role in filling pores, so the changes in pores after carbonization were not obvious. This is consistent with the results of the pore structure test. As can be seen from Figure 7c,d, carbonization led to a net volume increase and precipitation of $CaCO_3$ in the pore network of the PC0.65 group. After carbonization, the microstructure became denser and its porosity decreased, which is consistent with the test results of pore structure.

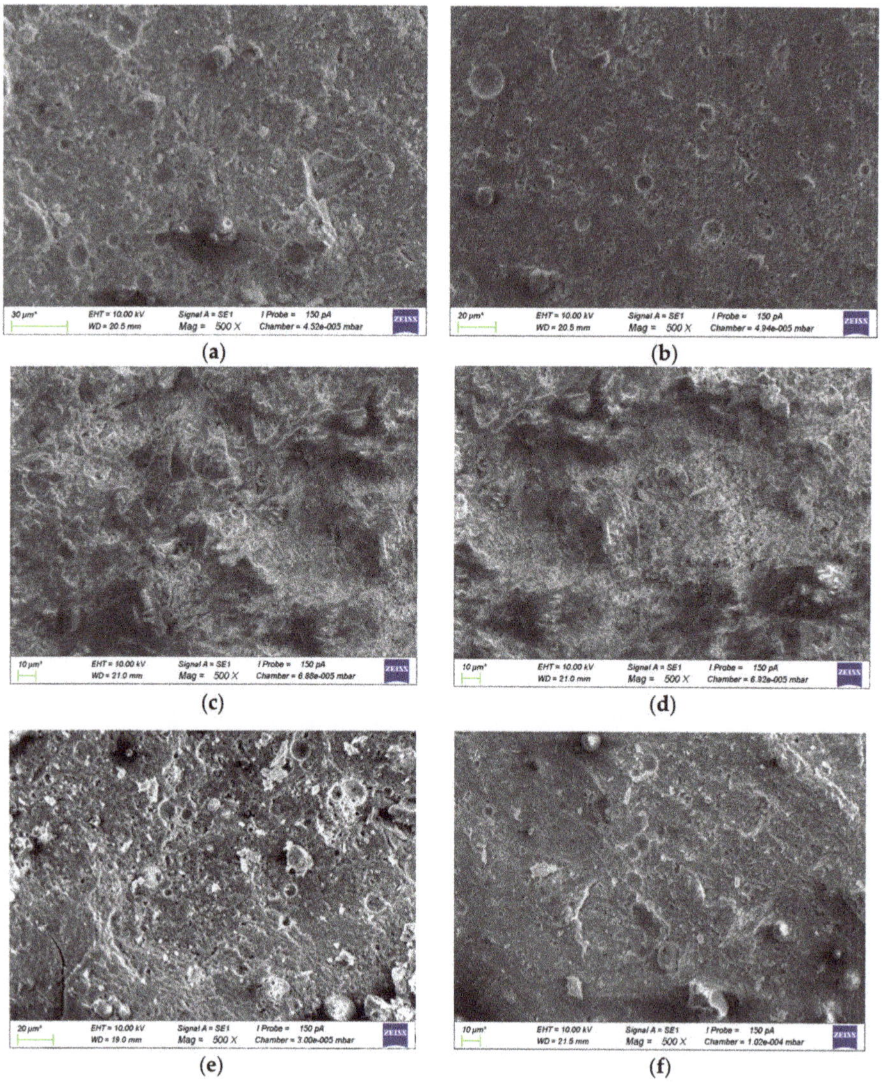

Figure 7. Micromorphology of geopolymers and cement before and after carbonization. (**a**) F2(pre-carbonization); (**b**) F2(28 d carbonation age); (**c**) PC0.65(pre-carbonization); (**d**) PC0.65(28 d carbonation age); (**e**) Ca10F2(pre-carbonization); (**f**) Ca10F2(28 d carbonation age).

Figure 7e,f show the micro-morphology of the Ca10F2 group before and after carbonization, respectively. As can be seen from Figure 7e, after the incorporation of $Ca(OH)_2$, various gels such as C–(A)–S–H and N–A–S–H appeared in the hydration products of the geopolymers, which formed a denser microstructure when filled with each other. At the same time, part of the unreacted $Ca(OH)_2$ was observed to be embedded in the gel. Therefore, the initial compressive strength of the geopolymers greatly improved after the addition of $Ca(OH)_2$. As can be seen from Figure 7f, the carbonized slurry developed many cracks, which may be because the $Ca(OH)_2$ in the gel generated a large amount of $CaCO_3$, resulting in volume expansion and cracks. At the same time, the added $Ca(OH)_2$ was not sufficient as a base reserve to resist the carbonization process, and a large number of calcium-containing gels were decalcified under CO_2 erosion. The cementation degree of the composite gel decreased, the gel structure became unstable, and the compressive strength decreased greatly after carbonization.

3.5. Phase Composition Analysis

3.5.1. XRD Phase Analysis

Figure 8 shows the XRD patterns of fly ash-metakaolin-based geopolymers and cement before and after carbonization. As can be seen from Figure 8, the hydration products of the F2 group were mainly quartz and mullite, and there was a small amount of muscovite. In addition, there was a dispersed steamed bun peak between $2\theta = 20°$ and $2\theta = 40°$, which is a typical spectrum characteristic of geopolymers. The results showed that the hydration products of fly ash-metakaolin-based geopolymers are mainly amorphous silicaluminate gels. The characteristic peaks of the Ca10F2 group after adding $Ca(OH)_2$ at $2\theta = 20°$~$40°$ were also characteristic amorphous peaks. The main hydration products of the Ca10F2 group were basically the same as the geopolymers without adding $Ca(OH)_2$. The main crystalline phases were quartz, mullite and a small amount of muscovite. At the same time, after $Ca(OH)_2$ was added to the system, an obvious $Ca(OH)_2$ diffraction peak appeared in the hydration products [39]. In addition, due to the introduction of calcium sources, some C–A–S–H and C–S–H gels formed in the hydration products, and the gels of the geopolymers existed in an amorphous form. For ordinary Portland cement, the hydration products of the PC0.65 group were mainly C–S–H gel and $Ca(OH)_2$.

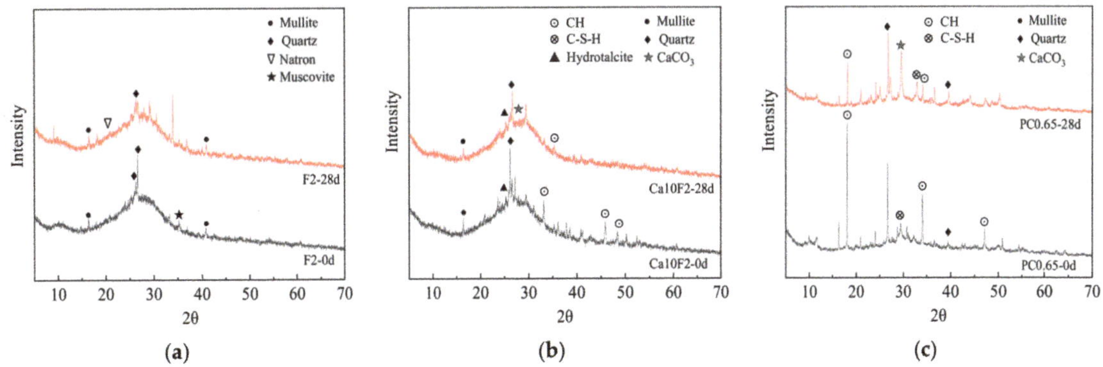

Figure 8. XRD patterns of geopolymers and cement before and after carbonization. (**a**) F2; (**b**) Ca10F2; (**c**) PC0.65.

As can be seen from Figure 8, the diffraction peak of $Ca(OH)_2$ in the PC0.65 group after carbonization significantly weakened, mainly because $Ca(OH)_2$ reacted with CO_2 to produce $CaCO_3$ during cement carbonization. However, the characteristic peak value of $CaCO_3$ did not appear after carbonization of the F2 group, and only a small amount of Natron was found. This is because the hydration products of fly ash-metakaolin-based geopolymers hardly contain substances that can react with CO_2. In the carbonization process, the development of carbonization was inhibited mainly by the alkalinity of its pore solution. However, the $Ca(OH)_2$ diffraction summit of Ca10F2 decreased significantly after carbonization, because the $Ca(OH)_2$ added into the system was consumed by the CO_2 reaction in the carbonization process to resist carbonization erosion, indicating that the incorporation of $Ca(OH)_2$ can significantly improve the carbonization resistance of fly ash-metakaolin-based geopolymers.

3.5.2. FT-IR Chemical Structure Test and Analysis

Figure 9 shows the FT–IR diagrams of fly ash-metakaolin-based geopolymers and cement before and after carbonization. Based on an infrared spectrum analysis test, qualitative analysis of material changes was conducted by studying fly ash-metakaolin-based geopolymers and specific functional groups in hydration products or carbonization products of cement. Figure 9 shows marked locations of major peaks, which are mainly functional groups corresponding to the structure of bonded water, carbonate, and gel. As can be seen from Figure 9, due to the stretching vibration of O–H in water, samples of F2, Ca10F2 and PC0.65 before carbonization appear absorption peaks at 3371 cm^{-1}, 3373 cm^{-1} and 3316 cm^{-1} respectively [40]. The absorption peak of the F2 group at 1654 cm^{-1} corresponded to the bending vibration of O–H, which was divided into bonded water in the gel after the hydration reaction. The crest at 960–970 cm^{-1} corresponds to the asymmetric stretching vibration of Si–O–T in the gel (where T is a tetrahedron of silicon or aluminum) [32,41], while the corresponding crest of group F2 is located at 983 cm^{-1}, which is due to the influence of the Si/Al ratio of raw materials; the crest was offset to a certain extent [42]. For the Ca10F2 group and PC0.65 group, the crest at 1403 cm^{-1}–1481 cm^{-1} represents the stretching vibration of the C–O bond in the CO_3^{2-} ion, and the crest at 954 cm^{-1}–990 cm^{-1} is caused by C–(A)–S–H gel [42]. However, this absorption peak did not appear in the F2 group without calcium, which indicates that the addition of $Ca(OH)_2$ into the fly ash-metakaolin-based geopolymers generated part of the C–(A)–S–H gel in the composite system, which is consistent with the XRD test results.

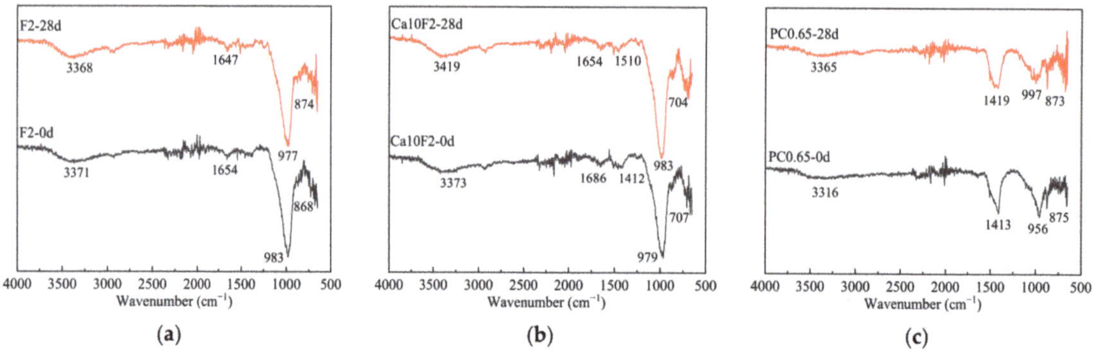

Figure 9. FT-IR images of geopolymers and cement before and after carbonization. (**a**) F2; (**b**) Ca10F2; (**c**) PC0.65.

As can be seen from Figure 9, since the fly ash-metakaolin-based geopolymers mainly rely on their own alkalinity to resist CO_2 erosion in the carbonization process, the corresponding gel peak value of group F2 was basically unchanged. However, the peak value of the 874 cm^{-1} position of group F2 after carbonization was due to the C–O bond in CO_3^{2-}, because some of the sodium crystals precipitated in the gel during the carbonization process. Carbonization led to the decalcification of calcium-containing gel and an increase in the degree of silica gel polymerization. Although there was no obvious change, in Ca10F2 and PC0.65 samples, the wave crest caused by the asymmetric stretching vibration of Si–O–T originally located at 960–970 cm^{-1} was shifted to a higher position after carbonization [32].

However, compared with cement, the calcium content of the fly ash-metakaolin-based geopolymers mixed with $Ca(OH)_2$ was still very low, so the peak deviation degree of the PC0.65 group was more obvious, which is consistent with the XRD test results.

4. Conclusions

Fly ash and metakaolin were used as raw materials to prepare fly ash-metakaolin-based geopolymers, and different dosages of $Ca(OH)_2$ were introduced to modify the carbonization resistance of the composite system. Through the accelerated carbonization test, the carbonization behavior and modification effect of the geopolymer was studied. The carbonization resistance of the geopolymer was compared with that of ordinary Portland cement, and the modification mechanism of the carbonization resistance of fly ash-metakaolin-based geopolymers was analyzed by adding $Ca(OH)_2$. The main conclusions are as follows:

(1) The initial compressive strength of the fly ash-metakaolin-based geopolymers mixed with $Ca(OH)_2$ increases between 3% and 40%, and the higher the mass proportion of fly ash in the composite system, the greater the increase. The variation of compressive strength of the unmodified geopolymers before and after carbonization was between −2.83% and 9.77%, while the reduction of compressive strength of the carbonized geopolymers with $Ca(OH)_2$ was more than 48%.

(2) The addition of $Ca(OH)_2$ can inhibit the development of the carbonation depth of geopolymers, and with the increase of $Ca(OH)_2$ content, the carbonation rate of geopolymers decreases greatly. When 10% $Ca(OH)_2$ was added, the carbonization rate of geopolymer was all below 2.43, while the carbonization rate of cement was 2.97, indicating that $Ca(OH)_2$ had an obvious modification effect on the carbonization resistance of fly ash-metakaolin-based geopolymers.

(3) The material alkalinity of fly ash-metakaolin-based geopolymers without $Ca(OH)_2$ decreased significantly during the carbonization process, and the material alkalinity of the 28 d carbonization stage decreased to 9.793. The addition of $Ca(OH)_2$ can play the role of an alkali reserve in the system, and the alkalinity of the material at the carbonation age of 28 d can be maintained at about 11, which plays a significant role in inhibiting the decrease of material alkalinity during the carbonization of geopolymers.

(4) Most of the pore sizes of the fly ash-metakaolin-based geopolymers and cement were distributed in the range of 10–100 nm; the addition of $Ca(OH)_2$ can improve the microstructure of the geopolymers, and the pore structure becomes denser. Furthermore, the pores of both geopolymers and cement developed to smaller pore sizes after carbonization.

(5) The hydration products of fly ash-metakaolin-based geopolymers were mainly amorphous silicaluminate gels, and after $Ca(OH)_2$ was added into the system, an obvious diffraction peak of $Ca(OH)_2$ appeared in the hydration products. The $Ca(OH)_2$ diffraction summit decreased significantly after carbonization, indicating that the addition of $Ca(OH)_2$ can significantly improve the carbonization resistance of fly ash-metakaolin-based geopolymers.

(6) A wave crest caused by C–(A)–S–H gel appeared at 954 cm^{-1}–990 cm^{-1} for the fly ash-metakaolin-based geopolymers mixed with $Ca(OH)_2$. However, carbonization led to decalcification of the calcium-containing gel, and increased silica gel polymerization to a degree, so that the peak value corresponding to Si–O–T after carbonization of fly ash-metakaolin-based geopolymers mixed with $Ca(OH)_2$ shifted to a higher position, which is consistent with XRD test results.

In this paper, the experiment successfully verified that $Ca(OH)_2$ admixture can obviously improve the carbonization resistance of fly ash-metakaolin-based geopolymers, which will greatly promote the application of geopolymers in engineering. However, the optimal dosage of $Ca(OH)_2$ in the system needs to be determined by further tests. In addition, the modification effect of nano-SiO_2, MgO and other admixtures on the carbonization resistance of fly ash-metakaolin-based geopolymers remains to be explored.

Author Contributions: Conceptualization, Y.L. and X.J.; data curation, J.Q. and B.P.; formal analysis, J.Q. and W.H.; investigation, W.H. and H.P.; Methodology, J.Q. and Y.L.; Supervision, H.P.; writing—original draft, W.H. and B.P.; writing—review & editing, J.Q. and X.J. All authors have read and agreed to the published version of the manuscript.

Funding: The work was supported by the Open Fund of Key Laboratory of Advanced Engineering Materials, Structure Behavior and Functional Control of the University of Hunan Province (Changsha University of Science & Technology, Grant No. 10KA10); the Innovation Fund of Characteristic Advantage Key Discipline of Civil Engineering of Changsha University of Science & Technology (Grant No. 17ZDXK07); the Natural Science Foundation of Hunan Province (Grant No. 2022JJ30600); the Natural Science Foundation of Changsha (Grant No. kq2202208); the Open Fund of Scientific Research and Innovation Platform of the Education Department of Hunan Province (Grant No. 20K005); the Open Fund of National Engineering Research Center of Highway Maintenance Technology (Changsha University of Science & Technology) (Grant No. kfj210102).

Institutional Review Board Statement: Not applicable.

Informed Consent Statement: Not applicable.

Data Availability Statement: Data is contained within the article.

Conflicts of Interest: The authors declare no conflict of interest.

References

1. Davidovits, J. Geopolymers: Inorganic polymeric new materials. *J. Therm. Anal. Calorim.* **1991**, *37*, 1633–1656. [CrossRef]
2. Cong, P.; Cheng, Y. Advances in geopolymer materials: A comprehensive review. *J. Traffic Transp. Eng. Engl. Ed.* **2021**, *8*, 283–314. [CrossRef]
3. Ahmaruzzaman, M. A review on the utilization of fly ash. *Prog. Energy Combust. Sci.* **2010**, *36*, 327–363. [CrossRef]
4. Zhang, B.; Zhu, H.; Chen, Y.; Huseien, G.; Shah, K. Shrinkage mechanisms and shrinkage-mitigating strategies of alkali-activated slag composites: A critical review. *Constr. Build. Mater.* **2022**, *318*, 125993. [CrossRef]
5. Gao, X.; Yu, Q.; Brouwers, H.J.H. Reaction kinetics, gel character and strength of ambient temperature cured alkali activated slag–fly ash blends. *Constr. Build. Mater.* **2015**, *80*, 105–115. [CrossRef]
6. Ding, Y.; Dai, J.G.; Shi, C.J. Mechanical properties of alkali-activated concrete: A state-of-the-art review. *Constr. Build. Mater.* **2016**, *127*, 68–79. [CrossRef]
7. Abdelli, K.; Tahlaiti, M.; Belarbi, R.; Oudjit, M.N. Influence of the origin of metakaolin on pozzolanic reactivity of mortars. *Energy Procedia* **2017**, *139*, 230–235. [CrossRef]
8. Imtiaz, L.; Rehman, S.; Memon, S.A.; Khan, M.K.; Javed, M.F. A review of recent developments and advances in eco-friendly geopolymer concrete. *Appl. Sci.* **2020**, *10*, 7838. [CrossRef]
9. Sitarz, M.; Hager, I.; Choińska, M. Evolution of mechanical properties with time of fly-ash-based geopolymer mortars under the effect of granulated ground blast furnace slag addition. *Energies* **2020**, *13*, 1135. [CrossRef]
10. Wu, X.; Shen, Y.; Hu, L. Performance of geopolymer concrete activated by sodium silicate and silica fume activator. *Case Stud. Constr. Mater.* **2022**, *17*, e01513. [CrossRef]
11. Wang, F.; Sun, X.; Tao, Z.; Zhu, P. Effect of silica fume on compressive strength of ultra-high-performance concrete made of calcium aluminate cement/fly ash based geopolymer. *J. Build. Eng.* **2022**, *62*, 105398. [CrossRef]
12. Yang, J.; Bai, H.; He, X.; Zeng, J.; Su, Y.; Wang, X.; Zhao, H.; Mao, C. Performances and microstructure of one-part fly ash geopolymer activated by calcium carbide slag and sodium metasilicate powder. *Constr. Build. Mater.* **2023**, *367*, 130303. [CrossRef]
13. Yang, T.; Zhu, H.; Zhang, Z. Influence of fly ash on the pore structure and shrinkage characteristics of metakaolin-based geopolymer pastes and mortars. *Constr. Build. Mater.* **2017**, *153*, 284–293. [CrossRef]
14. Elmesalami, N.; Celik, K. A critical review of engineered geopolymer composite: A low-carbon ultra-high-performance concrete. *Constr. Build. Mater.* **2022**, *346*, 128491. [CrossRef]
15. Gharzouni, A.; Joussein, E.; Samet, B.; Baklouti, S.; Rossignol, S. Effect of the reactivity of alkaline solution and metakaolin on geopolymer formation. *J. Non-Cryst. Solids* **2015**, *410*, 127–134. [CrossRef]
16. Hadi MN, S.; Al-Azzawi, M.; Yu, T. Effects of fly ash characteristics and alkaline activator components on compressive strength of fly ash-based geopolymer mortar. *Constr. Build. Mater.* **2018**, *175*, 41–54. [CrossRef]
17. Njimou, J.R.; Pengou, M.; Tchakoute, H.K.; Tamwa, M.S.; Tizaoui, C.; Fannang, U.; Lemougna, P.N.; Nanseu-Njiki, C.P.; Emmanuel Ngameni, E. Removal of lead ions from aqueous solution using phosphate-based geopolymer cement composite. *J. Chem. Technol. Biotechnol.* **2021**, *96*, 1358–1369. [CrossRef]
18. Babu, G.K.; Rao, K.V.; Dey, S.; Veerendra, G.T.N. Performance studies on quaternary blended Geopolymer concrete. *Hybrid Adv.* **2023**, *2*, 100019. [CrossRef]
19. Tang, Z.Q.; Sui, H.; De Souza, F.B.; Sagoe-Crentsil, K.; Duan, W.H. Silane-modified graphene oxide in geopolymer: Reaction kinetics, microstructure, and mechanical performance. *Cem. Concr. Compos.* **2023**, *139*, 104997. [CrossRef]

20. Duan, P.; Yan, C.; Zhou, W.; Luo, W.; Shen, C. An investigation of the microstructure and durability of a fluidized bed fly ash–metakaolin geopolymer after heat and acid exposure. *Mater. Des.* **2015**, *74*, 125–137. [CrossRef]
21. Singh, B.; Ishwarya, G.; Gupta, M.; Bhattacharyya, S.K. Geopolymer concrete: A review of some recent developments. *Constr. Build. Mater.* **2015**, *85*, 78–90. [CrossRef]
22. Provis, J.L.; Palomo, A.; Shi, C. Advances in understanding alkali-activated materials. *Cem. Concr. Res.* **2015**, *78*, 110–125. [CrossRef]
23. Pławecka, K.; Bazan, P.; Lin, W.T.; Korniejenko, K.; Sitarz, M.; Nykiel, M. Development of Geopolymers Based on Fly Ashes from Different Combustion Processes. *Polymers* **2022**, *14*, 1954. [CrossRef]
24. Bernal, S.A.; San Nicolas, R.; Myers, R.J.; Gutiérrez, R.M.; Puertas, F.; Deventer, J.S.J.; Provis, J.L. MgO content of slag controls phase evolution and structural changes induced by accelerated carbonation in alkali-activated binders. *Cem. Concr. Res.* **2014**, *57*, 33–43. [CrossRef]
25. Bakharev, T.; Sanjayan, J.G.; Cheng, Y.B. Resistance of alkali-activated slag concrete to carbonation. *Cem. Concr. Res.* **2001**, *31*, 1277–1283. [CrossRef]
26. Godwin, J.; Njimou, J.R.; Abdus-Salam, N.; Panda, P.K.; Tripathy, B.C.; Ghosh, M.K.; Basu, S. Nanoscale ZnO-adsorbent carefully designed for the kinetic and thermodynamic studies of Rhodamine B. *Inorg. Chem. Commun.* **2022**, *138*, 109287. [CrossRef]
27. Lv, Y.; Xiao, B.; Han, W.; Peng, H.; Qiao, J.; Wang, C.; Li, X. Research Progress on carbonization properties of alkali excited cementified materials. *Sci. Technol. Rev.* **2022**, *40*, 94–104.
28. Gao, K.; Lin, K.L.; Wang, D.Y.; Hwang, C.L.; Shiu, H.S.; Chang, Y.M.; Cheng, T.W. Effects SiO_2/Na_2O molar ratio on mechanical properties and the microstructure of nano-SiO_2 metakaolin-based geopolymers. *Constr. Build. Mater.* **2014**, *53*, 503–510. [CrossRef]
29. Park, S.M.; Jang, J.G.; Lee, H.K. Unlocking the role of MgO in the carbonation of alkali-activated slag cement. *Inorg. Chem. Front.* **2018**, *5*, 1661–1670. [CrossRef]
30. He, J.; Gao, Q.; Wu, Y.; He, J.; Pu, X. Study on improvement of carbonation resistance of alkali-activated slag concrete. *Constr. Build. Mater.* **2018**, *176*, 60–67. [CrossRef]
31. Glukhovsky, V.D. *Slag-Alkali Concretes Produced from Fine-Grained Aggregate*; Vishcha Shkola: Kiev, Ukraine, 1981. (In Russian)
32. Li, N.; Farzadnia, N.; Shi, C. Microstructural changes in alkali-activated slag mortars induced by accelerated carbonation. *Cem. Concr. Res.* **2017**, *100*, 214–226. [CrossRef]
33. *GB/T50082*; Standard of Test Methods for Long-Term Performance and Durability of Ordinary Concrete. China Academy of Building Research: Beijing, China, 2009.
34. Paudel, S.R.; Yang, M.; Gao, Z. pH level of pore solution in alkali-activated fly-ash geopolymer concrete and its effect on ASR of aggregates with different silicate contents. *J. Mater. Civ. Eng.* **2020**, *32*, 04020257. [CrossRef]
35. Natkunarajah, K.; Masilamani, K.; Maheswaran, S.; Lothenbach, B.; Amarasinghe, D.A.S.; Attygalle, D. Analysis of the trend of pH changes of concrete pore solution during the hydration by various analytical methods. *Cem. Concr. Res.* **2022**, *156*, 106780. [CrossRef]
36. Papadakis, V.G.; Vayenas, C.G.; Fardis, M.N. Fundamental modeling and experimental investigation of concrete carbonation. *Mater. J.* **1991**, *88*, 363–373.
37. Papadakis, V.G.; Vayenas, C.G.; Fardis, M.N. A reaction engineering approach to the problem of concrete carbonation. *AIChE J.* **1989**, *35*, 1639–1650. [CrossRef]
38. Papadakis, V.; Vayenas, C.; Fardis, M. Physical and chemical characteristics affecting the durability of concrete. *ACI Mater. J.* **1991**, *88*, 186–196.
39. Chang, P.H.; Chang, Y.P.; Chen, S.Y.; Yu, C.T.; Chyou, Y.P. Ca-rich Ca–Al-oxide, high-temperature-stable sorbents prepared from hydrotalcite precursors: Synthesis, characterization, and CO_2 capture capacity. *ChemSusChem* **2011**, *4*, 1844–1851. [CrossRef] [PubMed]
40. Kljajević, L.; Nenadović, M.; Ivanović, M.; Bučevac, D.; Mirković, M.; Nikolić, N.M.; Nenadović, S. Heat Treatment of Geopolymer Samples Obtained by Varying Concentration of Sodium Hydroxide as Constituent of Alkali Activator. *Gels* **2022**, *8*, 333. [CrossRef] [PubMed]
41. Lodeiro, I.G.; Macphee, D.E.; Palomo, A.; Fernández-Jiménez, A. Effect of alkalis on fresh C–S–H gels. FTIR analysis. *Cem. Concr. Res.* **2009**, *39*, 147–153. [CrossRef]
42. Tian, X.; Rao, F.; León-Patiño, C.A.; Song, S. Co-disposal of MSWI fly ash and spent caustic through alkaline-activation consolidation. *Cem. Concr. Compos.* **2021**, *116*, 103888. [CrossRef]

materials

MDPI

Article

Improved Tensile and Bond Properties through Novel Rod Constructions Based on the Braiding Technique for Non-Metallic Concrete Reinforcements

Anwar Abdkader [1], Paul Penzel [1,*], Danny Friese [1], Matthias Overberg [1], Lars Hahn [1], Marko Butler [2], Viktor Mechtcherine [2] and Chokri Cherif [1]

[1] Institute of Textile Machinery and High Performance Material Technology (ITM), Technische Universität Dresden, 01062 Dresden, Germany

[2] Institute of Construction Materials, Technische Universität Dresden, 01062 Dresden, Germany

* Correspondence: paul.penzel@tu-dresden.de

Abstract: Textile reinforcements have established themselves as a convincing alternative to conventional steel reinforcements in the building industry. In contrast to ribbed steel bars that ensure a stable mechanical interlock with concrete (form fit), the bonding force of smooth carbon rovings has so far been transmitted primarily by an adhesive bonding with the concrete matrix (material fit). However, this material fit does not enable the efficient use of the mechanical load capacity of the textile reinforcement. Solutions involving surface-profiled rods promise significant improvements in the bonding behavior by creating an additional mechanical interlock with the concrete matrix. An initial analysis was carried out to determine the effect of a braided rod geometry on the bonding behavior. For this purpose, novel braided rods with defined surface profiling consisting of several carbon filament yarns were developed and characterized in their tensile and bond properties. Further fundamental examinations to determine the influence of the impregnation as well as the application of a pre-tension during its consolidation in order to minimize the rod elongation under load were carried out. The investigations showed a high potential of the impregnated surface-profiled braided rods for a highly efficient application in concrete reinforcements. Hereby, a complete impregnation of the rod with a stiff polymer improved the tensile and bonding properties significantly. Compared to unprofiled reinforcement structures, the specific bonding stress could be increased up to 500% due to the strong form-fit effect of the braided rods while maintaining the high tensile properties.

Keywords: carbon-reinforced concrete; bond behavior; tensile test; braided rods; bond test; profiled roving; non-metallic reinforcement

Citation: Abdkader, A.; Penzel, P.; Friese, D.; Overberg, M.; Hahn, L.; Butler, M.; Mechtcherine, V.; Cherif, C. Improved Tensile and Bond Properties through Novel Rod Constructions Based on the Braiding Technique for Non-Metallic Concrete Reinforcements. *Materials* **2023**, *16*, 2459. https://doi.org/10.3390/ma16062459

Academic Editors: Nikolina Zivaljic, Hrvoje Smoljanović and Ivan Balić

Received: 27 February 2023
Revised: 14 March 2023
Accepted: 15 March 2023
Published: 20 March 2023

1. Introduction

In civil engineering, carbon-reinforced concrete (CRC) is of great interest to both research and industry in order to decrease the tremendous CO_2 emissions of up to 38% globally in the building industry significantly [1,2]. Due to corrosion resistance and a high load-bearing capacity compared to conventional steel bars, CRC allows up to 80% reduced concrete overlay and a lifespan of more than 200 years, making it a perfect material for a sustainable building of the future [3]. It is especially suitable in the use of thin-walled and lightweight concrete components with high load-bearing capacities as well as for reinforcing existing constructions [4].

In order to exploit the excellent properties of CRC, the bonding between the textile reinforcement and the surrounding concrete matrix (outer bond) as well as the bonding between the single filaments (internal bond) is of great importance. Especially when considering the energy-intensive production of the carbon fiber textiles, a high material utilization is needed for an efficient and resource-saving application of CRC for a sustainable building of the future.

In accordance with this, the bond mechanisms in textile reinforcements have long been the subject of extensive research [5–10]. A major prerequisite for the effective material utilization of the reinforcement potential in concrete is a strong internal bonding, since otherwise the core filaments of the rod will be significantly less or not at all involved in the load transmission, and the bond will fail accordingly early [5,9]. Furthermore, the strain (ε) at tensile failure of the reinforcement rod must be limited ($\varepsilon \leq 0.2\%$) in order to transmit the full load at small deformations of reinforced concrete structures and in order to ensure small crack widths in the composite as soon as the concrete matrix fails under tensile stress at app. 0.2% strain. Increased structural elongation of the reinforcement rod results in excessively wide cracks (bad durability) or high deflections (bad functionality) [7]. Therefore, a suitable rod structure (surface profiling) and consolidation as well as complete impregnation must ensure a good internal bonding [8].

As a result, various impregnations were developed to improve the internal bonding between the filaments [9]. Impregnations for rovings commonly used today are based on styrene-butadiene rubber (SBR), epoxy resin (EP), vinyl ester resin (VE), or polyacrylate (PA) [8,10]. All of these improve the internal bonding compared to unimpregnated rods and improve the structural stability as well [9]. SBR-impregnated reinforcements are more flexible and are therefore suitable for curved shapes and thus for retrofitting of existing buildings [11], whereas stiff EP-impregnated reinforcement structures are preferably used for precast concrete structures [12].

The outer bond between the rod and concrete matrix is based primarily on three mechanisms as shown in Figure 1 [13–16]. The Adhesive bond is based on adhesion (chemical bonding) between impregnation and hydrated cement. A relative displacement between the rod and the concrete overcomes the adhesion and activates the frictional bond. It primarily depends on the roughness in the boundary layer between rod and concrete matrix [15,17]. The highest outer bond is achieved through a form-fit-based shear bond (mechanical interlock) between a profiled rod structure and the surrounding concrete.

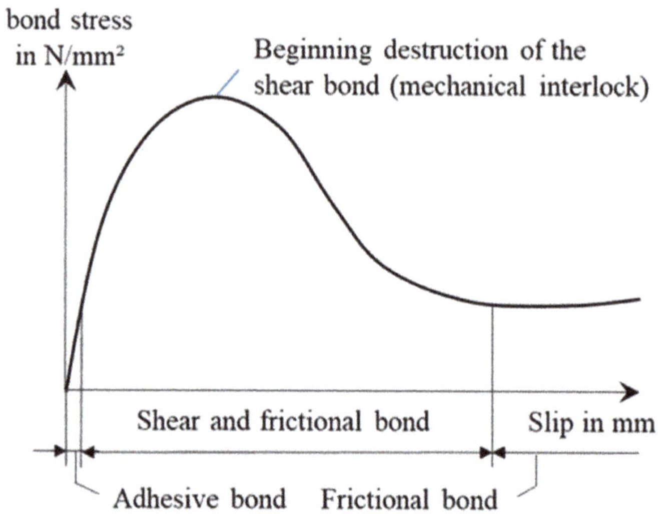

Figure 1. Schematic bond mechanisms of impregnated yarns for concrete reinforcements.

The first studies with continuously profiled carbon rovings showed that the bonding behavior could be adapted by surface profiling and variations in the profile depth (form fit) [18]; however, the exact mechanisms and principles of mechanical interlocking between the rod and concrete through profiling remain to be researched. For the analysis of the bond mechanisms, the bond behavior is characterized using the derived pull-out force–slip deformation relationships or the derived bond stress–crack-opening relationship [14,19].

An approach to increasing the bonding behavior of reinforcement yarns is the profiling of carbon rovings through a shaping process that was developed at the Institute of Textile Machinery and High Performance Material Technology (ITM). Hereby, the patented tetrahedral profile [20] is created by shaping the freshly polymer-impregnated roving with interlocking profile tools. The profiled yarn/roving is permanently stabilized through the cross-linking of the impregnation polymer via infrared (IR) radiation during the shaping process. This method allows the production of profiled rovings with different profile configurations. They show different and well defined bonding behaviors depending on the profile parameters [18,21].

Reinforcement structures with a large cross section compared to grid-like textile reinforcements can be produced by pultrusion from a single-fiber system [22]. Using polymeric matrix material in combination with thermal processes, fiber-reinforced plastics (FRP) profiled along the yarn axis can be generated by embossing [22]. The milling of rib-like structures into pultruded reinforcement bars (rebars) has also become a sufficient solution to produce rigid rebars with, in some cases, very high and reproducible bond parameters [23]. However, the methods are not suitable for reinforcements with smaller yarn cross-sections such as those required in flat components or for building reinforcements in combination with a low concrete layer thicknesses. Furthermore, the separated filament course of the milled structure results in shearing of the ribs and telescopic pull-out failures of concrete embedded rebars [24] in addition to a reduced efficiency with regard to material use and resource savings.

Other established and well-known methods for creating undulated yarn architectures in favor of a profiled surface and reproducible characteristic are, for example, braiding, twisting, and cabling [25]. With these technologies, the formation of a defined yarn geometry is realized by alternately inserting, intertwining, or winding of the individual threads. As a result, an improvement in the internal fiber bond and outer bond with the surrounding concrete matrix connection to the yarn surface is achieved [26].

This paper presents braiding technology for the development of novel rod structures from carbon rovings with defined yarn properties; e.g., a defined force–strain behavior; a stretched, shear-resistant anchored yarn structure; and surface profiling. The reinforcement structures made from these rods are expected to have significantly improved shear stress transmission in concrete with adjustable property profiles (Young's modulus, structural elongation, bond stress, and tensile strength) in combination with a high tensile strength.

The following Table 1 gives a brief overview of critical work and the state of the art this paper was based on.

Table 1. A brief overview of the literature and the state of the art.

Topic	State of the Art	References
Basic properties of carbon-reinforced concrete (CRC)	Use of carbon filament yarns with smooth surface structure; use of warp knitted grid-like structures with low rod count; high tensile properties; minimal structural elongation for initial load transfer with concrete matrix	[4,5,7,9–12,27]
Bond properties of CRC	Mainly adhesive bond; no dominant form fit or shear bond; yarn pull-out due to insufficient bonding properties; increased bond lengths for load transfer	[8,13–19,21–24,28]
Profiled textile reinforcement structures	Modified yarn structure by twisting and cabling; increased inner bond by fiber friction; no sufficient outer bond; surface profiling of thick rebars through additive/subtractive methods, reduced shear resistance and material efficiency; profiling through reshaping with defined bond properties	[18,20,21,23,25,26]

2. Materials and Methods

The focus of the study was the characterization of tensile and bond properties of different braided rods in comparison to unprofiled carbon rovings. In order to reduce the structural elongation of the braided rods, a purposeful impregnation and subsequent consolidation was investigated. Hereby, the influence of different impregnation agents and the solid content of the impregnation was studied to derive process recommendations. In the case of braided rods with a very high rod count and therefore a high predicted structural elongation, a pre-tension force was applied during the consolidation process in order to stretch out the rod and stabilize this condition for a further reduction in the structural elongation (improvement in the Young's modulus). For the characterization of the braided rods, tensile and pull-out test are performed. Furthermore, micrographic analyses were conducted to evaluate the impregnation quality and rod geometry.

2.1. Braided Rods Composed of Carbon Fiber Heavy Tows

For a broad characterization of the tensile and bonding properties of braided rods, different reinforcement rods consisting of several carbon fiber heavy tows (CFHT) from the Teijin Carbon Europe GmbH (Wuppertal, Germany) [29] were produced. An overview of the used fiber material is shown in Table 2. The tensile properties were characterized according to ISO 3341 [30] (Section 3.3) at the ITM. The tensile strength and Young's modulus refer to the filament area, which is determined by the density and count of the CFHT (for 3200 tex CFHT: 1.81 mm^2).

Table 2. Characteristics of the used carbon filament yarn [29].

Properties of the Carbon Fiber Heavy Tows						
Product Name	**Filaments in k**	**Count in Tex**	**Density in g/cm^3**	**Area in mm^2**	**Tensile Strength in MPa**	**Young's Modulus in Gpa**
Tenax®-E HTS45 E23	12	800	1.77	0.45	1632	219
Tenax®-E STS40 E23	48	3200	1.77	1.81	1687	232

A 4 × 4 variation braider VF 1/4-32-140 from Herzog GmbH (Oldenburg, Germany) was used as a basic braiding machine for targeted further development of the machine technology and for the development of braided rods (Figure 2).

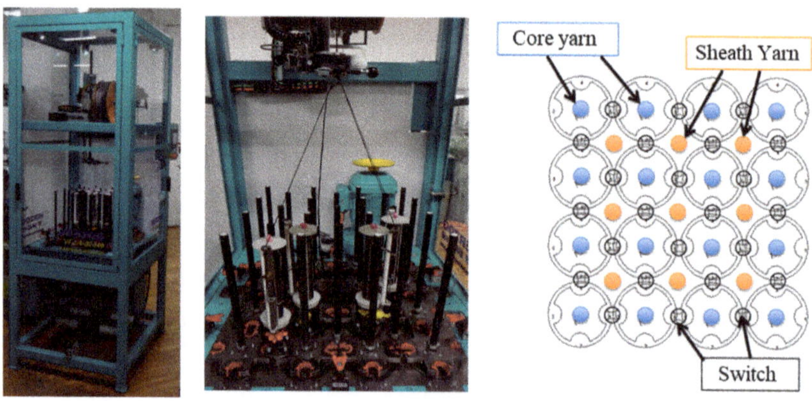

(a) (b) (c)

Figure 2. Illustration of the VF 1/4-32-140 braiding machine (**a**), the machine bed (**b**), and scheme (**c**).

The specifications of the variation braider were:

- 4 × 4 impellers to "park" several bobbins;
- Central control of the impellers;
- Up to 32 clappers and 24 pneumatic switching points;
- Clappers with closed ceramic yarn guides;
- 9 core inlet tubes Ø 8 mm between the impellers for core yarns;
- 16 drilled paddle wheel pillars for standing threads.

The variation braider was specifically further developed for gentle processing of brittle fibers (e.g., carbon filament yarn) with regard to the following points:

- Optimization of the bobbin design;
- Optimization of the deflection rollers with regard to surface properties and geometry;
- Optimization of the yarn tension compensation.

Using the modified VF 1/4-32-140 variation braider, continuous braided rods with different surface profiles were produced according to Table 3 with a 4-strand flat braid composed of 800 tex CFHTs (total rod count: 3200 tex) and a 6-strand flat braid composed of 3200 tex CFHTs (total rod count: 19,200 tex).

Table 3. Overview of the braided rod structures.

Overview of Dry Braided Rod					
Rod Type	**Sample Designation**	**Count in Tex**	**Braiding Structure**	**Pitch Length in mm**	**Illustration**
Braided rod	4_800	4 × 800	4-strand flat braid	120	
	6_3200	6 × 3200	6-strand flat braid	120	

2.2. *Impregnation and Consolidation*

In order to increase the internal bonding and for structural fixation, the produced braided rods were impregnated in an immersion bath method with water-based polymeric dispersions and consolidated using direct IR radiation. For the investigation, different impregnation agents were used (Table 4).

Table 4. Characteristics of the impregnation agents [31,32].

Overview of the Impregnation Agents				
Product Name	**Characteristics**	**Base Material**	**Solid Content in %**	**Linking Temperature in °C**
TECOSIT CC 1000 (CHT Germany GmbH)	Aqueous polymer dispersion	Polyacrylate (PA)	47 ± 1	160
Lefasol VL 90 (Lefatex Chemie GmbH)	Styrene-butadiene dispersion	Styrene-butadiene rubber (SBR)	50 ± 1	130–160

In a series of experiments, the resulting solid polymer content in the impregnated rods was varied (Table 4) by adding normal water to the polymeric dispersion. For the consolidation (drying and stabilization), the impregnated rods were positioned between two opposite positioned IR modules from Heraeus Noblelight GmbH (Kleinostheim, Germany) with the specification Typ MW Gold B 9755255 2500 W (230 V) with fast middle wave and 50% power (1.25 kW) and a distance of 10 mm to the roving. To minimize structural elongation, the braided rods were subjected to different pre-tension forces during the impregnation and consolidation process.

As a reference structure for the braided rods with a low count (3200 tex), an impregnated straight roving (with a circular cross section) consisting of a single 3200 tex CFHT was produced (called impregnated roving).

In the case of braided rods with a very high rod count and therefore a high predicted structural elongation, a pre-extension force was applied during the consolidation process in order to stretch out the rod and stabilize this condition for a further reduction in the structural elongation (improvement in the Young's modulus). The levels of the applied pre-extension force in order to reduce structural elongation and increase the Young's modulus were determined in tensile tests of dry braided rods and are further justified in Section 3.1.

In order to apply the pre-extension force during consolidation for minimizing the structural elongation of the braided rods with a high count, a test stand consisting of a frame, two opposite IR radiators, and a suspension was used. The schematic setup for the production of the samples composed of pre-extended braided rods with a high count is shown in Figure 3.

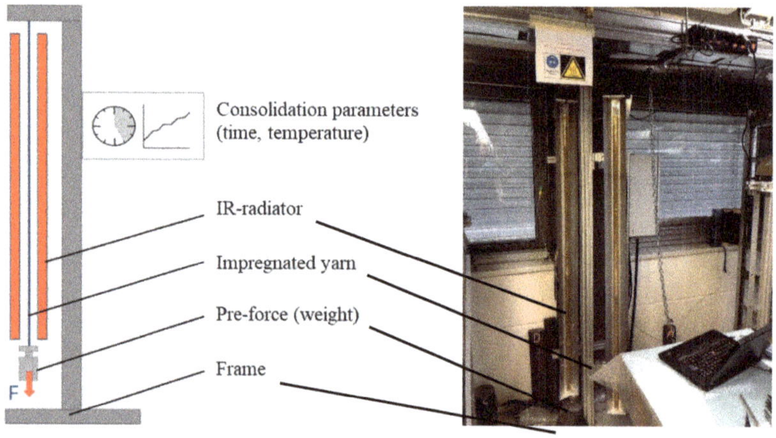

Figure 3. Schematic setup for consolidation with applied pre-extension force.

An overview of all test series of the different reinforcing rod architectures is given in Table 5.

Table 5. Overview of the tested rod architectures (common roving and braided rods) with impregnation.

Overview of Samples with Impregnation					
Material Type	**Sample Designation**	**Rod Count in tex**	**Impregnation Agent**	**Polymer Content in Mass %**	**Pre-Extension Force in N**
		Impregnated roving as reference			
Impregnated roving	1_3200_PA	1 × 3200	PA	~22	2
		Impregnated braided rods with low count			
Impregnated braided rod		*Variation in impregnation agent*			
	4_800_SBR	4 × 800	SBR	~22	2
	4_800_PA	4 × 800	PA	~22	2
		Variation in solid content			
	4_800_PA_18	4 × 800	PA	~18	2
	4_800_PA_15	4 × 800	PA	~15	2
	4_800_PA_12	4 × 800	PA	~12	2
	4_800_PA_10	4 × 800	PA	~10	2

Table 5. *Cont.*

Overview of Samples with Impregnation					
Material Type	**Sample Designation**	**Rod Count in tex**	**Impregnation Agent**	**Polymer Content in Mass %**	**Pre-Extension Force in N**
Impregnated braided rods with high count					
Impregnated braided rod	6_3200_PA	6 × 3200	PA	~22	20
	Variation in Pre-Extension force				
	6_3200_PA_1000	6 × 3200	PA	~22	1000
	6_3200_PA_2000	6 × 3200	PA	~22	2000
	6_3200_PA_3000	6 × 3200	PA	~22	3000

The three different investigated braided rods for concrete reinforcement are illustrated in Table 6.

Table 6. Illustrations of the impregnated braided rods and references.

Overview of Braided Rods		
Sample Designation	**Rod Count in Tex**	**Illustration**
1_3200_PA	1 × 3200	
4_800_PA	4 × 800	
6_3200_PA	6 × 3200	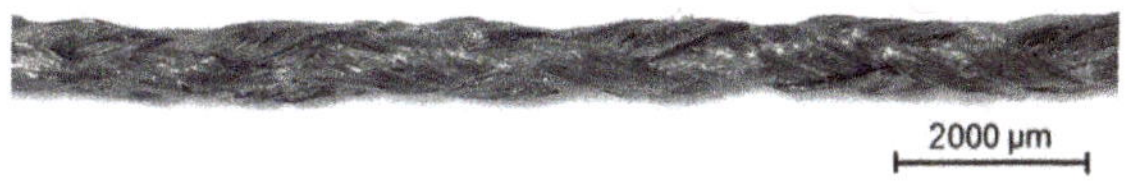

2.3. Concrete Matrix

Fiber-based reinforcements are very often embedded in cementitious matrices with a small maximum grain size of 1–2 mm [6].

For the characterization of the bond and tensile behaviors of concrete-embedded braided rods, two different fine concretes were used. The used high-performance concrete for the pull-out tests consisted of a fine concrete dry-mix called BMK 45-220-2. The detailed composition is presented in [21].

For the tensile test on carbon-reinforced concrete (CRC) specimens, the fine concrete TF 10 CARBOrefit® was used; it consisted of the fine concrete dry mix TF10 by the company PAGEL Spezial-Beton GmbH & Co. KG (Essen, Germany). The plastic consistency of concrete is suitable for laminating in layers and therefore for production of sample plates of CRC tensile test specimens.

The basic mechanical properties of the concrete were determined on prisms (dimensions $40 \times 40 \times 160$ mm^3) according to DIN EN 196-1 [33]; the mean values are listed in Table 7.

Table 7. Characteristics of the concrete composition.

Characteristics of the Concrete Matrix			
Product Name	**Maximum Grain Size in mm**	**Compressive Strength in MPa**	**Flexural Strength in MPa**
BMK 45-220-2 [21]	2	≥105	≥11.5
TF 10 CARBOrefit® fine concrete [27]	1	≥80	≥6

2.4. Characterization of the Rods' Tensile Properties and Bond Behavior in Concrete as Well as Tensile Properties of the Composite

For the determination of the tensile properties, the different dry and impregnated braided rods were tested at the Institute of Textile Machinery and High Performance Material Technology (ITM), TU Dresden. The dry braided rods were tested according to ISO 3341 [30] with wrap clamps. Due to the stiffness of the impregnated braided rods, which resulted in filament damage when wrapped around the clamps, the ends of the rods were resinated, clamped by metal clamps with a free clamping length of 200 mm, and tested acc. to DIN EN ISO 10618 [34]. The detailed test setup and testing parameters are presented in [18,21]. All tensile tests were performed with a Zwick Z100 testing machine. The tests were performed in a normal climate according to DIN EN ISO 139. A minimum of 10 specimens for each series were tested.

In order to determine the quality of the impregnation and the cross-sectional geometry of the different impregnated reinforcement rods, a micrographic analysis was conducted. Hereby, resin-embedded samples of the reinforcement rods were cut through the circumference and were analyzed with the AXIOImager.M1m microscopic workstation by the company Carl Zeiss AG (Germany) at the Institute of Textile Machinery and High Performance Material Technology (ITM), TU Dresden, with a magnification factor of 200:1 in a bright field and a reflective light microscopy method.

For the computer-supported analysis of the impregnation agent amount, its distribution, and the presence of air voids, the open source image-processing software ImageJ Version 1.54b was used to color the different components (filaments, polymer, embedment resin, and air gaps) by assigning new colors to detected color ranges. Hereby, the picture was transformed into a four-color image with one assigned color for each component. Once calibrated by manual selection of the color values for each component, the software allowed batch processing for similar images. The open source software ilastik enabled the image classification and segmentation of the colored images by counting the different pixel values of the colored microsection in a histogram. The diameter of the cross-section was determined by manual evaluation of the microsection dimension. In case of elliptical deformed geometries, its major axis and minor axis (the longest and shortest dimension) were measured and averaged.

For the tensile behavior composite composed of concrete and embedded braided rods with low fineness, tensile tests of composite specimens (reinforcement rod and concrete matrix) were performed at the Institute of Concrete Structures (IMB), TU Dresden. All tests took place at 20 °C exactly 28 days after casting according to the recommendations in [15]. Detailed information about the test setup and specimen production is given in [21]. Three composite samples (each reinforced by three braided rods) were tested.

For the characterization of the bonding behavior of the single reinforcement rod, pull-out tests acc. to [28] were conducted at the Institute of Construction Materials (IfB), TU Dresden. The specimen for the pull-out testing consisted of an upper concrete block to cause the pull-out of the reinforcement rod with an embedding length of 50 mm (bond length) and a lower concrete block for the rod fixation with an anchoring length of 90 mm. The test setup and specimen production are described in [21].

The test setup comprised a tensile strength testing machine by Instron 8802 with one upper, smaller specimen holder composed of metal and one lower specimen holder (Figure 4). The specimens had a free length of 120 mm. The test velocity was 1 mm per minute. The concrete-embedded specimens were tested at 20 °C, 65% relative humidity, and a standard ambient pressure of 1 bar. At least five specimens for each series were tested.

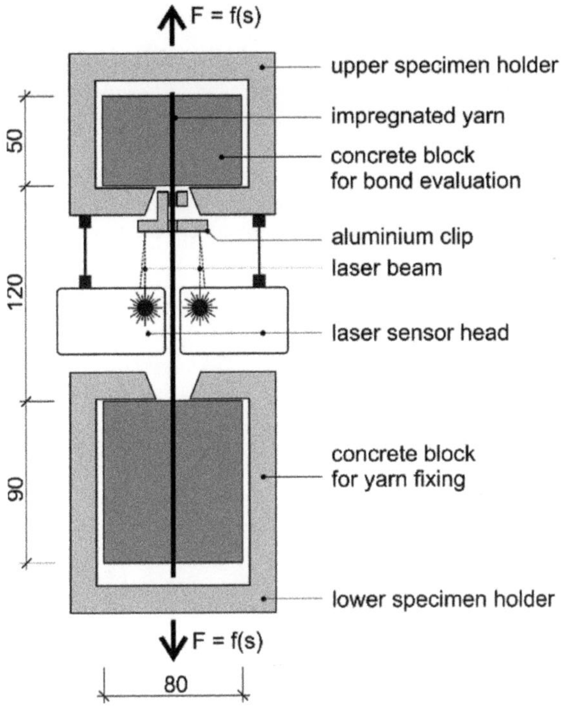

Figure 4. Schematic pull-out test setup of concrete-embedded braided rods (measured in mm) [8].

2.5. Specimens Manufacturing

For the microscopic examinations, 10 mm-long roving sections were placed in cylinders with a 20 mm diameter and fully resinated. After one day of drying, the front side was ground with sandpaper several times from rough to fine and finally polished with a polishing agent.

For the rod tensile tests, 450 mm-long rod sections were cut to size. Then, the rods were stretched and clamped in a frame. With the help of metal molds, the ends were embedded in epoxy resin at 125 mm each.

For the determination of the tensile strength of rods embedded in TF10 CARBOrefit© fine concrete, six tensile specimens with three profiled rods each and a concrete cover of 5 mm were produced for each rod series by laminating. This was done in a formwork in which the individual rods were fixed and aligned with a rod spacing of 13 mm. Then, the fine-grained concrete was filled in; first a bottom layer was created that was subsequently slightly compressed. In a second step, the top concrete layer was filled in and smoothed. The 120 cm-long, 1 cm-thick, and 33 cm-wide plate was then covered with damp cloths. From the 2nd to 7th days, the plate was stored in water. From day 8 to day 28, the plate was stored in a climate chamber at 20 °C and 65% relative humidity. Before the tensile tests, the plate was sawn into 5.2 cm-wide strips containing three rods each.

Specimens for the yarn pull-out (YPO) tests were created by embedding single braided rods as well as rovings with no profile (reference) in self-compacting, fine-grained concrete (BMK 45-220-2) in a cube formwork. The specimens consisted of two centered concrete blocks at the rod ends and a free rod segment of 120 mm in between the blocks. This clearly defined the area in which composite failure could occur. The specimens were stored for 7 days under water and stored for additional 21 days in a climate chamber (20 °C and 65% relative humidity).

3. Results and Discussions

3.1. Tensile Properties

The diagrams in Figures 5 and 6 illustrate the mean values of the tensile test results of the different series of reinforcement rod configurations with their standard deviations. For each of the following series, at least 10 single specimens were tested according to DIN EN ISO 10618 [34] (impregnated rods) and ISO 3341 [30] (dry braided rods). In order to analyze the influence of the braiding as well as the impregnation parameters, an important part of the study was a comparison between the tensile properties of dry braided rods (4_800, compare Table 5), impregnated braided rods with different impregnation agents (4_800_SBR, 4_800_PA), and impregnated rovings with a circular geometry (1_3200_PA) as well as impregnated braided rods with reduced polymer content (4_800_X).

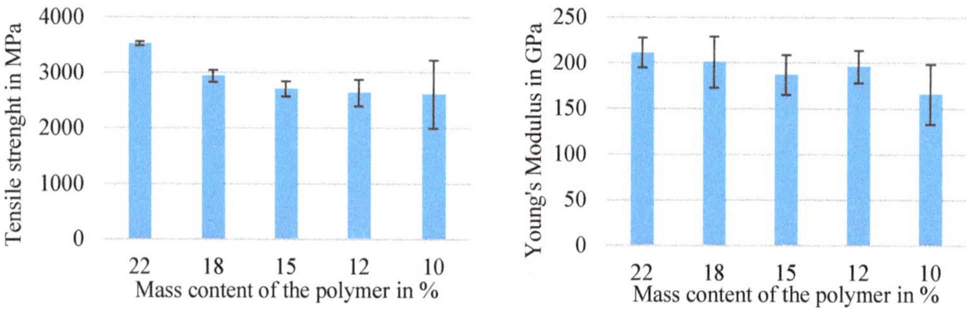

Figure 5. Influence of the polymer content on the tensile properties of 4 × 800 tex braided rods.

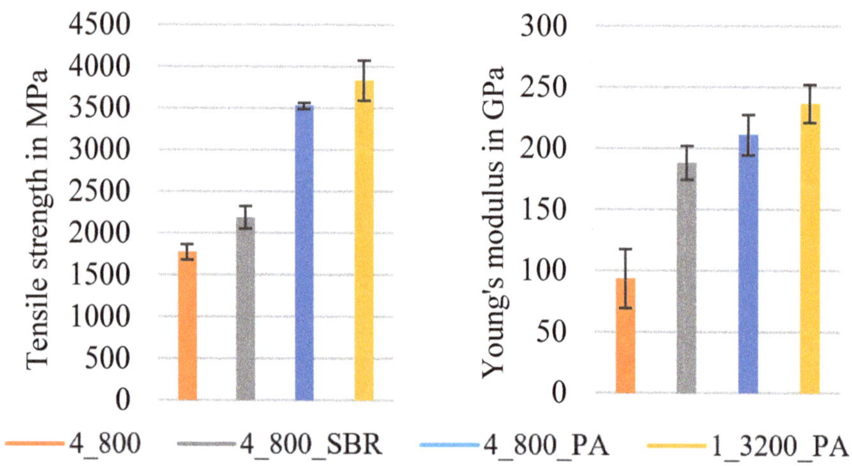

Figure 6. Tensile properties of impregnated 4 × 800 tex braided rods.

The determined tensile strength (in N/mm^2) refers in all tests (dry and impregnated rods) to the measured force (in N) divided by the net filament area (actual area of reinforcement). For the 3200 tex single roving and the braided rods with a total rod count of 3200 tex (4 × 800 tex), the compact filament area was 1.81 mm^2. The braided rods with a high fineness of 6 × 3200 tex had a compact filament area of 10.86 mm^2 (6 × 1.81 mm^2).

The following Figure 5 shows the influence of a varied polymer content of the rod on the tensile properties of 4 × 800 tex PA-impregnated braided rods.

The study's results showed that a reduced polymer content in the impregnated rod resulted in reduced tensile properties (tensile strength and Young's modulus). The braided rods with 22% mass content of the impregnation (4_800_PA) showed a result of 3524 MPa, which was the highest tensile strength, as well as the highest Young's modulus at 211 GPa in addition to the smallest standard deviation, indicating a good and even impregnation quality (complete impregnation) and an improved internal bond between the filaments that resulted in a more even load distribution. Furthermore, the tensile properties were in the range of approved textile reinforcements for concrete applications according to the general building approval [27] (values according to [27]: tensile strength ≥2700 MPa; Young's modulus ≥170 GPa) with an additional safe limit of up to 20%. Further reducing factors were regarded in the design of the reinforcement structure as described in [27].

With reduced mass content of the impregnation down to 10 M.-% (4_800_PA_10), the tensile properties dropped almost 25%, and the standard deviation increased significantly to almost ±20%. As deduced from that, the reduced solid content of the impregnation resulted in an uneven and incomplete impregnation of the braided rods and therefore an uneven load distribution between the filaments when stressed. The reduced Young's modulus especially indicated an increased structural elongation under load. The thesis was that a higher impregnation mass-content impregnation and therefore a higher polymer content in the braided rod resulted in a polymer accumulation in the gaps of the braided structure between the single fibers and reduced or limited the transverse contraction of the braided rod under load depending on the polymer proportion and stiffness, resulting in a higher tensile stiffness (Young's modulus) and a reduced structural elongation.

Figure 6 shows the mean values of the tensile properties with a single standard deviation of four different braided rods with a total count of 3200 tex. All impregnated rods in Figure 6 had an averaged polymer content of ~22 M.-%, which was determined as favorable in order to achieve high tensile properties (see Figure 5).

The diagrams in Figure 6 show that the impregnated heavy tow roving (1_3200_PA) had the highest tensile strength at app. 3830 MPa and Young's modulus at app. 236 GPa and thus only typical structural elongation under load. The impregnated braided rod with PA as an impregnation agent (4_800_PA) had very high tensile properties as well: an app. 10% decrease compared to the impregnated roving (3524 MPa tensile strength and 211 GPa Young's modulus), which resulted from the slight deviation in the individual filaments in the braided rods compared to the perfectly uniaxial orientated filaments in the straight roving. Therefore, no complete uniform load distribution was possible, and lateral forces were introduced in the filaments due to the transverse contraction of the braided structure, thereby resulting in a slightly premature failure compared to the straight impregnated roving. However, no structural elongation under load was recognizable. The SBR-impregnated braided rod (4_800_SBR) had (at a 2190 MPa tensile strength and a 188 GPa Young's modulus) significantly lower tensile properties compared to the PA-impregnated braided rod and an evident structural elongation of app. 0.3% due to the soft SBR impregnation. The lowest tensile properties show unimpregnated, dry braided rods (4_800) with a tensile strength of 1773 MPa, a Young's modulus of 93 GPa, and a significant structural elongation under load of ca. 0.5% resulting from the undulated braided rod structure.

The results showed that impregnation increased the tensile behavior of the braided rods significantly and that the PA impregnation was much stiffer compared to the SBR impregnation and therefore more suitable in the use of reinforcement rods with reduced structural elongation. A PA-impregnated braided rod showed no evident structural elongation under load and similar tensile properties compared to straight impregnated rovings.

Figure 7 shows the averaged force–strain behavior of the dry braided rods with a high rod count of 6 × 3200 tex as well as the linear behavior (tangent), the structural elongation under load $\Delta\varepsilon$ (non-linear behavior, red dotted line) and the determined pre-extension forces for stretching out of the braided rod during consolidation (dotted lines).

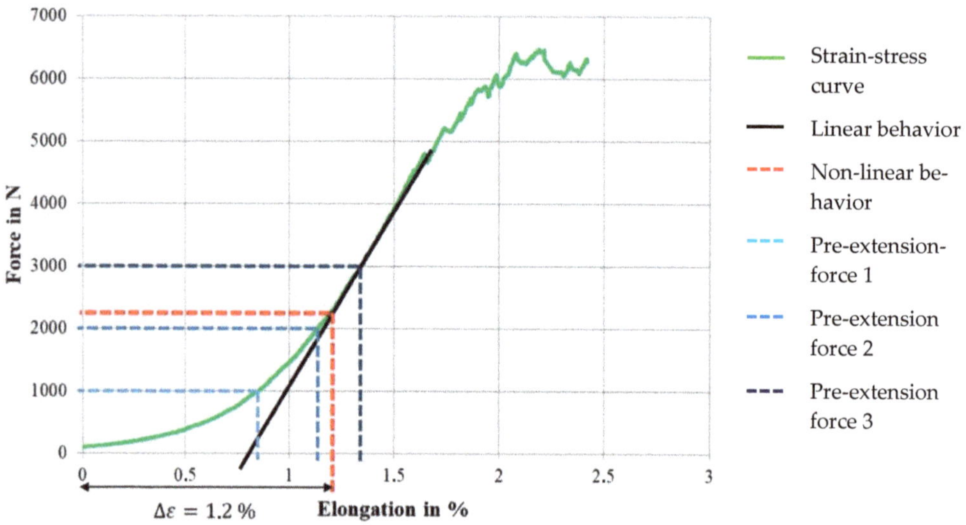

Figure 7. Structural elongation and pre-tension levels of dry 6 × 3200 tex braided rods.

In contrast to the braided rods with a relatively low count of 4 × 800 tex (total rod count 3200 tex), the dry braided rods with 6 × 3200 tex (total rod count 19,200 tex) had very low tensile properties with a tensile strength at 684 MPa (6420 N breaking strength), a Young's modulus of about 17 GPa, and a structural elongation of app. 1.2% (elongation at failure app. 2.4% composed of 1.2% structural and 1.2% material elongation under load). Compared to the braided rod with a low count, the braided rod with a high count had a 60% lower tensile strength but a 250% higher structural elongation. This inefficient use of the performance potential was due to the strongly undulated braided structure of the six braided 3200 tex CFHTs with significant voids between the single CFHTs compared to a more dense braided structure of the braided rod with a low count. The undulated structure increased the transverse contraction of the braided rod under load, resulting in higher structural elongation, lateral forces, and therefore premature failure compared to a roving with a comparable count but straight filament orientation.

Such a high structural elongation under load would be unsuitable for use as reinforcements in concrete structures because after the failure of the concrete matrix at ~0.2% strain, a further structural elongation of the reinforcement rod resulted in increased crack openings and deflection of the structure. For a reduction in the structural elongation of the impregnated braided rod with a high count, a pre-extension force was applied during consolidation (after impregnation) according to the load levels measured in the force–strain curve (1000 N/2000 N/3000 N). The thesis was that the applied load would stretch out the 1.2% structural elongation of the dry braided rod according to the load level as shown in Figure 7 and that the consolidation process stabilized the structure in the outstretched condition and therefore reduced the structural elongation under further load application. Hereby, a pre-extension force of 1000 N would reduce the structural elongation by 0.8% down to 0.4%, 2000 N would reduce elongation to 0.1%, and 3000 N would reduce elongation completely to 0%.

The diagram in Figure 8 shows the averaged stress–strain curve of the impregnated and outstretched braided rods with a high rod count in comparison to the dry braided rod.

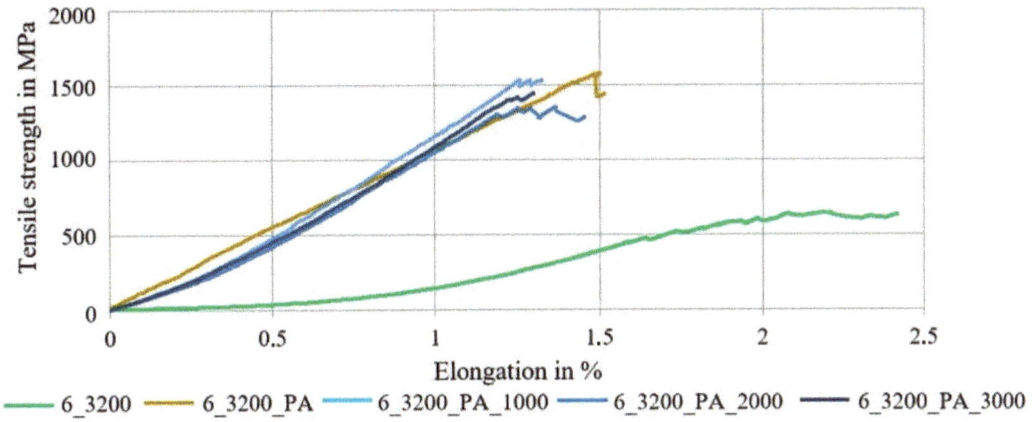

Figure 8. Influence of the pre-tension on tensile properties of impregnated 6 × 3200 tex braided rods.

It was evident that the impregnation of the braided rods with a high count increased the tensile properties and reduced the structural elongation under load significantly, which was similar to the braided rods with a low count. Figure 9 compares the averaged tensile properties.

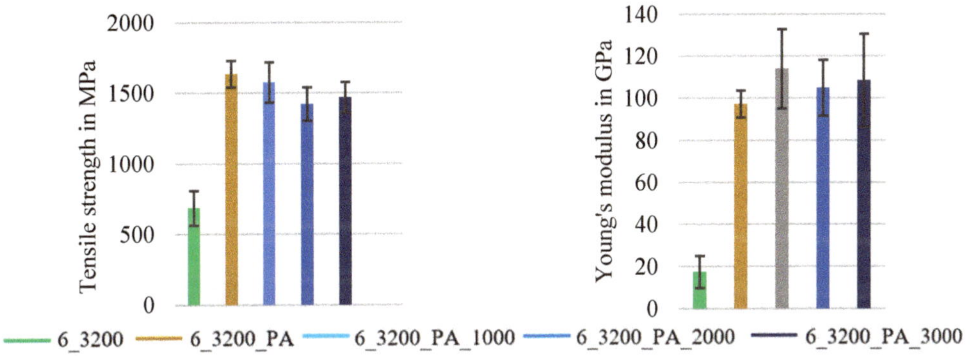

Figure 9. Tensile properties of 6 × 3200 tex braided rods with and without pre-tension.

Hereby, the impregnation with the stiff PA impregnation agent and a high polymer content increased the tensile strength by almost 250% to 1634 MPa, and the Young's modulus was at about 97 GPa (five times higher). The structural elongation was almost eliminated.

The application of a pre-extension force resulted in an up to 10% increase in the Young's modulus (ca. 110 GPa) and a similar tensile strength at around 1574 MPa compared to the impregnated braided rods. Therefore, the tensile stiffness was increased and the structural elongation could be reduced. Higher pre-extension forces resulted in a slightly lower tensile strength at 1420–1470 MPa and an increased single standard deviations with up to 15% (probably due to increased fiber damage) but a higher Young's modulus at ~110 GPa.

In summary, the investigations showed that an impregnation with high polymer content of at least 22 mass-% (see Figure 5) and a stiff impregnation agent on a PA basis (see Figure 6) were favorable to achieve the highest tensile properties of braided rods. In the case of braided rods with a high rod count (e.g., 6 × 3200 tex), the application of a limited pre-extension force during consolidation increased the Young's modulus (see Figure 9), making them suitable for transmission of high initial forces with reduced structural elongation.

3.2. Micrographic Analysis

In order to validate the quality of the impregnation, the ratio of the polymer content as well as to examine the dimensions of the impregnated reinforcement rods, microsection analyses were conducted as described in Section 3.4 in addition to a gravimetric analysis for the determination of the polymer content. The microscopic tests were performed at the textile physical testing laboratory at the Institute of Textile Machinery and High Performance Material Technology (ITM), TU Dresden. At least four cross-sections of each following series were analyzed. Hereby, the following colors of the original microsections indicate different components: white—filament; grey (outside)—embedment resin; light grey (inside)—impregnation/polymer; and black—air gaps (see Table 8). For a computer-supported analysis of the impregnation ratio (filament/polymer/air content), image-classification programs were used for a coloring of the components (see Section 3.3).

Table 8. Micrographic analysis and geometry parameters of different reinforcement rods.

Micrographic Analysis and Geometry Parameters of Different Reinforcement Rods			
Rod Type	**Properties**	**Values**	**Illustration**
1_3200_PA	Total rod count in tex	3200	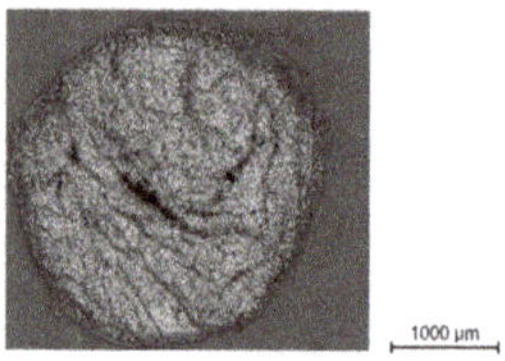
	Comparable diameter in mm	2.0	
	Circumference in mm	6.3	
	Polymer content in mass.-%	~22	
4_800_PA	Total rod count in tex	3200	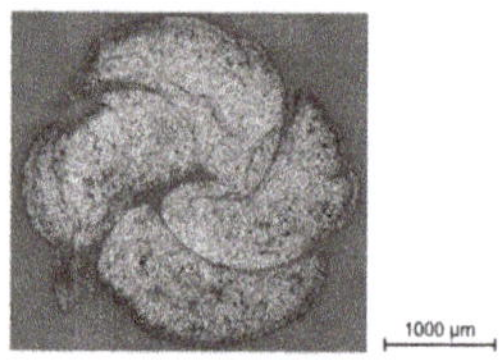
	Comparable diameter in mm	2.0	
	Circumference in mm	6.3	
	Polymer content in mass.-%	~22	
6_3200_PA	Total rod count in tex	19,200	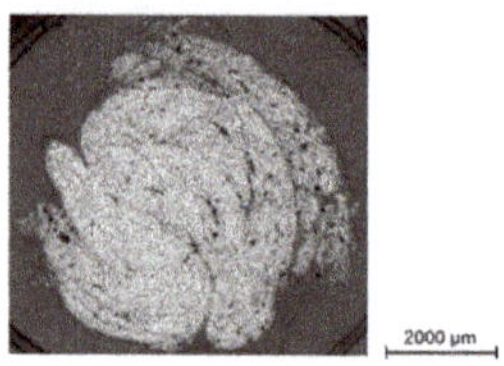
	Comparable diameter in mm	4.5	
	Circumference in mm	14.1	
	Polymer content in mass.-%	~22	

Table 9 shows an exemplary comparison of an original and colored microsection of a braided rod and the component assignment.

The results of the micrographic and gravimetric analysis showed that all rod types had at ca. 22 mass.-% a similar polymer content. In contrast to the straight, impregnated roving with no evident polymer accumulations and a comparatively dense fiber arrangement, due to linear orientation of the filaments without undulation, the braided rods, which consisted of several braided filament yarns, showed an increased accumulation of the polymer in the voids between the single filaments in the internal yarn structure. Furthermore, the micrographic analysis showed minor air voids of the impregnated rod structure ($\leq$1% for braided rods with low count) with up to 2% for braided rods with a high count. Such air voids or inclusions could destabilize the rod structure and result in reduced tensile properties.

Table 9. Micrographic analysis of an impregnated reinforcement rod (left—original; right—colored).

Computer-Supported Micrographic Analysis		
Real Microsection	**Components**	**Colored Microsection**
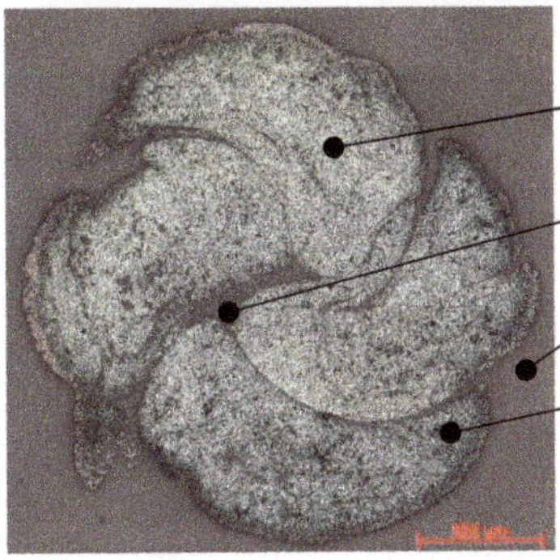	Filament (white/yellow) Polymer (light grey/pink) Embedment resin (dark gray/white) Air void (black/green) 1000 µm	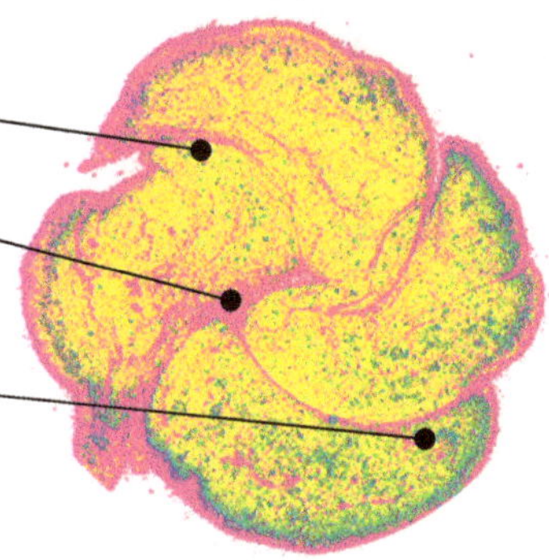

The micrographic analysis presented in this study used a computer-supported evaluation of the quality of the impregnation (polymer content, distribution, air voids, etc.). It allowed a fast and exact investigation of the rods' material and structural properties such as polymer accumulations and air voids as well as the dimensions and filament arrangement, which were needed to verify a high impregnation quality and for calculation of the bond stress on the basis of the rod dimensions. Especially when deriving correlations between rod structures with defined surface-profiles and the resulting tensile and bonding properties, such investigations are needed for further understanding of the inter-relations.

3.3. Tensile Properties of Concrete-Embedded Braided Rods

Figure 10 shows the force–strain behavior of three individual composite test specimens, which consisted of three concrete-embedded braided rods with a low count and PA impregnation (4_800_PA) with a very low scatter.

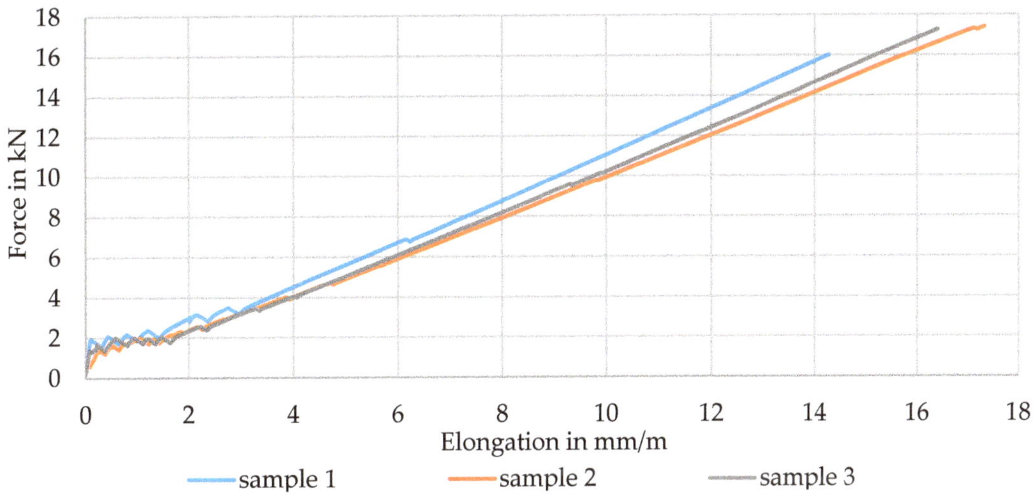

Figure 10. Tensile behavior of CRC-specimens (left) and tested specimen after failure (right).

The evaluation of the CRC tensile test specimens is compared in Table 10.

Table 10. Evaluation of the tensile tests on CRC specimens.

Sample	First Crack in kN	Crack-Count	Splitting/ Cracking	Longitudinal Crack	Failure Force in kN	Tensile Strength in MPa	Failure Pattern
1	1.91	7	Cracking at failure	at failure	16.00	2946	complete spalling
2	1.37	9			17.40	3204	
3	1.49	9			17.27	3174	
Mean	1.59	8			16.89	3110	

A mean failure force of 16.9 kN or a mean rod tensile strength of 3110 MPa, referring to the dry and compacted filament cross section of 1.81 mm^2 for each rod, was determined. These values were in the range of the tensile strength of the single rod tensile tests (see Section 3.1) and therefore confirmed a high material utilization due to an almost uniform load distribution. According to the general building approval [27], the concrete-embedded rod required a tensile strength $\geq$2250 MPa. The braided rods achieved up to 40% higher tensile properties and therefore had a very high safety factor and great suitability for use as concrete reinforcement.

In all tests, rod rupture occurred, indicating a full load transfer between the textile and concrete matrix. The braided structure was still visible after testing. The rods were not pulled out from the load transfer areas, indicating a sufficient outer bond between the braided rods and the concrete matrix. After appearance of several cracks (an average of eight cracks before failure) in an evenly distributed pattern during the test, speaking for an even load distribution, the concrete was completely spalled in the measuring area at the moment of rod failure (see Table 10). During the tests, no splitting or delamination cracks could be observed.

In conclusion, the force–strain behavior curve or the tensile strength was similar to that of the reference expansion specimens composed of construction-approved reference textiles (e.g., CARBOrefit® variant 3 with 3200 tex CFHT and PA impregnation [27]).

3.4. Bond Properties

The following Figure 11 shows the averaged pull-out force–slip-deformation curves of the PA- and SBR-impregnated braided rods with a low count (4_800_PA/SBR) in comparison to the impregnated roving (1_3200_PA).

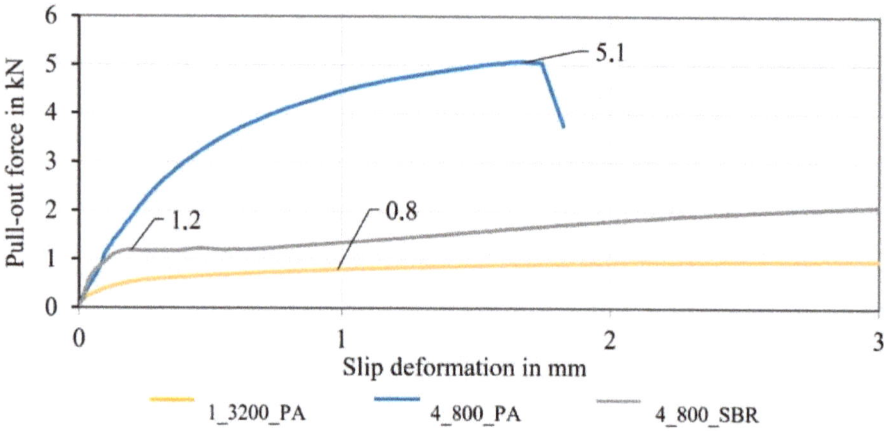

Figure 11. Bond behavior of impregnated braided rods (3200 tex) and different impregnation agents.

Due to the improved mechanical interlock of the braided rods, the resulting bond strengths were well above that of the impregnated rovings with no profile. The PA-impregnated braided rods (4_800_PA) transmitted with approximately 5.1 kN, which was almost five times higher than the pull-out forces for impregnated rovings with no profile (0.8 kN), and showed a distinct shear bond. In addition to that, they showed a steep increase in the pull-out force, emphasizing a high bond stiffness and strong anchoring (mechanical interlock and dominant shear bond). The sudden drop in the bond force was due to the filament failure at 5.1 kN, indicating that the embedment length of 50 mm was sufficient for a complete anchoring and load transmission of the braided rod in the concrete matrix. In contrast to that, the impregnated roving with no profile (1_3200_PA) was pulled out completely due to a lack of a sufficient shear bond and mainly an adhesive or frictional bond. The SBR-impregnated braided rod (4_800_SBR) initially showed a steep increase in the pull-out force due to an initial shear bond, but the mechanical interlock failed prematurely at about 1.2 kN, and the braided rod was pulled out and showed only frictional bond. A possible reason for the premature destruction of the mechanical interlock was a deformation of the braided structure due to the "soft" SBR impregnation. The results showed that the stiff PA impregnation was suitable for creating profiled reinforcement rods with a strong mechanical interlock, a distinct shear bond, and significantly increased bond performance.

Due to the different rod count and therefore the different load distribution among the filaments, the transferable pull-out force of the rods with a total count of 3200 tex were not comparable to the braided rods with a total rod count of 19,200 tex (6_3200_PA). For a direct comparison, the bond stress (in N/mm^2) had to be determined by dividing the measured pull-out force by the rods' outer surface area. The outer surface area was hereby calculated by the product of the initial bond length (50 mm) and the averaged rod circumference of the different reinforcement rods. For simplification of the complex surface geometry of the braided rods, the circumference was calculated based on a circular cross-section with the comparing diameter of the rods, which was determined in the micrographic analysis in Section 3.2 (see also [23] for the comparing diameter) multiplied by pi.

Figure 12 shows the averaged bond stress–slip deformation curves of the different reinforcement rods, in which the comparison of the standardized bond stress of the braided rods between low and high rod count can be seen.

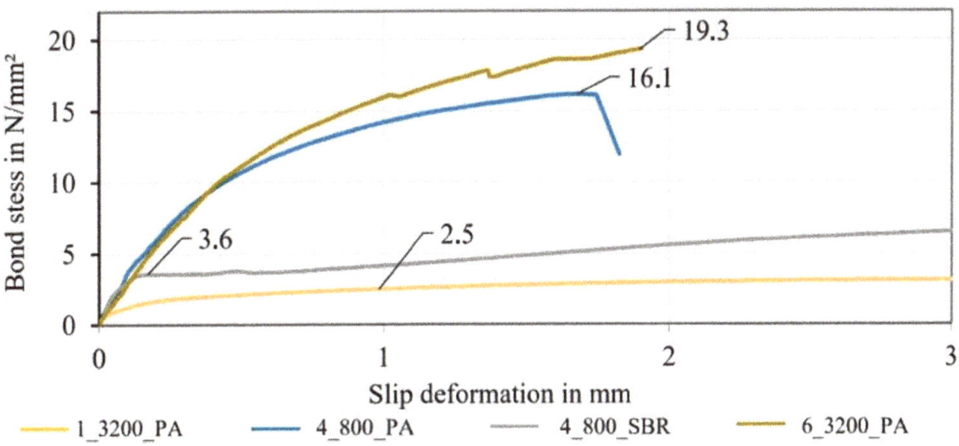

Figure 12. Bond stress of different braided rods.

Hereby, the braided rod with a low count (4_800_PA) and a high count (6_3200_PA) had an almost identical bond stress–slip deformation behavior. The initial bond stiffness was identical, yet the braided rod with a high count showed, at 19.3 N/mm^2, an almost 20% higher bond stress than braided rods with a low count, which possibly resulted from the rougher and more undulated braided structure and therefore strong mechanical interlock. In comparison with the impregnated roving without profile (2.5 N/mm^2), the braided rod with a low count (4_800_PA) showed a 544% higher bond stress behavior, and the one with a high count (6_3200_PA) showed a 672% higher bond stress behavior. The bond stress of the braided rod with a low count and SBR impregnation agent (4_800_SBR) was 44% higher compared to the impregnated roving without profile.

4. Conclusions

The results showed that the developed braided rods were able to transmit much higher pull-out loads than common carbon fiber architectures without profile and demonstrated a significantly improved bond-slip behavior with up to five times the maximum bond strength compared to straight carbon rovings with no profile, yet maintained high tensile properties with almost no structural elongation and properties above the required values in the general building approval with sufficient safe limits, making them particularly suitable as concrete reinforcement. Hereby, the investigation of different coating agents and coating agent contents showed that a complete impregnation with a stiff polymeric impregnation resulted in very high tensile and bond properties and a strong mechanical interlock due to a reduced rod deformation under load. Braided rods with a high count showed similar results to braided rods with a low count, yet the tensile performance potential was not fully used because the high rod count resulted in an more undulated braided structure. Nonetheless, a complete impregnation eliminated the structural elongation under load, and the rough structure increased the bond behavior. As expected, the application of a pre-extension force increased the modulus of elasticity but had no significant effect on the tensile strength of the braided rods with a high count.

As a result of the bond investigations, fundamental knowledge was gained about the force transmission between the textile reinforcement and the fine concrete matrix. These are required for the subsequent design and dimensioning of the anchorage lengths of the new carbon concrete. As a major result, the braided rod designs developed were able to significantly increase pull-out forces while maintaining the concrete composite's tensile strength.

5. Outlook

The developed braided rods with a low count are perfectly suitable for the use in rollable, grid-like textile reinforcement structures, whereas the braided rods with high count can be used for rigid reinforcement mats known from conventional steel reinforcements or for individual reinforcement bars used as additional reinforcement elements.

For a production of profiled, grid-like textile reinforcement structures for concrete applications in form of rollable or rigid reinforcement mats, the conventional textile manufacturing processes such as the multiaxial warp knitting process as well as robot supported rod placement will be adapted and further developed. Further research is planned on braided rods with various braiding structures for specific surface profiles and defined tensile and bond properties for application specific designs of CRC-structures and highest material efficiency. Furthermore epoxy-resin impregnated braided rods with a high rod count will be investigated for further improvement in the tensile behavior.

Such carbon fiber reinforcements with high tensile and bonding properties will clearly increase the material efficiency of carbon-reinforced concrete in the future, especially in the areas of new construction and strengthening. In the case of component strengthening, for example, shortened end anchorage and overlap lengths will improve handling. Additionally, the lower material consumption will reduce costs, which increases competitiveness compared to other reinforcement methods.

Author Contributions: Conceptualization, A.A., L.H. and C.C.; methodology, P.P., A.A., M.O. and D.F.; software, P.P. and D.F.; validation, A.A., P.P. and D.F.; formal analysis, A.A., P.P. and D.F.; investigation, A.A., P.P., D.F. and M.O.; resources, A.A.; data curation, P.P., D.F. and M.O.; writing—original draft preparation, A.A., P.P. and D.F.; writing—review and editing, M.O., L.H., M.B., V.M. and C.C.; visualization, A.A., P.P. and D.F.; supervision, L.H., M.B., V.M. and C.C.; project administration, C.C.; funding acquisition, C.C. All authors have read and agreed to the published version of the manuscript.

Funding: This article presents parts of the results achieved within the Project C^3 V2.4 and the ZIM research project ZF4008344KI9, which were supported by the Federal Ministry for Economic Affairs and Energy (BMWi) on the basis of a decision by the German Bundestag. The IGF research project 21375 BR of the Forschungsvereinigung Forschungskuratorium Textil e. V. was funded through the AiF within the program for supporting the "Industriellen Gemeinschaftsforschung (IGF)" from funds of the Federal Ministry for Economic Affairs and Climate Action on the basis of a decision by the German Bundestag. The Article Processing Charge (APC) was funded by the joint publication fund of the TU Dresden, the Medical Faculty Carl Gustav Carus, and the SLUB Dresden. The funding is gratefully acknowledged.

Institutional Review Board Statement: Not applicable.

Informed Consent Statement: Not applicable.

Data Availability Statement: The data presented in this study are available on request from the corresponding author.

Conflicts of Interest: The authors declare no conflict of interest. The funders had no role in the design of the study; in the collection, analyses, or interpretation of data; in the writing of the manuscript; or in the decision to publish the results.

References

1. Witsch, K. Klimakiller Beton: So will die Zementbranche Klimaneutral Werden. *Handelsblatt* 25 November 2020. Available online: https://www.handelsblatt.com/unternehmen/energie/klimaschutz-klimakiller-beton-so-will-die-deutsche-zementindustrie-co2-neutral-werden-/26652040.html (accessed on 29 April 2022).
2. vdz gGmbH: CO_2 Capturing in Cement Production. Available online: https://www.vdz-online.de/fileadmin/Forschung/01_Ruppert_2019_08_22_CO2_capturing_in_cement_production_EN.pdf (accessed on 22 September 2022).
3. Seifert, W.; Lieboldt, M. Ressourcenverbrauch im globalen Stahlbetonbau und Potenziale der Carbonbetonbauweise. *Beton Stahlbetonbau* **2020**, *115*, 469–478. [CrossRef]
4. Friese, D.; Scheurer, M.; Hahn, L.; Gries, T.; Cherif, C. Textile reinforcement structures for concrete construction applications—A review. *J. Compos. Mater.* **2022**, *56*, 002199832211271. [CrossRef]
5. Hahn, L.; Rittner, S.; Nuss, D.; Ashir, M.; Cherif, C. Development of Methods to Improve the Mechanical Performance of Coated Grid-Like Non-Crimp Fabrics for Construction Applications. *Fibres Text. East. Eur.* **2019**, *27*, 51–58. [CrossRef]
6. Hahn, L. Entwicklung einer In-situ-Beschichtungs- und Trocknungstechnologie für multiaxiale Gelegestrukturen mit hohem Leistungsvermögen. Ph.D. Thesis, Technische Universität Dresden, Dresden, Germany, 2020.
7. Jesse, F. Tragverhalten von Filamentgarnen in zementgebundener Matrix. Ph.D. Thesis, Technische Universität Dresden, Dresden, Germany, 2004.
8. Kruppke, I.; Butler, M.; Schneider, K.; Hund, R.-D.; Mechtcherine, V.; Cherif, C. Carbon Fibre Reinforced Concrete: Dependency of Bond Strength on Tg of Yarn Impregnating Polymer. *Mater. Sci. Appl.* **2019**, *10*, 328–348. [CrossRef]
9. Curbach, M.; May, S.; Egbert, M.; Schumann, A.; Schütze, E.; Wagner, E. Verstärken mit Carbonbeton. In *BetonKalender 2022*; John Wiley & Sons, Ltd: Hoboken, NJ, USA, 2022; pp. 761–804.
10. Köckritz, U. In-Situ-Polymerbeschichtung zur Strukturstabilisierung offener nähgewirkter Gelege. Ph.D. Thesis, Technische Universität Dresden, Dresden, Germany, 2007.
11. Curbach, M.; Hauptenbuchner, B.; Ortlepp, R.; Weiland, S. Textilbewehrter Beton zur Verstärkung eines Hyparschalentragwerks in Schweinfurt. *Beton Stahlbetonbau* **2007**, *102*, 353–361. [CrossRef]
12. Hegger, J.; Kulas, C.; Horstmann, M. Realization of TRC Façades with Impregnated AR-Glass Textiles. *Polym. Concr.* **2011**, *466*, 121–130. [CrossRef]
13. Schütze, E.; Curbach, M. Zur experimentellen Charakterisierung des Verbundverhaltens von Carbonbeton mit Spalten als maßgeblichem Versagensmechanismus/Experimental characterisation of the bond behaviour of carbon reinforced concrete with concrete splitting as significant failure mode. *Bauingenieur* **2019**, *94*, 133–141. [CrossRef]
14. Lorenz, E. Endverankerung und Übergreifung textiler Bewehrungen in Betonmatrices. Ph.D. Thesis, Fakultät Bauingenieurwesen, Dresden, Germany, 2014.

15. Bielak, J.; Spelter, A.; Will, N.; Claßen, M. Verankerungsverhalten textiler Bewehrungen in dünnen Betonbauteilen. *Beton Stahlbetonbau* **2018**, *113*, 515–524. [CrossRef]
16. Preinstorfer, P.; Kromoser, B.; Kollegger, J. Kategorisierung des Verbundverhaltens von Textilbeton/Categorisation of the bond behaviour of textile reinforced concrete. *Bauingenieur* **2019**, *94*, 416–424. [CrossRef]
17. Banholzer, B. Bond Behaviour of a Multi-Filament Yarn Embedded in a Cementitious Matrix. Ph.D. Thesis, RWTH Aachen University, Aachen, Germany, 2004.
18. Penzel, P.; May, M.; Hahn, L.; Cherif, C.; Curbach, M.; Mechtcherine, V. Tetrahedral Profiled Carbon Rovings for Concrete Reinforcements. *Text. Mater.* **2022**, *333*, 173–182. [CrossRef]
19. Li, Y.; Bielak, J.; Hegger, J.; Chudoba, R. An incremental inverse analysis procedure for identification of bond-slip laws in composites applied to textile reinforced concrete. *Compos. Part B Eng.* **2018**, *137*, 111–122. [CrossRef]
20. Waldmann, M.; Rittner, S.; Cherif, C. DE 10 2017 107 948 A1. Bewehrungsstab zum Einbringen in eine Betonmatrix Sowie Dessen Herstellungsverfahren, ein Bewehrungssystem aus Mehreren Bewehrungsstäben Sowie ein Betonbauteil. 12.04.17. Available online: https://patentscope.wipo.int/search/de/detail.jsf?docId=WO2018189345 (accessed on 27 February 2023).
21. Penzel, P.; May, M.; Hahn, L.; Scheerer, S.; Michler, H.; Butler, M.; Waldmann, M.; Curbach, M.; Cherif, C.; Mechtcherine, V. Bond Modification of Carbon Rovings through Profiling. *Materials* **2022**, *15*, 5581. [CrossRef] [PubMed]
22. Böhm, R.; Thieme, M.; Wohlfahrt, D.; Wolz, D.; Richter, B.; Jäger, H. Reinforcement Systems for Carbon Concrete Composites Based on Low-Cost Carbon Fibers. *Fibers* **2018**, *6*, 56. [CrossRef]
23. Schumann, A.; May, M.; Schladitz, F.; Scheerer, S.; Curbach, M. Carbonstäbe im Bauwesen. *Beton Stahlbetonbau* **2020**, *115*, 962–971. [CrossRef]
24. Schumann, A. Experimentelle Untersuchungen des Verbundverhaltens von Carbonstäben in Betonmatrices. Ph.D. Thesis, Technische Universität Dresden, Institut für Massivbau, Dresden, Germany, 2020.
25. Kolkmann, A.; Wulfhorst, B. Garnstrukturen für den Einsatz in textilbewehrtem Beton (Lehrstuhl für Textilmaschinenbau und Institut für Textiltechnik). Available online: http://publications.rwth-aachen.de/record/102203 (accessed on 27 February 2023).
26. Kolkmann, A. Methoden zur Verbesserung des Inneren und äußeren Verbundes Technischer Garne zur Bewehrung zementgebundener Matrices. Ph.D. Thesis, RWTH Aachen University, Aachen, Germany, 2008.
27. CARBOCON GMBH: CARBOrefit®—Verfahren zur Verstärkung von Stahlbeton mit Carbonbeton: Allgemein bauaufsichtliche Zulassung/Allgemeine Bauartgenehmigung. 2021. Available online: https://cloud.carborefit.de/index.php/s/PgMiAwyYMTz8Gse?dir=undefined&path=%2FCARBOrefit%C2%AE%20Zulassung&openfile=27808 (accessed on 27 February 2023).
28. Schneider, K.; Michel, A.; Liebscher, M.; Mechtcherine, V. Verbundverhalten mineralisch gebundener und polymergebundener Bewehrungsstrukturen aus Carbonfasern bei Temperaturen bis 500 °C. *Beton Stahlbetonbau* **2018**, *113*, 886–894. [CrossRef]
29. Teijin Limited. *Teijin Carbon Europe GmbH: Tenax Filament: YarnProduct Datasheet*; Teijin Limited: Tokyo, Japan, 2020.
30. *ISO 3341:2000-05*; Textilglas-Garne-Bestimmung der Reißkraft und Bruchdehnung. Beuth Publishing Din: Berling, Germany.
31. *CHT Germany GmbH: TECOSIT CC 1000: Technical Datasheet*; CHT Germany GmbH: Tübingen, Germany, 2022.
32. *Lefatex Chemie GmbH: Lefasol VL 90. Technical Datasheet*; Lefatex Chemie GmbH: Brüggen, Germany, 2017.
33. *DIN EN 196-1:2016*; Prüfverfahren für Zement-Teil 1: Bestimmung der Festigkeit. Beuth Publishing Din: Berling, Germany.
34. *DIN EN ISO 10618:2004-11*; Kohlenstofffasern—Bestimmung des Zugverhaltens von harzimprägnierten Garnen. Beuth Publishing Din: Berling, Germany.

materials

MDPI

Article

Preparation and Hydration Properties of Steel Slag-Based Composite Cementitious Materials with High Strength

Zhiming Xu [1], Ying Ma [1,*], Jiahao Wang [1] and Xiaodong Shen [1,2]

1 College of Materials Science and Engineering, Nanjing Tech University, Nanjing 211816, China
2 State Key Laboratory of Materials-Oriented Chemical Engineering, Nanjing Tech University, Nanjing 211816, China
* Correspondence: maying@njtech.edu.cn

Abstract: Steel slag (SS) has been largely discharged but little utilized, causing an environmental problem in China. In this paper, SS-based composite cementitious materials with high strength were prepared by the high volume of SS ($\geq$40%), granulated blast-furnace slag (GBFS), fly ash (FA), flue gas desulfurization gypsum (FGDG) and cement to promote the effective utilization of SS. The hydration and hardening properties were studied through setting time, compressive strength, length change, isothermal calorimetry (IC), X-ray diffraction (XRD), mercury intrusion porosimetry (MIP), and scanning electron microscopy equipped with energy dispersive spectroscopy (SEM-EDS) tests. The results show that SS-based composite cementitious material exhibited a lower hydration heat release, an appropriate setting time, and volume stability. The SS cementitious material with 40% SS could obtain high strength of over 65 MPa at 28 days and 80 MPa at 90 days. The strength value of > 60 MPa is present in the binder, with 50% SS at 56 days. GBFS promotes hydration reactions and the formation of AFt and C-(A)-S-H gel, thus enhancing compressive strength. FA has a beneficial effect on later strength. The small and fine pore structures contribute to the development of strength. The main hydration products of SS composite cementitious materials are C-(A)-S-H gel, and ettringite (AFt), with less $Ca(OH)_2$. The C-(A)-S-H gel with a lower Ca/Si ratio and a higher Al/Ca ratio in cementitious material, promotes mechanical properties.

Keywords: steel slag; solid wastes; compressive strength; hydration; microstructure

Citation: Xu, Z.; Ma, Y.; Wang, J.; Shen, X. Preparation and Hydration Properties of Steel Slag-Based Composite Cementitious Materials with High Strength. *Materials* **2023**, *16*, 2764. https://doi.org/10.3390/ma16072764

Academic Editors: Nikolina Zivaljic, Hrvoje Smoljanović and Ivan Balić

Received: 28 February 2023
Revised: 25 March 2023
Accepted: 27 March 2023
Published: 30 March 2023

1. Introduction

Steel slag (SS) is an industrial solid waste discharged during steelmaking. Its emissions were 160 million tons, and the utilization rate was less than 30% in China. A great part of SS is randomly discarded, resulting in land occupation and environmental pollution. The improvement of efficient utilization of SS is conducive to ecological environment protection and achieving sustainable development of the steel industry.

The mineral composition of SS contains C_3S, γ-C_2S, C_4AF, C_2F, RO phase (CaO–FeO–MnO–MgO solid solution), free-CaO, and free-MgO [1]. SS has been mainly used as aggregate substitution [2–4] and supplementary cementitious materials [5,6] in building materials. Some mechanical properties and durabilities of concrete with complete or partial substitution of SS aggregate were examined. The compressive strength of concrete would increase as the substitution ratio of SS aggregate replacing traditional aggregate increased [4,7–9]. There was a reduction in the compressive strength of concrete containing SS aggregates after freezing-thawing cycles due to high porosity [10]. It was proposed that the replacement ratio of SS aggregate at 15% and 30% are beneficial to the compressive strength [11].

SS has a low reactivity with respect to the high amount of γ-C_2S resulting from the slow and natural cooling during SS production. Low reactivity calcium silicate in SS participates in cement hydration to generate hydrated calcium silicate. SS as supplementary cementitious materials could improve the workability, decrease hydration heat and decrease compressive strength [5]. The compressive strength decreases with the increase in the SS replacement ratio [12–14]. The 28-day compressive strength of mortar is about 37 MPa when the SS content reaches 30% [15].

It should be noted that the use of high-volume SS (≥40%) in steel slag-based cementitious material can efficiently promote the utilization of SS. Low compressive strength was reported in cementitious materials with high-volume SS. A low 28-day compressive strength of 30 MPa was exhibited in cement paste with 50% SS [16] and with the replacing ratio increase to 70%, the strength of cement paste was only 20.8 MPa [17]. He et al. [18] reported that 80% SS-20% cement paste has a 28-day compressive strength of about 35 MPa and decreases by 60 MPa compared with that of pure cement paste.

Due to the low activity of SS, activation is necessary to promote the activity and hydration properties of SS-based composite binders. Alkali and calcium sulfur activation were primarily used in low-activity cementitious materials. Potassium and sodium compounds were commonly used as activators in alkali-activated cementitious materials. Under alkali conditions, the silicate and aluminate minerals dissolve and hydrate to produce C-S-H, C-A-S-H and N-A-S-H. A typical calcium sulfur-activated cementitious material (supersulfated cement) is composed of 70% to 90% slag, 0% to 5% Portland cement (clinker) or calcium hydroxide, and 10% to 20% sulfate activators (hemihydrate and gypsum dihydrate). The activity of SiO_2 and Al_2O_3 is activated by calcium and sulfate and undergoes a hydration reaction to form C-(A)-S-H and AFt.

The steel slag-based composite cementitious material with a high volume of SS could be prepared based on the activity matching between the aluminate and silicate phases in SS, GBFS, FA with the calcium and sulfate phases in cement (clinker) and gypsum. It hydrates to generate C-S-H, Aft, or AFm, and potentially results in high cementitious properties. It was proposed that the addition of SS of less than 20% could slightly improve the compressive strength of SS-GBFS-FGDG mortar [19]. But a higher content of SS would have a negative impact on the compressive strength [19,20]. The hydration degree of the composite cementitious materials was improved by GBFS with higher activity, thereby improving the compressive strength. [16,21,22]. The compressive strength of the SS-FA-FGDG ternary system decreases with the substitution ratio of FA replacing SS. The strength of 60% SS-20%FA-20%FGDG paste is 30 MPa, and the strength decreases to 15 MPa while the replacing ratio of FA increases to 40% [23]. The previous studies indicate that composite cementitious materials can possess a higher strength with a lower volume of SS (≤20%), and for a high volume SS, low compressive strength occurred for SS-based composite cementitious material [24].

This study aims to prepare a high-strength SS-based composite cementitious material using high volume SS (≥40%) along with GBFS, FA, FGDG and less Portland cement. The hydration and hardening properties of SS-based composite cementitious materials were analyzed through hydration heat evolution, setting time, compressive strength, length change, XRD, MIP, and SEM-EDS tests. This study will further promote the effective utilization of steel slag, sustainable development of resources and environmental protection.

2. Experimental

2.1. Raw Materials

Fine powders of SS from Jigang Group Co., Ltd., Shandong, GBFS from Jintaicheng Environmental Resources Co., Ltd., Hebei, and low calcium FA were used in the experiment. Flue gas desulfurization gypsum was treated through a 200-mesh sieve in size of 75 μm. PC-II 52.5 Portland cement from Jiangnan Onoda was used. Particle size distributions of these fine materials were measured and shown in Figure 1. The medium diameter (D50) of SS, GBFS, PC, FGDG and FA were 11.51, 6.58, 11.56, 29.02, and 8.36 μm, respectively.

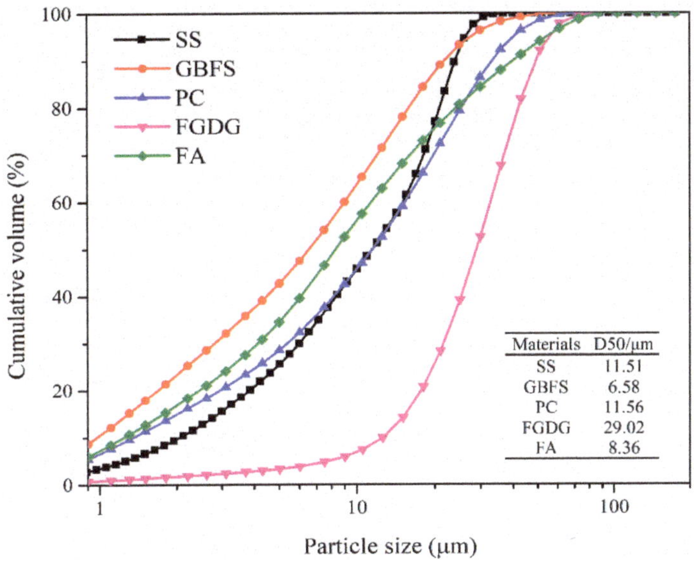

Figure 1. Particle size distributions of raw materials.

The chemical and mineral compositions of raw materials were measured by X-ray fluorescence (XRF) and XRD as shown in Table 1 and Figure 2. The main chemical composition of SS is CaO, followed by Fe_2O_3 and SiO_2, MgO, and MnO. Its mineral composition includes C_2S, $Ca_2Fe_2O_5$, RO and $Ca(OH)_2$ phases. The alkalinity coefficient of the SS is 1.94 regarded as medium alkalinity according to Mason's theory [25]. GBFS is composed by CaO and SiO_2, followed by Al_2O_3 as well as the amorphous phase. Fly ash consists of SiO_2, $CaSO_4$, $Ca_3Al_2O_6$, etc. The main mineral phases of FGDG were dihydrate ($CaSO_4 \cdot 2H_2O$), anhydrite ($CaSO_4$) along with calcium carbonate ($CaCO_3$).

Table 1. Chemical composition of raw materials (wt. %).

Materials	CaO	SiO_2	Al_2O_3	SO_3	MgO	Fe_2O_3	Na_2O	K_2O	MnO	P_2O_5	LOI
SS	36.46	16.64	5.71	0.24	7.42	21.03	0.29	0.09	6.06	2.14	1.06
GBFS	40.38	30.42	16.74	1.34	7.56	1.24	-	0.43	0.23	0.14	-
PC	62.13	21.75	5.21	1.97	2.09	2.91	-	0.63	0.05	0.16	2.40
FGDG	32.06	1.65	0.80	42.46	0.70	0.20	-	0.08	0.02	0.02	21.59
FA	7.95	46.41	31.18	1.91	1.30	4.97	0.60	0.86	0.08	0.60	2.40

2.2. Preparation and Curing Procedures

The experimental flowchart of this study is shown in Figure 3. The mixture proportions of SS-based composite cementitious materials are shown in Table 2. The water-to-binder ratio of 0.3 was used. The fluidity of SS-based composite cementitious material is a range of 190–250 mm. The moulds in sizes of 20 mm × 20 mm × 20 mm and 20 mm × 20 mm × 80 mm were used for compressive strength and length change measurements, respectively. The raw materials were premixed, and the paste was mixed by a mixer for 4 min. The paste with the mould was vibrated by a vibrating machine. After casting, that was placed in a curing box (20 ± 2 °C and RH 95 ± 2%) for 24 h. Then, the cubes were demoulded and cured in water at 20 ± 1 °C until certain periods. At the ages of 3, 7, 14, 28, 56, and 90 days, parts of the samples were taken out and prepared for compressive strength and length change tests. Moreover, other parts of the samples were broken into pieces and soaked in anhydrous ethanol for 24 h [26] to stop the hydration. The sample pieces were dried in a vacuum oven at 40 °C for 24 h. The prepared sample pieces were used for MIP and

SEM-EDS analysis. Parts of the sample pieces were ground and passed through a 200-mesh sieve for the XRD test.

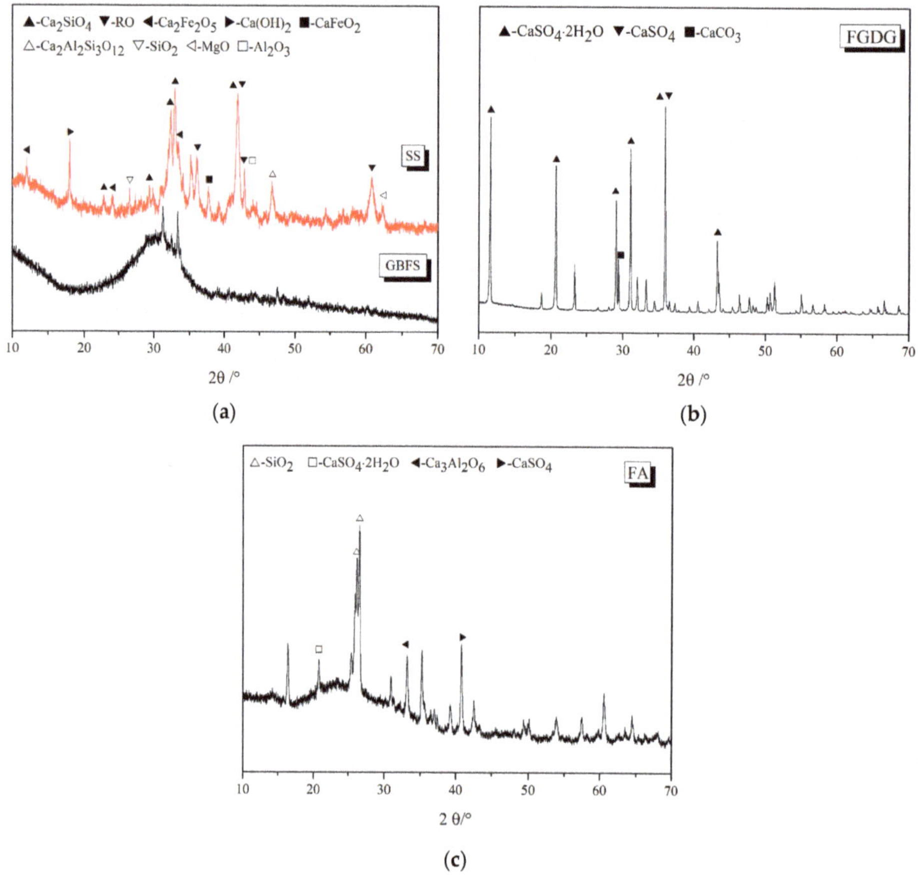

Figure 2. XRD patterns of (**a**) SS and GBFS, (**b**) FGDG and (**c**) FA.

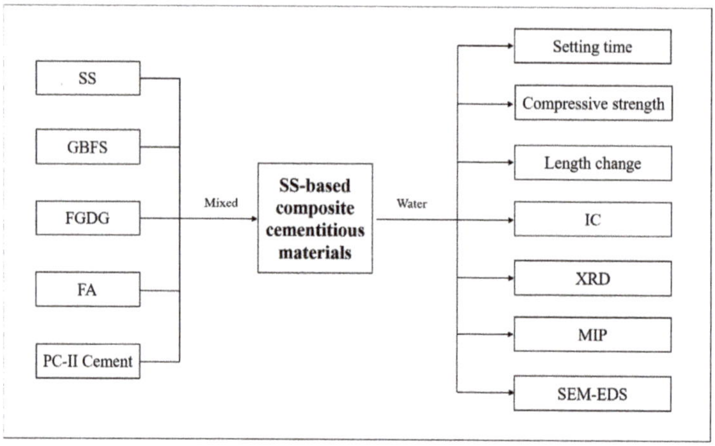

Figure 3. Experimental flowchart of this study.

Table 2. Mixture proportions (wt. %) of SS-based composite cementitious materials.

Binders	SS	FGDG	PC	FA	GBFS
SS40	40	13	15	0	32
SS43	43	13	15	0	29
SS40-FA7	40	13	15	7	25
SS40-FA12	40	13	15	12	20
SS50-FA2	50	13	15	2	20

2.3. Test Methods

The hydration heat evolution of SS-based composite cementitious materials was tested and recorded for 72 h using an isothermal calorimeter (TAM Air). The weight of 1.8 g of deionized water and 6 g of the binder were used. The slurry was mixed outside and immediately loaded into the isothermal calorimeter. The temperature was set and stable at 20 °C.

The fluidity of binders was tested according to GBT/2419-2016.

The initial and final setting times were conducted by using a Vicat apparatus according to the Chinese standard of GB/T 1346-2011.

The compressive strength was measured by a pressure testing instrument (AEC-201 type) with 2.4 KN/s according to JGJ/T 70-2009. The strength value was calculated by taking the average of at least four cubes. The length change was measured according to GB/T 29417-2012. The data on length change was collected by a digital comparator (precision of 0.001 mm). The average value of three cubes was used. The linear expansion ratio was obtained by calculating the length value at specific periods corresponding to the initial length value.

The hydration products of pastes were identified by XRD tests. XRD analysis was performed by a Rigaku-Smart-lab 3000A X-ray diffractometer with Cu Kα radiation at 35 mA and 40 kV. The scanning range is between 5° and 70° at 5° per minute with a step of 0.01°.

The pore parameters of samples were measured by using a MIP (Auto Pore V 9600, Micromeritics). The pressure was 0.1 to 33,000 psi, and the pore size was in a range of 5.48 to 10,000 nm.

The samples were coated with a gold layer for SEM-EDS measurement. The micro-morphology of samples was observed by a ZEISS instrument (Ultra 55 FESEM) at the age of 90 days. Elements in the selected micro-regions were determined by the EDS at an accelerated voltage of 10 kV.

3. Results

3.1. Hydration Heat Evolution and Setting Time

The hydration heat flow and cumulative hydration heat of cementitious materials are shown in Figure 4. Three exothermic peaks were observed on heat flow curves for pastes except for the SS50-FA2 group. The first exothermic peak which corresponds to the dissolution of mineral ions and initial hydration of pastes is not accurate and fully displayed because of the limitation of the device and mixing way. The first exothermic peak was followed by an induction period with relatively small amounts of heat release and a short duration. The duration time shortens with the increase in SS. The heat flow then begins to enter the acceleration period, and the starting time is ordered by SS50-FA2 < SS43 and SS40 < SS40-FA7 < SS40-FA12.

A high exothermic peak II occurred in the SS50-FA2 paste. Other cementitious materials show a relatively lower exothermic peak II at around 10 to 14 h. The occurrence time of peak II is prolonged with the rise of FA content in SS40-FA7 and SS40-FA12 pastes compared with the SS40 group. The cement is replaced by SS with medium alkalinity, and the second exothermic peak appears around 10 h [27]. It is inferred that the second exothermic peak would be caused by the hydration of SS particles based on the high heat release at peak II for SS50-FA2 paste with a high content of SS.

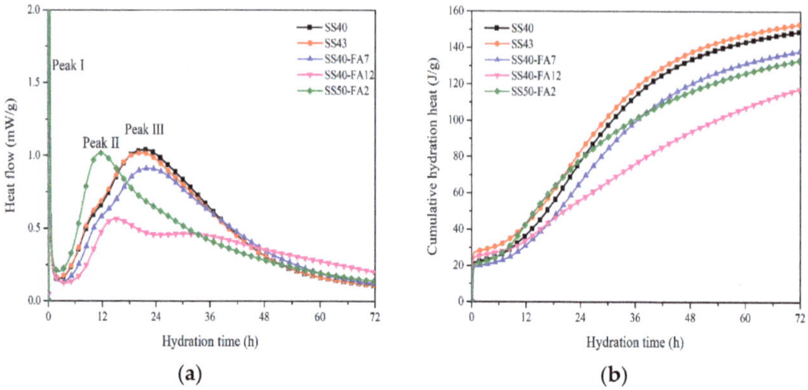

Figure 4. (**a**) Hydration heat flow and (**b**) cumulative hydration heat of SS-based composite cementitious materials.

The peak III appeared in SS40, SS43, and SS40-FA7 pastes after 17 h. SS40-FA12 paste presents a small peak III after 28 h resulting by a high addition of FA. The third exothermic peak is considered in relation to the main hydration of active mineral phases of binders as a result of the formation of C-S-H gel and $Ca(OH)_2$ [28].

The cumulative hydration heat presents about 117.52–152.92 J/g for SS-based composite cementitious materials, which is lower than Portland cement systems. The cumulative hydration heat of SS43 paste at 1 and 3 days is slightly higher than that of SS40 paste. The cumulative hydration heat of SS40-FA7 and SS40-FA12 paste decreased with the increase of FA compared to SS40 at 1 and 3 days. It has been recognized that FA leads to a lower cumulative heat release than GBFS and Portland cement [29]. The fact that SS50-FA2 paste has a larger cumulative hydration heat than SS40-FA12 paste shows that SS releases more heat than FA. It is indicated that the activity of SS is higher than FA in cementitious materials at an early age.

The initial and final setting times are important indicators of early hydration. The initial and final setting times of pastes are shown in Figure 5. The initial and final setting times of pastes are at ranges of 282–348 and 460–512 min respectively, which meets the requirement of the standard of GB/T 1346–2011 [30]. Compared with SS40 paste, the initial and final setting times of SS40-FA7 and SS40-FA12 pastes were prolonged by the addition of FA. SS40-FA12 and SS50-FA2 pastes exhibited similar initial and final setting times. The proper setting time of steel slag paste facilitates the placement and handling.

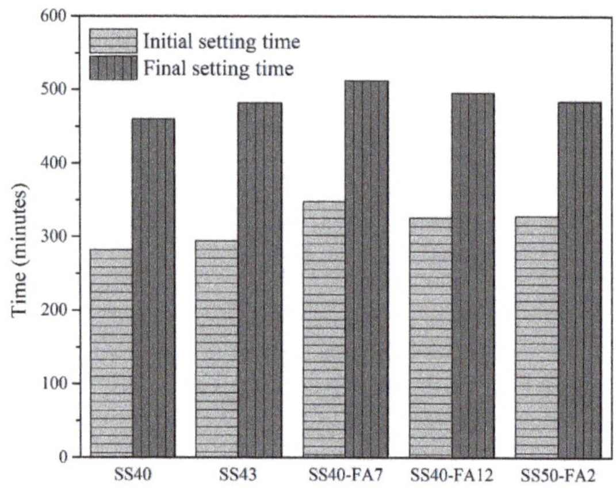

Figure 5. Initial and final setting times of SS-based composite cementitious materials.

3.2. Compressive Strength

Results of the compressive strength of cementitious materials at different ages are shown in Figure 6. It can be seen that the compressive strength increases as hydration time increases. The compressive strength of SS40 paste exceeds 60 MPa at 28 days and 80 MPa at 90 days, while the compressive strength of other pastes exceeds 70 MPa at 90 days. The SS40 specimen had higher compressive strength values. The continuous hydration of SS and GBFS increases the amount of C-(A)-S-H gel, which greatly improves the pore structure of pastes and results in the development of compressive strength.

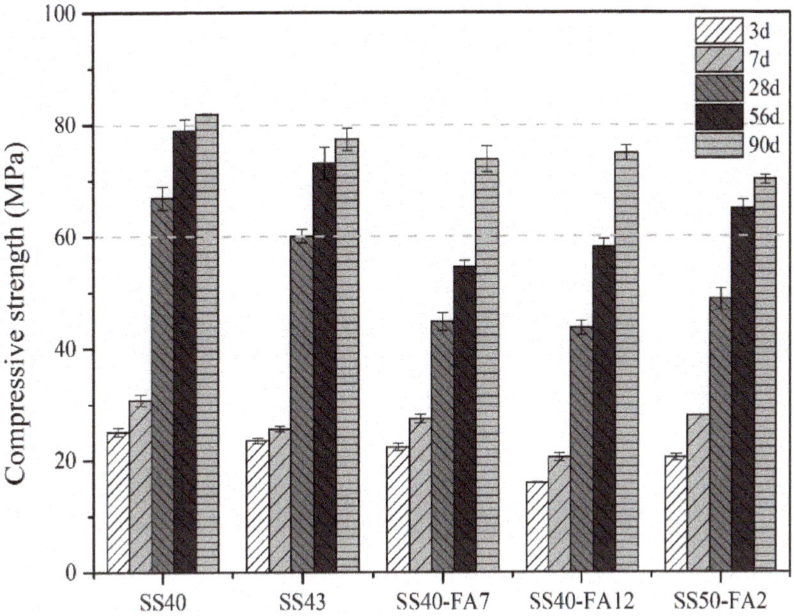

Figure 6. Compressive strength of SS-based composite cementitious materials at different ages.

Compared to the SS40 specimen, SS40-FA7 and SS40-FA12 pastes possessed lower compressive strength at all test ages because of the addition and the lower activity of FA in binders [31]. A high rise in compressive strength occurred for SS40-FA7 and SS40-FA12 pastes after 56 days, and the increase ratio is 35% and 29% from 56 to 90 days. It indicates that the FA plays an advantageous role in the development of compressive strength, due to the activation of the pozzolanic activity of the FA by the $Ca(OH)_2$ formed in the paste at later ages. The C-S-H gel formed from the hydration of FA can fill the pores [32], resulting in a rapid rise in compressive strength at 90 days. The compressive strength of SS50-FA2 paste was higher than that of SS40-FA12 within 56 days, indicating that SS is superior to FA in the development of compressive strength.

3.3. Hydration Products

The hydration products of all pastes at different ages (up to 90 days) determined by XRD are shown in Figure 7. It is observed that the phases, of AFt, $Ca(OH)_2$, and various unhydrated phases including RO, γ-C_2S, and gypsum, are produced in cementitious materials. A stronger intensity of AFt and a relatively weaker intensity of $Ca(OH)_2$ are exhibited in SS cementitious materials, while amorphous C-S-H cannot be detected.

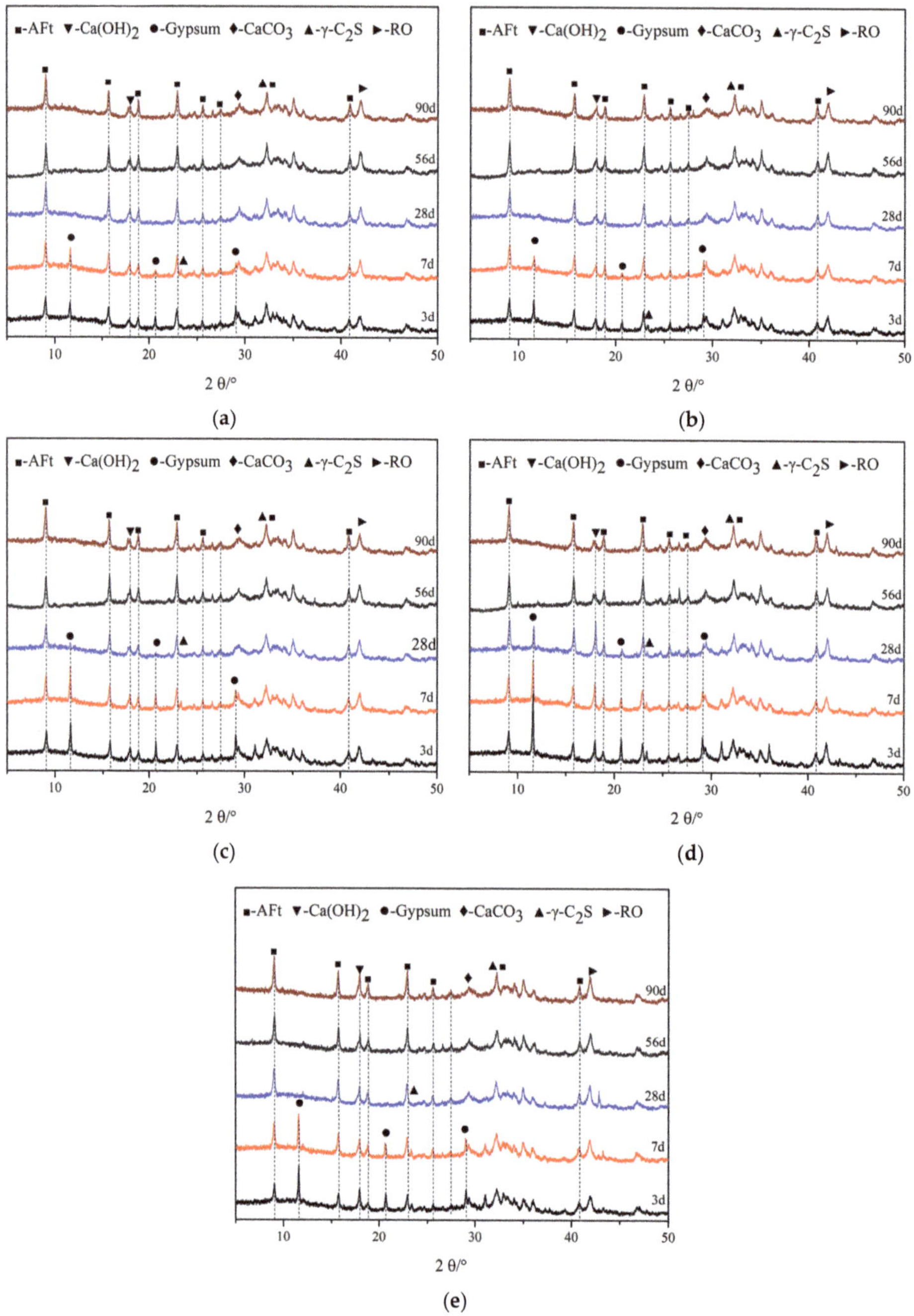

Figure 7. XRD patterns of (**a**) SS40, (**b**) SS43, (**c**) SS40-FA12, (**d**) SS40-FA12 and (**e**) SS50-FA2 at different ages.

When the binders were mixed with water, cement minerals hydrated first and produced C-S-H gel and $Ca(OH)_2$. The Si-O and Al-O bonds in GBFS, FA [33] and SS particles are decomposed in an alkaline environment to produce Ca^{2+}, SiO_4^{4-} and AlO_4^{5-}, respectively. These ions can participate in hydration reactions to form hydrated calcium silicate and hydrated calcium aluminate according to Equations (1) and (2). The latter can react with SO_4^{2-} released from FGDG to produce AFt [20], as shown in Equation (3). The skeleton structure formed by AFt is conducive to the formation of early strength, and the continuous formation of C-S-H gel and AFt promotes the development of compressive strength.

$$Ca^{2+} + xSiO_4^{4-} + yH_2O \rightarrow CaO{\cdot}xSiO_2{\cdot}yH_2O \quad (1)$$

$$5Ca^{2+} + 2AlO_4^{5-} + 8H_2O \rightarrow 3CaO{\cdot}Al_2O_3{\cdot}6H_2O + 2Ca(OH)_2 \quad (2)$$

$$3CaO{\cdot}Al_2O_3{\cdot}6H_2O + 3CaSO_4 + 26H_2O \rightarrow 3CaO{\cdot}Al_2O_3{\cdot}3CaSO_4{\cdot}32H_2O \quad (3)$$

It can be seen that the diffraction peak intensity of AFt increases and the diffraction peak of gypsum gradually decreases with the development of hydration time, due to the consumption of gypsum during the formation of AFt. The disappearance of the gypsum diffraction peak can be observed in SS40 and SS43 pastes after 28 days. After 56 days, there were no obvious signs of a gypsum diffraction peak in other cementitious materials. The higher amount of GBFS in SS40 and SS43 pastes results in faster consumption of gypsum, which would be caused by a greater hydration reaction between the aluminate of GBFS and gypsum to produce AFt [34,35]. The gypsum diffraction peak of SS50-FA2 is lower than that of SS40-FA12 indicating that more steel slag leads to faster consumption of gypsum.

The diffraction peak of γ-C_2S decreases with the hydration time. The presence of $CaCO_3$ would be caused by carbonization, but most of it comes from FGDG. The diffraction peak of the RO phase remained and exhibited insignificant change up to 90 days.

3.4. Length Variation

The linear expansion ratio of SS-based composite cementitious materials is presented in Figure 8. All pastes expended and the values of the expansion ratio are less than 0.46% and cured in water for up to 90 days. There is a rapid increase in the linear expansion ratio before 28 days and then a steady increase afterward.

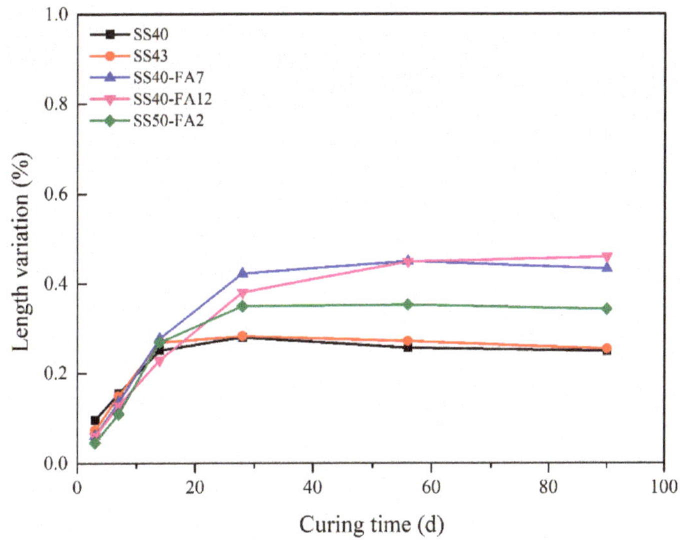

Figure 8. The linear expansion ratio of SS-based composite cementitious materials under water curing up to 90 days.

SS40 and SS43 pastes showed similar length changes, and the linear expansion ratios are less than 0.25%. It cannot be observed an increase in the expansion ratio up to 90 days when the amount of SS increases from 40% in SS40-FA12 to 50% in SS50-FA2. The potential volume instability has been proposed because the reaction between f-CaO and f-MgO in the SS and water produce $Ca(OH)_2$ and $Mg(OH)_2$ as a result of expansion [36]. The volume stability of SS cementitious materials exhibited a relatively weak variation with the rise of SS content up to 50% in this paper.

It can be seen that cementitious materials incorporating FA occurred a larger length change at 90 days. SS40-FA12 paste exhibited a higher expansion ratio than SS40-FA7 and SS50-FA2 specimens. The linear expansion ratio rises with the increase in FA content. The expansion development is dominated by AFt formation in the expansion performance of cement pastes with FA at early ages [37].

3.5. Pore Parameters

The pore size distribution curves and pore volume fractions of cementitious materials are shown in Figure 9. Pore sizes smaller than 5.48 nm were not detected due to the limited method of testing. To gain more insights into the pore structure, the measured pore size distribution was divided into three size ranges: gel pores (<10 nm), medium pores (10–50 nm) and large pores (>50 nm) [38]. Table 3 summarizes the pore parameters including average and median pore diameter as well as the total porosity of specimens at 90 days. In general, the smaller the average pore size, the denser the pore structure, and the higher the strength [39]. Previous studies have shown that compressive strength is inversely proportional to the size of the porosity [40,41].

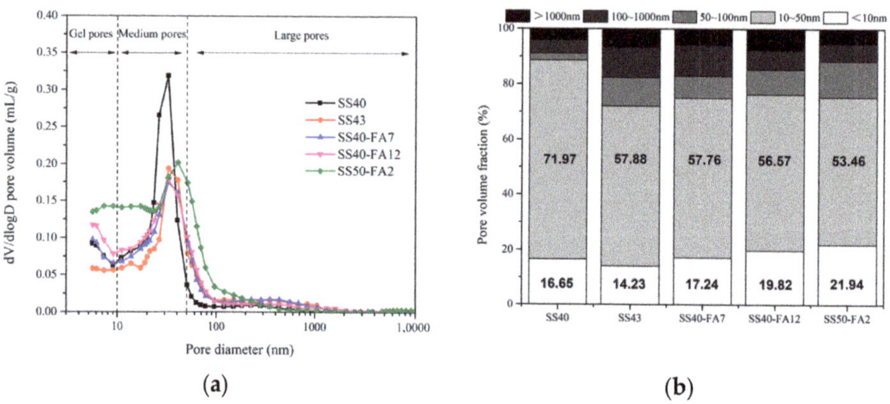

Figure 9. (**a**) Pore size distribution curves and (**b**) pore volume fractions of SS-based composite cementitious materials at 90 days.

Table 3. Pore parameters of SS-based composite cementitious materials at 90 days.

Groups	Average Pore Diameter (nm)	Median Pore Diameter (Volume) (nm)	Porosity (%)
SS40	19.07	27.77	23.61
SS43	23.10	34.35	20.91
SS40-FA7	20.60	31.00	22.84
SS40-FA12	19.19	28.91	23.37
SS50-FA2	18.35	27.29	30.28

The gel and medium pores possess roughly 75%–89% of the volume fractions, meaning most occurrence pores in specimens (Figure 9). SS40 paste possesses a high-volume fraction of up to about 72% in the medium pore range, and the proportion of <50 nm pores accounts for 89%. The SS40 specimen exhibited a lower average pore diameter, and its porosity is 23.61%. It is predicted that its small and fine pores contribute to its high compressive strength (Figure 6). A large proportion of large pores and small porosity occurred in the SS43 paste. SS40-FA7 and SS40-FA12 specimens present similar pore parameters. A high fraction of gel pores but a low proportion of medium pores were observed in the SS50-FA2 paste. SS50-FA2 specimen has the smallest average pore size, but the highest porosity than other pastes. The lowest compressive strength would be mainly attributed to its high porosity.

3.6. Micromorphology

SEM micrographs of pastes at 90 days are shown in Figure 10. The presence of AFt needles with a width of around 200 nm and C-(A)-S-H gels was found in all pastes. The dot-like substances, which may be the initial C-S-H gel [28] were observed in the SS40 paste. The structure of AFt interlaced with C-(A)-S-H gels occurred in other samples. It can be observed the spots of net structure with regard to the interconnection between AFt and C-(A)-S-H in SS40-FA12 and SS50-FA2 samples.

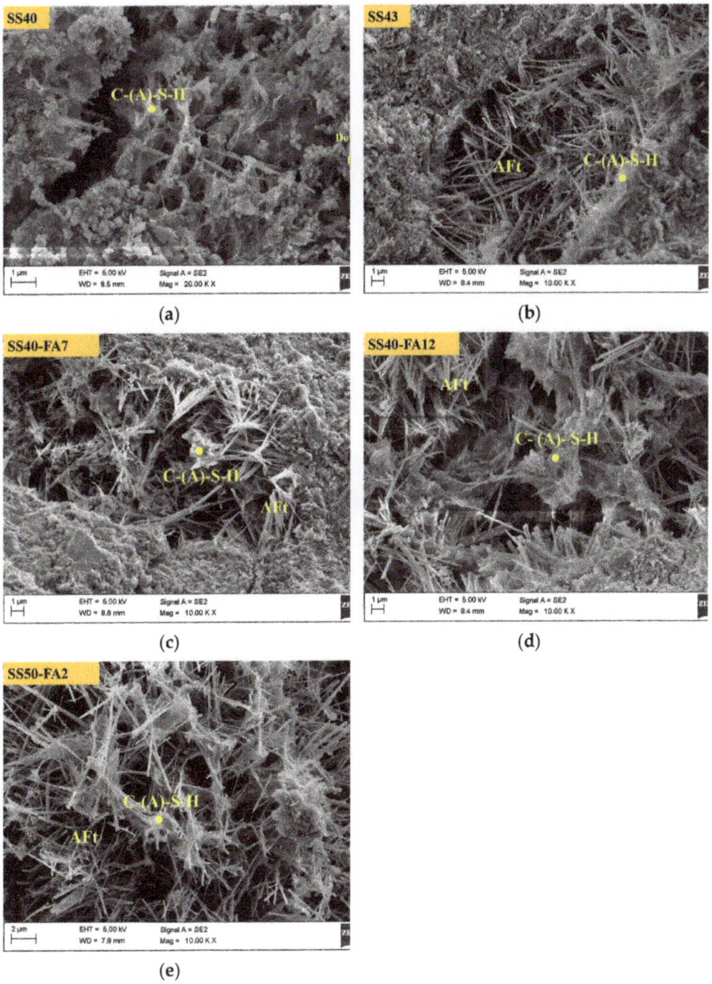

Figure 10. SEM images of SS-based composite cementitious materials at 90 days.

The percentages of Ca and Si atoms as well as Ca/Si and Al/Ca atomic ratios in C-(A)-S-H of pastes were explored by EDS analysis at 90 days, as shown in Figure 11. The Ca/Si ratio of C-(A)-S-H gel in cementitious materials is in the range of 1.32–1.84. Relevant works of literature [42–44] suggest that the Ca/Si of C-S-H for cement-based materials ranges from 1.50 to 2.58. SS cementitious materials exhibit a lower Ca/Si ratio of C-(A)-S-H compared to cement-based materials. The average silicate chain length of C-(A)-S-H increases with the decrease in the Ca/Si ratio [45]. Longer silicate chains contribute to the development of mechanical properties [43].

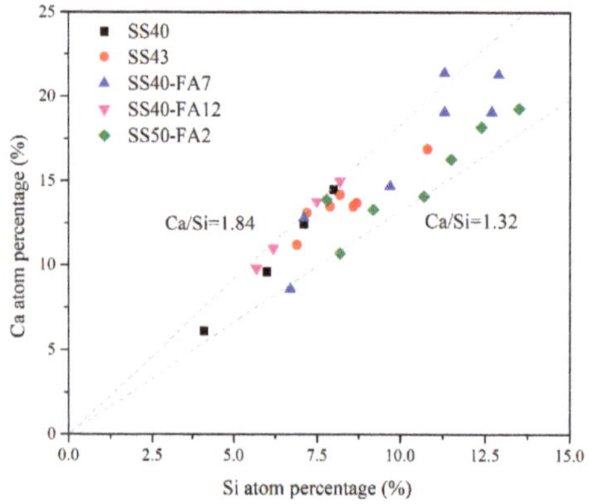

(a)

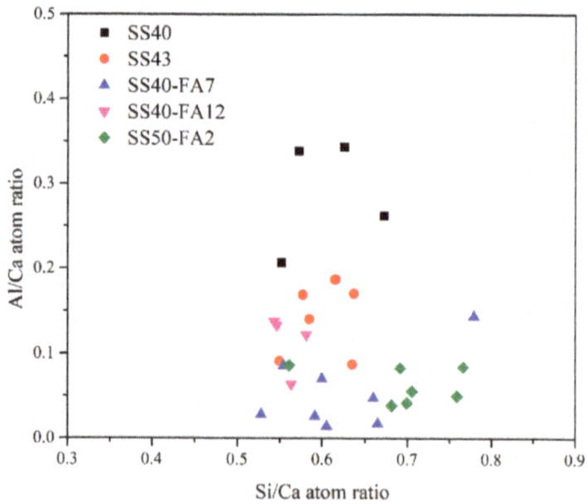

(b)

Figure 11. (**a**) Ca and Si atoms percentage, (**b**) Si/Ca and Al/Ca atomic ratios of C-(A)-S-H.

The Al/Ca ratio of pastes ranges from 0.02 to 0.34, and the average Al/Ca is 0.12 (Figure 11b). Al enters C-S-H mainly through silicic acid chain bridging sites [46], forming dense C-A-S-H gels [47]. The Al substitutes Si in C-S-H and high Al/Ca provide excellent mechanical properties [48].

4. Conclusions

In this paper, the high-strength SS-based composite cementitious materials were prepared by high volume SS (≥40%) along with GBFS, FA, FGDG and less Portland cement.

SS-based composite cementitious materials exhibited a low hydration heat release. The second exothermic peak before the main peak on the heat flow curves is caused by the hydration of SS particles. The increase in SS or FA prolongs the initial and final setting times. The setting times of SS-based composite binders satisfy the requirement.

The SS-based composite cementitious material (40% SS) could obtain a high compressive strength of over 65 MPa at 28 days and 80 MPa at 90 days. The strength value of >60 MPa is present in the binder with 50% SS at 56 days. GBFS with a higher activity promotes the hydration reaction and the formation of AFt and C-(A)-S-H gel, thus facilitating the compressive strength of the binder. The FA performs an advantageous effect on the strength at later periods. Meanwhile, the SS-based composite binders exhibited good volume stability.

The main hydration products of SS-based composite binders are AFt and C-(A)-S-H gel, with less $Ca(OH)_2$. The network structure of the interconnection between the needles AFt and C-(A)-S-H gel formed, which is conducive to the dense structure. A lower Ca/Si ratio and a higher Al/Ca ratio of C-(A)-S-H gel formed, which promotes mechanical properties.

Small and fine pores contribute to the high compressive strength of paste with 40% SS, and a higher SS volume (50%) would lead to an increase in porosity, resulting in a reduction in the strength of the paste (SS50-FA2).

The results could provide a reference for the preparation and application of high-strength steel slag-based composite cementitious materials, which facilitate the efficient utilization of steel slag.

Author Contributions: Z.X.: Investigation, Methodology, Writing-original draft. Y.M.: Writing-review, editing, Supervision. J.W.: Investigation, Methodology. X.S.: Writing—review, editing. All authors have read and agreed to the published version of the manuscript.

Funding: This research was jointly funded by the State Key Laboratory of Materials-Oriented Chemical Engineering, the National Natural Science Foundations of China (No. 51802151) and the Priority Academic Program Development of Jiangsu Higher Education Institutions (PAPD).

Data Availability Statement: Data will be made available on request.

Conflicts of Interest: The authors declare no conflict of interest.

References

1. Shi, C. Characteristics and cementitious properties of ladle slag fines from steel production. *Cem. Concr. Res.* **2002**, *32*, 459–462. [CrossRef]
2. Lee, J.-I.; Choi, S.-J. Effect of replacement ratio of ferronickel slag aggregate on characteristics of cementitious mortars at different curing temperatures. *Case Stud. Constr. Mat.* **2023**, *18*, e01882. [CrossRef]
3. Huang, Y.; Yang, X.; Wang, S.; Liu, Z.; Liu, L.; Xu, B. Evaluating Cement Treated Aggregate Base Containing Steel Slag: Mechanical Properties, Volume Stability and Environmental Impacts. *Materials* **2022**, *15*, 8277. [CrossRef] [PubMed]
4. Guo, Y.; Xie, J.; Zhao, J.; Zuo, K. Utilization of unprocessed steel slag as fine aggregate in normal- and high-strength concrete. *Constr. Build. Mater.* **2019**, *204*, 41–49. [CrossRef]
5. Han, X.; Feng, J.; Shao, Y.; Hong, R. Influence of a steel slag powder-ground fly ash composite supplementary cementitious material on the chloride and sulphate resistance of mass concrete. *Powder Technol.* **2020**, *370*, 176–183. [CrossRef]
6. Zhao, Q.; Pang, L.; Wang, D. Adverse Effects of Using Metallurgical Slags as Supplementary Cementitious Materials and Aggregate: A Review. *Materials* **2022**, *15*, 3803. [CrossRef]
7. Gencel, O.; Karadag, O.; Oren, O.H.; Bilir, T. Steel slag and its applications in cement and concrete technology: A review. *Constr. Build. Mater.* **2021**, *283*, 122783. [CrossRef]
8. Adegoloye, G.; Beaucour, A.L.; Ortola, S.; Noumowé, A. Concretes made of EAF slag and AOD slag aggregates from stainless steel process: Mechanical properties and durability. *Constr. Build. Mater.* **2015**, *76*, 313–321. [CrossRef]
9. Maslehuddin, M.; Sharif, A.M.; Shameem, M.; Ibrahim, M.; Barry, M.S. Comparison of properties of steel slag and crushed limestone aggregate concretes. *Constr. Build. Mater.* **2003**, *17*, 105–112. [CrossRef]
10. Manso, J.M.; Polanco, J.A.; Losañez, M.; González, J.J. Durability of concrete made with EAF slag as aggregate. *Cem. Concr. Compos.* **2006**, *28*, 528–534. [CrossRef]
11. Qasrawi, H.; Shalabi, F.; Asi, I. Use of low CaO unprocessed steel slag in concrete as fine aggregate. *Constr. Build. Mater.* **2009**, *23*, 1118–1125. [CrossRef]
12. Zhu, H.; Ma, M.; He, X.; Zheng, Z.; Su, Y.; Yang, J.; Zhao, H. Effect of wet-grinding steel slag on the properties of Portland cement: An activated method and rheology analysis. *Constr. Build. Mater.* **2021**, *286*, 122823. [CrossRef]
13. Kourounis, S.; Tsivilis, S.; Tsakiridis, P.; Papadimitriou, G.; Tsibouki, Z. Properties and hydration of blended cements with steelmaking slag. *Cem. Concr. Res.* **2007**, *37*, 815–822. [CrossRef]
14. Rosales, J.; Cabrera, M.; Agrela, F. Effect of stainless steel slag waste as a replacement for cement in mortars. Mechanical and statistical study. *Constr. Build. Mater.* **2017**, *142*, 444–458. [CrossRef]
15. Pang, L.; Liao, S.; Wang, D.; An, M. Influence of steel slag fineness on the hydration of cement-steel slag composite pastes. *J. Build. Eng.* **2022**, *57*, 104866. [CrossRef]
16. Liu, Y.; Zhang, Z.; Hou, G.; Yan, P. Preparation of sustainable and green cement-based composite binders with high-volume steel slag powder and ultrafine blast furnace slag powder. *J. Clean. Prod.* **2021**, *289*, 125133. [CrossRef]
17. Hao, X.; Liu, X.; Zhang, Z.; Zhang, W.; Lu, Y.; Wang, Y.; Yang, T. In-depth insight into the cementitious synergistic effect of steel slag and red mud on the properties of composite cementitious materials. *J. Build. Eng.* **2022**, *52*, 104449. [CrossRef]
18. He, W.; Zhao, J.; Yang, G.; Liu, Y. Investigation on the Role of Steel Slag Powder in Blended Cement Based on Quartz Powder as Reference. *Adv. in Civ. Eng.* **2021**, *2021*, 1–15. [CrossRef]
19. Niu, F.; An, Y.; Zhang, J.; Chen, W.; He, S. Synergistic Excitation Mechanism of CaO-SiO_2-Al_2O_3-SO_3 Quaternary Active Cementitious System. *Front. Mater.* **2021**, *8*, 514. [CrossRef]
20. Cui, X.; Ni, W.; Ren, C. Hydration Mechanism of All SolidWaste Cementitious Materials Based on Steel Slag and Blast Furnace Slag. *Chin. J. Mater. Res.* **2017**, *31*, 687–694. [CrossRef]
21. Lizarazo-Marriaga, J.; Claisse, P.; Ganjian, E. Effect of Steel Slag and Portland Cement in the Rate of Hydration and Strength of Blast Furnace Slag Pastes. *J. Mater. Civ. Eng.* **2011**, *23*, 153–160. [CrossRef]

22. Wang, Q.; Yan, P.; Mi, G. Effect of blended steel slag–GBFS mineral admixture on hydration and strength of cement. *Constr. Build. Mater.* **2012**, *35*, 8–14. [CrossRef]
23. Duan, S.; Liao, H.; Cheng, F.; Song, H.; Yang, H. Investigation into the synergistic effects in hydrated gelling systems containing fly ash, desulfurization gypsum and steel slag. *Constr. Build. Mater.* **2018**, *187*, 1113–1120. [CrossRef]
24. Xu, C.; Ni, W.; Li, K.; Zhang, S.; Xu, D. Activation mechanisms of three types of industrial by-product gypsums on steel slag–granulated blast furnace slag-based binders. *Constr. Build. Mater.* **2021**, *288*, 123111. [CrossRef]
25. Mason, B. The constitution of some open-heart slag. *J. Iron Steel Inst.* **1994**, *11*, 69–80.
26. Li, Z.; Zhang, J.; Li, S.; Lin, C.; Gao, Y.; Liu, C. Feasibility of preparing red mud-based cementitious materials: Synergistic utilization of industrial solid waste, waste heat, and tail gas. *J. Clean. Prod.* **2021**, *285*, 124896. [CrossRef]
27. Chen, L.; Wang, H.; Zheng, K.; Zhou, J.; He, F.; Yuan, Q. The mechanism of basic oxygen furnace steel slag retarding early-age hydration of Portland cement and mitigating approach towards higher utilization rate. *J. Clean. Prod.* **2022**, *362*, 132493. [CrossRef]
28. Zhuang, S.; Wang, Q. Inhibition mechanisms of steel slag on the early-age hydration of cement. *Cem. Concr. Res.* **2021**, *140*, 106283. [CrossRef]
29. Wu, M.; Sui, S.; Zhang, Y.; Jia, Y.; She, W.; Liu, Z.; Yang, Y. Analyzing the filler and activity effect of fly ash and slag on the early hydration of blended cement based on calorimetric test. *Constr. Build. Mater.* **2021**, *276*, 122201. [CrossRef]
30. Federation, C.B.M. *Test Methods for Water Requirement of Normal Consistency, Setting Time and Soundness of the Portland Cement*; Peoples R China; China Standard Press: Beijing, China, 2011. (In Chinese)
31. Nedunuri, S.S.S.A.; Sertse, S.G.; Muhammad, S. Microstructural study of Portland cement partially replaced with fly ash, ground granulated blast furnace slag and silica fume as determined by pozzolanic activity. *Constr. Build. Mater.* **2020**, *238*, 117561. [CrossRef]
32. Moon, G.D.; Oh, S.; Choi, Y.C. Effects of the physicochemical properties of fly ash on the compressive strength of high-volume fly ash mortar. *Constr. Build. Mater.* **2016**, *124*, 1072–1080. [CrossRef]
33. Fu, X.; Hou, W.; Yang, C.; Li, D.; Wu, X. Studies on high-strength slag and fly ash compound cement. *Cem. Concr. Res.* **2000**, *30*, 1239–1243. [CrossRef]
34. Sun, Z.; Zhou, J.; Qi, Q.; Li, H.; Zhang, N.; Mu, R. Influence of Fly Ash on Mechanical Properties and Hydration of Calcium Sulfoaluminate-Activated Supersulfated Cement. *Materials* **2020**, *13*, 2514. [CrossRef] [PubMed]
35. Zhong, S.; Ni, K.; Li, J. Properties of mortars made by uncalcined FGD gypsum-fly ash-ground granulated blast furnace slag composite binder. *Waste Manag.* **2012**, *32*, 1468–1472. [CrossRef] [PubMed]
36. Wang, G.; Wang, Y.; Gao, Z. Use of steel slag as a granular material: Volume expansion prediction and usability criteria. *J. Hazard. Mater.* **2010**, *184*, 555–560. [CrossRef]
37. Chen, X.; Gao, J.; Yan, Y.; Liu, Y. Investigation of expansion properties of cement paste with circulating fluidized bed fly ash. *Constr. Build. Mater.* **2017**, *157*, 1154–1162. [CrossRef]
38. Dhandapani, Y.; Santhanam, M. Assessment of pore structure evolution in the limestone calcined clay cementitious system and its implications for performance. *Cem. Concr. Compos.* **2017**, *84*, 36–47. [CrossRef]
39. Chen, X.; Wu, S.; Zhou, J. Influence of porosity on compressive and tensile strength of cement mortar. *Constr. Build. Mater.* **2013**, *40*, 869–874. [CrossRef]
40. Choi, Y.C.; Kim, J.; Choi, S. Mercury intrusion porosimetry characterization of micropore structures of high-strength cement pastes incorporating high volume ground granulated blast-furnace slag. *Constr. Build. Mater.* **2017**, *137*, 96–103. [CrossRef]
41. Saleh, F.K.; Teodoriu, C.; Sondergeld, C.H. Investigation of the effect of cement mixing energy on cement strength and porosity using NMR and UPV methods. *J. Nat. Gas Sci. Eng.* **2019**, *70*, 102972. [CrossRef]
42. Qiang, W.; Mengxiao, S.; Jun, Y. Influence of classified steel slag with particle sizes smaller than 20 μm on the properties of cement and concrete. *Constr. Build. Mater.* **2016**, *123*, 601–610. [CrossRef]
43. Li, Y.; Qiao, C.; Ni, W. Green concrete with ground granulated blast-furnace slag activated by desulfurization gypsum and electric arc furnace reducing slag. *J. Clean. Prod.* **2020**, *269*, 122212. [CrossRef]
44. Richardson, I. The nature of C-S-H in hardened cements. *Cem. Concr. Res.* **1999**, *29*, 1131–1147. [CrossRef]
45. He, Y.; Lu, L.; Struble, L.J.; Rapp, J.L.; Mondal, P.; Hu, S. Effect of calcium–silicon ratio on microstructure and nanostructure of calcium silicate hydrate synthesized by reaction of fumed silica and calcium oxide at room temperature. *Mater. Struct.* **2013**, *47*, 311–322. [CrossRef]
46. Yu, P.; Kirkpatrick, R.J.; Poe, B.; McMillan, P.F.; Cong, X. Structure of Calcium Silicate Hydrate (C-S-H): Near-, Mid-, and Far-Infrared Spectroscopy. *J. Am. Ceram. Soc.* **1999**, *82*, 742–748. [CrossRef]
47. Barbhuiya, S.; Mukherjee, S.; Nikraz, H. Effects of nano-Al2O3 on early-age microstructural properties of cement paste. *Constr. Build. Mater.* **2014**, *52*, 189–193. [CrossRef]
48. Qomi, M.J.A.; Ulm, F.-J.; Pellenq, R.J.-M. Evidence on the Dual Nature of Aluminum in the Calcium-Silicate-Hydrates Based on Atomistic Simulations. *J. Am. Ceram. Soc.* **2012**, *95*, 1128–1137. [CrossRef]

materials

MDPI

Article

Simulation of Tetrahedral Profiled Carbon Rovings for Concrete Reinforcements

Paul Penzel [1,*], Tobias Georg Lang [1], Philipp Benjamin Weigel [1], Thomas Gereke [1], Lars Hahn [1], Arthur Hilbig [2] and Chokri Cherif [1]

1 Institute of Textile Machinery and High Performance Material Technology (ITM), Technische Universität Dresden, 01069 Dresden, Germany
2 Institute of Machine Elements and Machine Design, Technische Universität Dresden, 01069 Dresden, Germany
* Correspondence: paul.penzel@tu-dresden.de

Abstract: Textile reinforcements are increasingly establishing their position in the construction industry due to their high tensile properties and corrosion resistance for concrete applications. In contrast to ribbed monolithic steel bars with a defined form-fit effect, the conventional carbon rovings' bond force is transmitted primarily by an adhesive bond (material fit) between the textile surface and the surrounding concrete matrix. As a result, relatively large bonding lengths are required to transmit bond forces, resulting in inefficient material utilization. Novel solutions such as tetrahedral profiled rovings promise significant improvements in the bonding behavior of textile reinforcements by creating an additional mechanical interlock with the concrete matrix while maintaining the high tensile properties of carbon fibers. Therefore, simulative investigations of tensile and bond behavior have been conducted to increase the transmittable bond force and bond stiffness of profiled rovings through a defined roving geometry. Geometric and material models were thus hereby developed, and tensile and pullout tests were simulated. The results of the simulations and characterizations could enable the optimization of the geometric parameters of tetrahedral profiled rovings to achieve better bond and tensile properties and provide basic principles for the simulative modeling of profiled textile reinforcements.

Keywords: carbon-reinforced concrete; bond behavior; bond simulation; profiled rovings; pullout test; 3D scan

Citation: Penzel, P.; Lang, T.G.; Weigel, P.B.; Gereke, T.; Hahn, L.; Hilbig, A.; Cherif, C. Simulation of Tetrahedral Profiled Carbon Rovings for Concrete Reinforcements. *Materials* **2023**, *16*, 2767. https://doi.org/10.3390/ma16072767

Academic Editors: Nikolina Zivaljic, Hrvoje Smoljanović and Ivan Balić

Received: 4 March 2023
Revised: 21 March 2023
Accepted: 24 March 2023
Published: 30 March 2023

1. Introduction

In recent years, the textile concrete sector has increasingly established itself as an innovation driver in both the construction and textile industries [1]. According to the forecasts of leading German research institutes, around 20% of today's reinforced concrete will be replaced by carbon concrete within the next ten years [2]. Since the start of research on textile-reinforced concrete in the mid-1990s [3], its development has been pushed forward continuously and intensively [4,5], and rapid transfer to practice has been ensured through close cooperation with companies. Interest is growing steadily and networks for textile-reinforced concrete, such as Composites United-Bau e.V. and Carbon Concrete Composite (C^3) e.V., belong to over 122 companies. The primary areas of application are plate- and shell-like constructions [6,7], so flat structures made of carbon fibers with a primarily biaxial structure are of particular relevance.

The bond between rovings and concrete is—similarly to steel-reinforced concrete—mainly based on three mechanisms: adhesion, mechanical interlock and friction [8–10].

Due to the high load-bearing behavior and small cross-sections required for textile-reinforced concrete, the bond between the textile reinforcement and concrete matrix is of great importance. In order to effectively transfer the forces between concrete and textile, a permanently rigid and non-slip bond is essential, which was severely limited using previous impregnation technologies. So far, the bonding forces have only been transmitted via material-to-material solutions based on the impregnation agents used.

These solutions have been extensively examined and further developed through numerous research projects and the limits of this technology have been worked out, for example, within the SFB 528 project [11] and the BMBF project C^3 [12]. With the latest impregnation systems, these limits have a transferable shear flow of up to 40 N/mm. However, the high impregnating agent content of around 30% by volume results in very stiff structures. In addition, the bond parameters are still insufficient and require overlapping lengths of up to 1 m. The results of national and international research groups, as well as our own preparatory work, show that the profiling of textile reinforcements is essential for a sufficient mechanical bond and load transmission between textile and concrete [12–16].

The modification of the yarn structure necessary for profiling influences yarn characteristics, particularly in terms of the stiffness and strength properties of the yarn. Due to deviation or interruption in a linear filament course, tensile properties are reduced by using common yarn profiling techniques. Moreover, for effective use in concrete structures, textile reinforcement must absorb the tensile forces to be dissipated even with expansions of around 0.1–0.2%. Known processes for generating a profiled yarn surface are cabling, twisting, conventional braiding or the Kemafil process. In all cases, the principle is the formation of a defined yarn geometry by alternately inserting, intertwining or winding the individual threads. As a result, this leads on the one hand to an improvement in the fiber structure inside the yarn and on the other hand, to an improved matrix connection on the yarn surface [17]. However, the principle-related disadvantage of all of the above methods is the associated structural expansion when a load is applied for the first time, which is often in the order of at least 1% elongation and thus represents a clear exclusion criterion for use as concrete reinforcement. In addition, without special consolidation measures, such profiled yarns usually have an insufficient dimensional stability in their cross-section. Especially in the case of core-sheath structures, such as in conventional braiding or Kemafil technology, this causes the shear structure to slide open and thus leads to a telescoping structural failure in the concrete bond, which usually occurs early on.

An alternative is a method in which, after yarn and/or textile fabric formation, contour-shaped fiber polymer composites are produced in a downstream processing step using polymeric matrix material in combination with thermal processes in the pultrusion process [18]. Strong and rigid rods are created, which achieve high bond values by the profiled geometrics, but can no longer be processed in terms of their textile technology.

For the development of carbon-reinforced concrete (CRC) structures with controllable and predictable bond behavior based on a defined form-fit effect, profiled rovings with a patented tetrahedral geometry were developed [19]. Therefore, a new profiling unit for a continuous production of tetrahedral profiled rovings was developed. The tetrahedral roving geometry is illustrated in Figure 1.

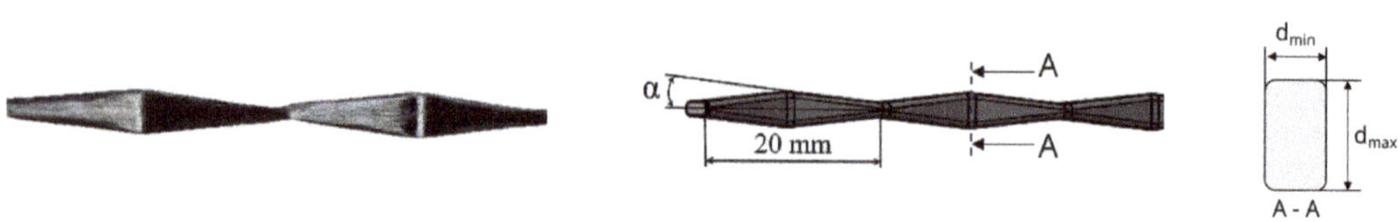

(a) (b) (c)

Figure 1. Profiled carbon roving with a tetrahedral geometry: (**a**) photography; (**b**) schematic illustration; (**c**) schematic cross-section (acc. to [20,21]).

A detailed presentation of the profiling unit as well as the general production process and the tetrahedral roving geometry is given in [20,21].

Preliminary investigations have shown that due to their profiling technology and roving geometry, tetrahedral profiled rovings transmit up to 500% higher bond forces [20,21] compared to impregnated rovings without profiling while maintaining high tensile properties. An extensive presentation of the profiling technology and roving properties is given in [20,21].

The numerical modeling and simulation of the structural–mechanical properties of textile structures based on the finite element method (FEM) enables their computer-aided design and design with significantly reduced experimental effort. However, the prediction quality of an FEM analysis of a textile structure depends heavily on the model, i.e., its geometry, the material model and the associated boundary conditions as well as the contact conditions. For this purpose, a large number of models for homogeneous textile structures has been successfully developed and validated. With macroscopic approaches, the textile is modeled as a material with homogenized properties [22–25]. Discrete approaches are based on the representation by yarns, individual fibers or filaments [26–31].

To realize composite-optimized profiled reinforcement yarns, a simulation model for the geometric and material description of tetrahedral profiled rovings was developed in this study. It enables the prediction of the structural–mechanical and composite-side properties in simulated yarn pull and concrete pullout tests. Simulation allows for the selection and further development of promising roving configurations. In contrast to profile-free rovings, tetrahedral profile rovings have an inhomogeneous structure, which must be taken into account when modeling their geometric and material properties. This paper presents simulative approaches for the generation of realistic geometric and material behavior of tetrahedral profiled rovings.

2. Materials and Methods

2.1. *Rovings with Different Configurations*

A carbon fiber heavy tow (CFHT) from Teijin Carbon Europe GmbH (Wuppertal, Germany) called Teijin Tenax-E STS 40 F13 48 K 3200 tex was selected to investigate the influence of profiling on mechanical properties. All of the different rovings in this study were produced with this CFHT. Tensile strength was determined in single yarn tensile tests according to the ISO 3341 [32,33]. The manufacturers' specifications are given in [34].

The rovings were impregnated using a polymeric dispersion on an acrylate base, i.e., TECOSIT CC 1000 from CHT Germany GmbH (Tübingen, Germany) as an impregnation agent. The properties of the impregnation agent are given in [21].

The mass content of the impregnation agent (polymer) of the impregnated rovings, $M_{impregnated}$ (in mass %), was determined by a weight comparison of dry and impregnated rovings according to

$$M_{impregnated} = \frac{m_{impregnated} - m_{dry}}{m_{impregnated}} \cdot 100\% \quad (1)$$

where m_{dry} is the mass of the dry roving and $m_{impregnated}$ is the mass of the impregnated roving (in g).

The fiber–volume content of the impregnated rovings V_{roving} (in vol. %) was determined by transforming the mass content of the impregnated rovings into a volume considering the density of single roving components as follows:

$$V_{roving} = \frac{m_{dry} \cdot \rho_{fiber} - m_{impregnated} \cdot \rho_{polymer}}{m_{impregnated}} \cdot 100\% \quad (2)$$

where ρ_{fiber} is the density of the fiber of the dry roving and $\rho_{polymer}$ is the density of the polymer of the impregnation agent (in g/cm^3).

For the development of the material and geometric model of the profiled rovings, impregnated rovings with different mass contents and, therefore, different fiber–volume contents and cross-sections, as well as tetrahedral profiled rovings with different profile configurations, were produced on a continuous working laboratory profiling unit. The characteristics of the different rovings are detailed in Table 1.

Table 1. Profile characteristics of different rovings.

Roving Configuration	Roving Geometry	Specimen Illustration	Schematic Cross-Section	Dimension of Cross-Section
Dry yarn	Band-shaped		—	Variable (no internal bond)
Impregnated roving	Circular		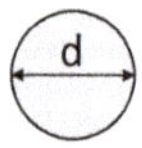	d = 1.8–2.4 mm
Tetrahedral profiled roving	Light profile	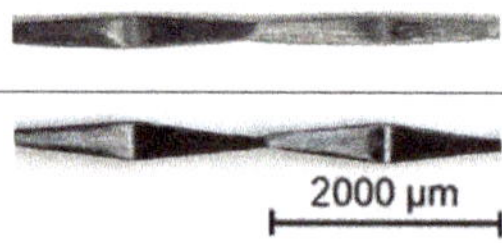	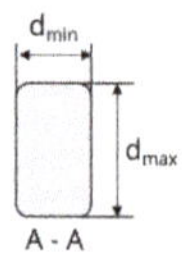	d_{diff} = 0.8 mm α = 4°
	Strong profile			d_{diff} = 1.0 mm α = 5°

The profile of the tetrahedral-shaped rovings is characterized by the difference between the minimum and maximum diameter in a profile indentation (smallest cross-section) and the angle of the filament orientation. The angle α is determined as the tangent between the distance of two neighboring profile indentations (profile spacing t in mm) of the vertical and horizontal plane (set to 10 mm) and the difference between the minimum and maximum diameter (d_{min} and d_{max} in Figure 1), d_{diff}, as follows:

$$\alpha = \arctan\left(\frac{d_{diff}}{t/2}\right) \quad (3)$$

An impregnated roving with no profile has a circular shape. Due to the different mass contents and, therefore, different fiber–volume contents, the diameters of the impregnated rovings without profiles varied between 1.8 mm and 2.4 mm.

To investigate the influence of the fiber–volume content, the consolidation parameters and profiling on tensile properties for the development of a material model, different test specimens of impregnated rovings with and without profiling were produced on the laboratory unit (see Table 2). Therefore, impregnated rovings without profiles and different fiber–volume contents were produced using different squeezing devices made of silicon-defined circular punctures (1.5 mm/2.0 mm/2.5 mm/3.0 mm). To investigate the influence of the consolidation parameters, profiled rovings with intensified consolidation were produced by reducing the production speed and therefore increasing the consolidation time using IR radiation according to the setup described in [21]. The tetrahedral profile varied (light or strong) by changing the vertical distance between the upper and lower profiling chain.

Table 2. Properties of different rovings.

Roving Configuration	Parameters			
	Roving Geometry	Impregnation Agent	Fiber–Volume Content (%)	Consolidation Time (min)
Rovings without profiles				
Dry roving	Band-shaped	–	100	–
Impregnated roving	Circular tiny	Tecosit	~78	4
	Circular small		~76	
	Circular middle		~69	
	Circular large		~44	

Table 2. *Cont.*

<table>
<tr><th rowspan="2">Roving Configuration</th><th colspan="4">Parameters</th></tr>
<tr><th>Roving Geometry</th><th>Impregnation Agent</th><th>Fiber–Volume Content (%)</th><th>Consolidation Time (min)</th></tr>
<tr><td></td><td colspan="4">Profiled rovings from laboratory unit with different profiles and impregnation agents</td></tr>
<tr><td rowspan="3">Tetrahedral profiled roving</td><td>Light profile</td><td rowspan="3">Tecosit</td><td rowspan="3">~74</td><td>4</td></tr>
<tr><td rowspan="2">Strong profile</td><td>4</td></tr>
<tr><td>10</td></tr>
</table>

2.2. *Concrete Matrix*

Fiber-based reinforcements are often embedded in cementitious matrices with a small maximum grain size, also called mortars [1,35]. For such fine concrete matrices, compressive strength and flexural tensile strength are determined according to the DIN EN 196-1 [36] after 28 days.

For the investigation of bond behavior, a fine concrete dry mix called BMK 45-220-2 was used for the pullout tests at the Institute of Construction Materials (IfB) of TU Dresden. The detailed composition and properties of the concrete are presented in [21].

2.3. *Test Program and Test Setups*

To analyze the roving geometry, the impregnated 100 mm long rovings with and without tetrahedral profiles were digitized by an optical measuring machine using the GOM ScanCobot in connection with the structured-light 3D scanner ATOS Q 12M MV100 from Carl Zeiss GOM Metrology GmbH (Braunschweig, Germany). This measurement setup allowed for the accurate and repeatable acquisition of a measurement uncertainty of 1 μm, a maximal divergence of 20 μm and a minimum measuring point distance of 29 μm per single capturing view. All capturing views were automatically registered through 0.8 mm reference points, while the rovings were surface-treated by an air-dissolving scanning spray as a matting agent to cover up reflective and transparent surface parts. For the exact evaluation of the roving geometry, the software Geomagic Wrap from 3D Systems, Inc. (Rock Hill, SC, USA) and ANSYS SpaceClaim from ANSYS, Inc. (Canonsburg, PA, USA) were used.

The tensile tests were conducted based on the DIN EN ISO 10618 [32] (see also [33,36]). The test setup and testing parameters are described in [20,21].

The characterization of the bond between textile reinforcement and concrete is possible in different ways. However, there is no standardized test method yet. Therefore, a preliminary test method suitable for single yarn pullout tests was examined, mainly to understand whether this test method was suitable for investigating profiled yarns.

Single yarn pullout (YPO) tests were conducted at the Institute of Construction Materials (IfB) at TU Dresden to analyze the characteristic bond–slip behavior of single rovings with different profile properties (see, e.g., [37]). In this type of experiment, individual rovings were embedded in cubic concrete blocks. The upper block provided an embedment length of 50 mm in the top roving section. The lower block possessed an increased embedment length of 90 mm in the bottom roving section for a defined roving fixation. The concrete cover was 40 mm. The specimens were fixed to an upper and lower specimen holder and the pullout force–slip deformation curve was measured by a single-sided pullout in the upper concrete block with a controlled quasi-static load. The pullout (slip) deformation was measured by an optical system consisting of laser sensors and aluminum clips that were fixed to the yarn. The detailed test setup is presented in [21].

2.4. *Specimen Manufacture*

For microscopic examinations, resonated short roving sections of about 10 mm were placed in cylinders of 20 mm in diameter. After one day of drying, the front side was grinded with sandpaper and finally polished.

For the yarn tensile tests, 450 mm long rovings were cast on each end in a 125 mm long epoxy resin embedment. Further details are given in [21].

Specimens for the YPO tests were made by embedding single profiled carbon rovings as well as rovings with no profile in the self-compacting fine-grained concrete BMK 45-220-2 in a cube formwork according to [20,21].

2.5. Modeling and Simulation of Tetrahedral Profiled Rovings

A geometric model and a material model were developed to simulate the tensile and bond tests of the tetrahedral profiled rovings. The challenge was the realistic representation of the complex structure of the roving, taking into account the geometric parameters and the type of impregnation and consolidation. A parametric geometric model was therefore derived from 3D scans of real profiled rovings.

Due to the profiling and changing roving cross-section with slight deviations in its linear filament orientation, an inhomogeneous material model was implemented. Therefore, a correlation between the roving area with respect to fiber–volume content and the tensile properties was derived by analyzing and testing impregnated rovings with different fiber–volume contents.

2.5.1. Geometric Model of Tetrahedral Profiled Rovings

In a first step, 3D scans of real impregnated rovings without a profile and tetrahedral profiled rovings were made. Exemplary illustrations of the different roving configurations are shown in Table 3.

Table 3. 3D scans of different rovings as the basis for the geometric models.

Roving Geometry	3D Scan
No profile	
Light profile	
Strong profile	

Thus, the strong profiled roving shows a dominant tetrahedral geometry, whereas the impregnated roving without a profile shows a circular cross-section and a smooth surface structure. In order to validate the quality of the impregnation as well as the roving cross-section and roving area, microscopic analyses of different roving sections were made. For this purpose, cross- and longitudinal sections in profile and transition segments between neighboring profile indentations were made according to Figure 2.

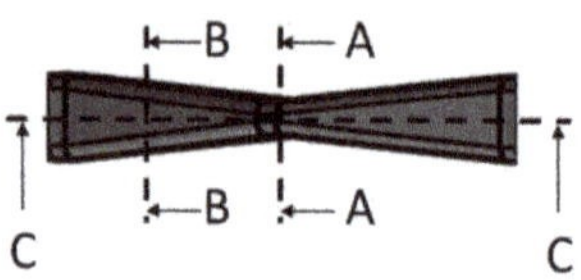

A:A—Cross-section in profile segment
B:B—Cross-section in transition segment
C:C—Longitudinal section

Figure 2. Schematic cross-sections for the microscopic analysis.

The comparison of the cross-sections of tetrahedral profiled rovings is illustrated in Table 4. There is a distinctive difference between the roving cross-sections of the profile segments of light and strong profiled rovings. The strong profiled rovings show a dominant rectangular cross-section due to the profiling, whereas the light profiled rovings show a more quadratic cross-section. The strong profiled rovings have a more significant difference between their minimum and maximum diameter (~1.0 mm), and according to Equation (3), a greater angle α (~5°). Both profiled rovings show a circular cross-section in the transition area between the neighboring profile segments. The longitudinal cross-section illustrates the transition and changing diameter along the transition between the horizontal and vertical profiles of the rovings. The different components of the microsection analysis (filament—white, polymer—light grey, embedment resin—dark grey, air void—black) are highlighted in the longitudinal sections.

Table 4. Comparison of the microscopic analysis of different cross-sections of tetrahedral profiled rovings.

Profiled Roving—Cross Section		
Profile Segment (Light Profile)	**Profile Segment (Strong Profile)**	**Transition Segment**
	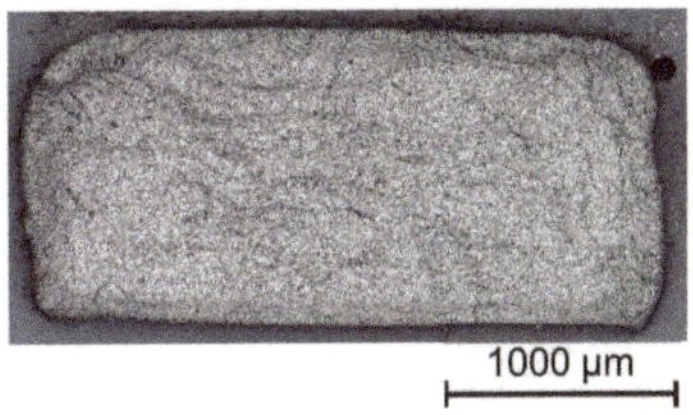	
Profiled Roving—Longitudinal Section (profile and transition segment)		
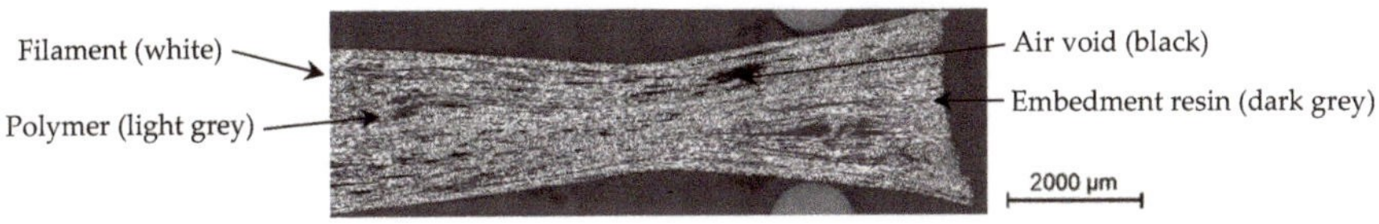		

For the purpose of modeling the yarn geometry, a parametric geometry model was determined, in which the tetrahedral geometry could be adapted by defined parameters. To approximate the geometric model parametrization by the 3D scans, a superelliptical function [38] was chosen as a base for the surface modeling of the periodic geometric model as follows:

$$\begin{pmatrix} x(s,t)\\ y(s,t)\\ z(s,t)\end{pmatrix} = \begin{pmatrix} \pm\left(\frac{a}{2}\right)\cdot\cos(t)^{\left(\frac{2}{n}\right)} \\ \pm\left(\frac{b}{2}\right)\cdot\sin(t)^{\left(\frac{2}{n}\right)} \\ s\end{pmatrix} \; for \; 0\le s\le z_{per};\; 0\le t\le\frac{\pi}{2} \tag{4}$$

Therefore, a and b describe the maximum axis dimensions of the superellipse in the x- and y-directions, respectively, while the exponent n influences the general shape. By varying the exponent n, the rectangular cross-sections of the tetrahedral profiled yarns can be reproduced as can be seen in Figure 3.

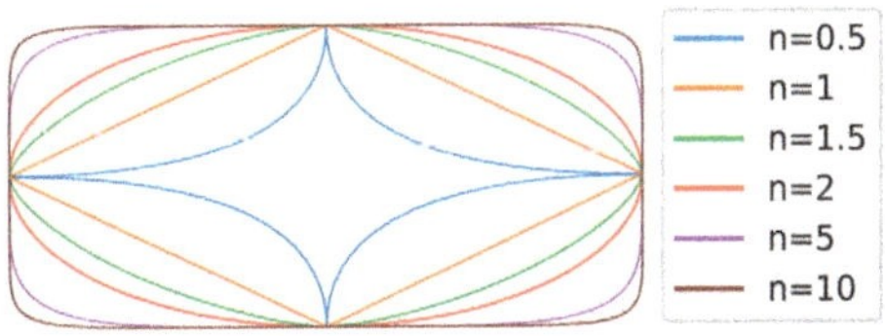

Figure 3. Superelliptical curves with varying curvatures (depicted by exponent n, aspect ratio 2).

To consider a profile progression in dimension with the variable s, trigonometrically based functions were selected for a and b accordingly:

$$\text{Profile dimension along y-axis}: \ a(s) = a_m + \sin\left(\frac{2\pi \cdot s}{a_p} + a_o\right) \cdot a_h \tag{5}$$

$$\text{Profile dimension along x-axis}: \ b(s) = b_m + \cos\left(\frac{2\pi \cdot s}{b_p} + b_o\right) \cdot a_h \tag{6}$$

Therefore, the parameter a_m describes the offset, a_h, the amplitude, a_p, the period and a_o, the phase shift of the function (without units). The parameters b_m, b_h, b_p and b_o are used analogously for the profile dimensions along the x-axis. In order to determine the variation in the profile progression over a period length, the parameters were fitted with the measured 3D scans. The yarn profiles were determined from equidistant cross-sections (distance 0.5 mm) along the yarn axis. For each cross-section, a minimum bounding box was computed to determine the yarn profile in the x- and y-directions. Equations (5) and (6) were adjusted to the consecutive profile dimensions using a non-linear least squares optimization. The exponent n was determined by adapting the superelliptical function to contours of multiple cross-sections. The optimizations were carried out using the toolbox available in scipy [39]. Finally, the geometric surface was modeled by using Equation (4) (see Figure 4a) and meshed with hexahedral elements (Figure 4b) for further use in LS-DYNA.

Figure 4. Parametric geometry models of the profiled rovings. (**a**) Parametric surface model, (**b**) Hexahedral mesh of geometry.

2.5.2. Material Model of Tetrahedral Profiled Rovings

Due to the profile-dependent and thus changing geometric and material properties along the yarn axis, an inhomogeneous material model for the profile yarns with varying material parameters along the yarn axes was created. The material parameters are based on roving tensile tests of impregnated rovings with varied yarn cross-sections and fiber–volume contents, to approximate the profile-dependent changes in each cross-section. In Table 5, the microscopic analysis of impregnated rovings with different fiber–volume contents is compared. For this purpose, the impregnated rovings were produced with different punctured silicon squeezers (squeezing gap 1.5/2.0/2.5/3.0 mm), defining the fiber–volume content of the rovings.

Due to the defined squeezing of the impregnated rovings, the fiber–volume contents of the rovings and, respectively, the roving areas varied. The results presented in Table 5 are mean values of five test specimens together with exemplary illustrations. As expected, the fiber–volume content decreased and, respectively, the roving area increased with greater squeezing gaps from approximately 78 vol. % and 2.3 mm^2 (squeezing gap 1.5 mm) to 44 vol. % and 4.1 mm^2 (3.0 mm). The impregnated rovings with 1.5 mm and 2.0 mm squeezing gaps thereby show a relatively circular cross-section with a dense fiber arrangement, whereas the impregnated rovings with a 3.0 mm gap show a deformed cross-section and relatively loose fiber arrangement with great polymeric accumulation.

Table 5. Microscopic analysis of impregnated rovings with different fiber–volume contents.

Squeezing gap: 1.5 mm Fiber–volume content: 78.2% Roving area: 2.31 mm^2	Squeezing gap: 2.0 mm Fiber–volume content: 76.2% Roving area: 2.37 mm^2
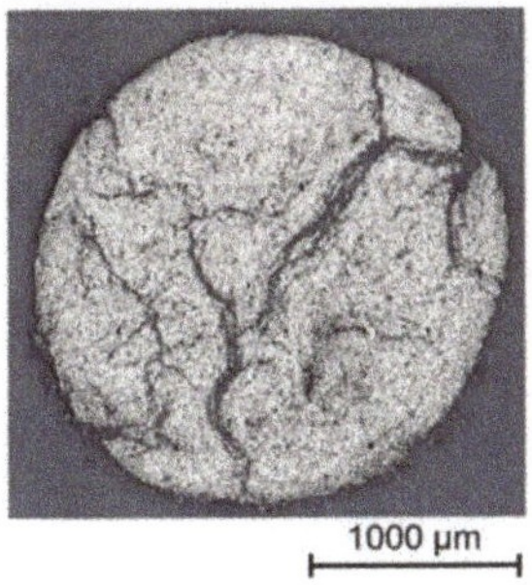	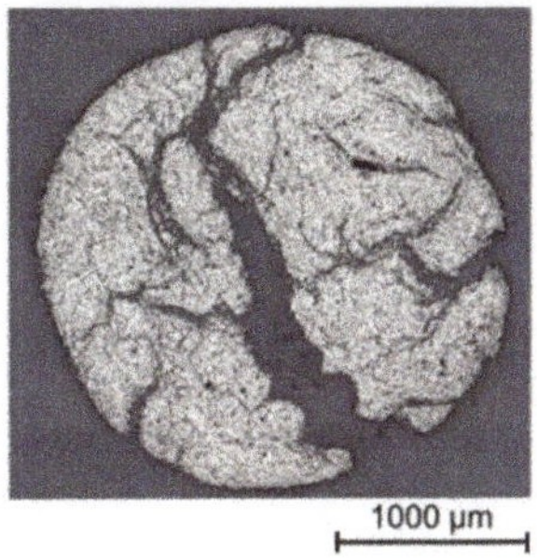
Squeezing gap: 2.5 mm Fiber–volume content: 68.6% Roving area: 2.63 mm^2	Squeezing gap: 3.0 mm Fiber–volume content: 43.8% Roving area: 4.1 mm^2
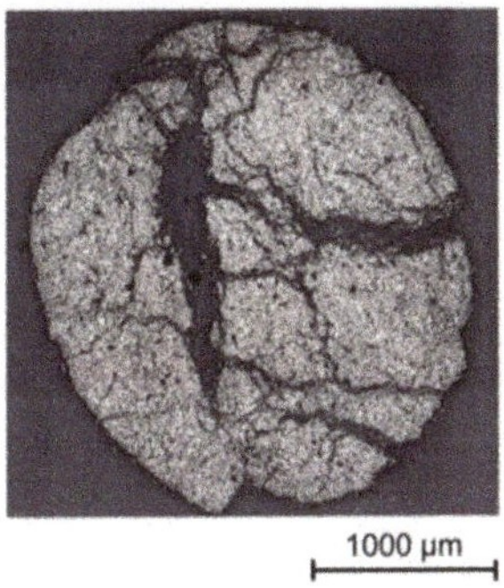	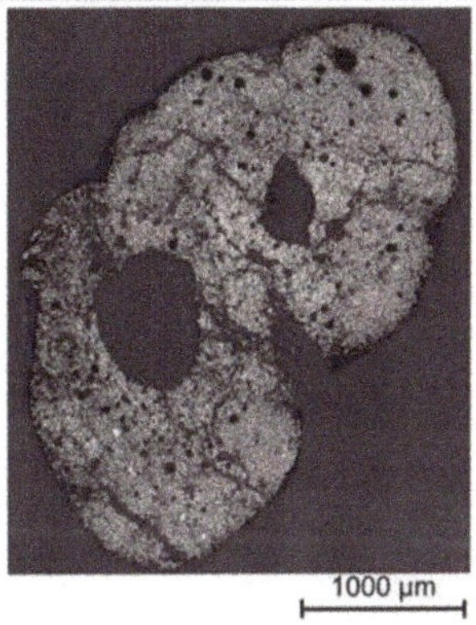

In order to derive a material model that considers the dependency of tensile properties and fiber–volume content, the tensile properties of the impregnated rovings with different fiber–volume contents were determined. Figure 5 shows the resulting correlation between Young's modulus based on the roving area and the fiber–volume content of the dry and impregnated carbon rovings (dots) with a single standard deviation (error bar) and a linear trend line (dotted line).

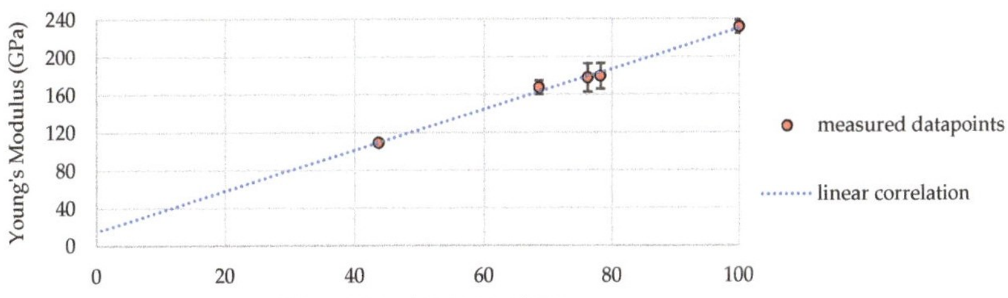

Figure 5. Tensile properties of the dependence of the fiber–volume content of impregnated rovings (3.200 tex with Tecosit as the impregnation agent) tested acc. to the DIN EN ISO 10618.

Dry rovings had the highest Young's modulus of 235 GPa. They furthermore had, at 100% fiber–volume content, a dry filament area of about 1.81 mm^2 and therefore the smallest area resulting in the highest Young's modulus based on the roving area. With decreasing fiber–volume content, the roving area increases, resulting in a reduced Young's modulus. Therefore, a linear correlation could be approximated between the data points, resulting in the following function:

$$E_{fiber-volume-content} \approx 2.15 * V_{roving} + 15.36 \quad (7)$$

This correlation was considered in the developed material model of the tetrahedral profiled rovings by cutting the profiled rovings into 0.1 mm long segments along the roving axis (see Figure 6a), determining the averaged roving area and resulting fiber–volume content in each segment and assigning each segment to an averaged fiber–volume content based on Young's modulus according to the trend line in Figure 5.

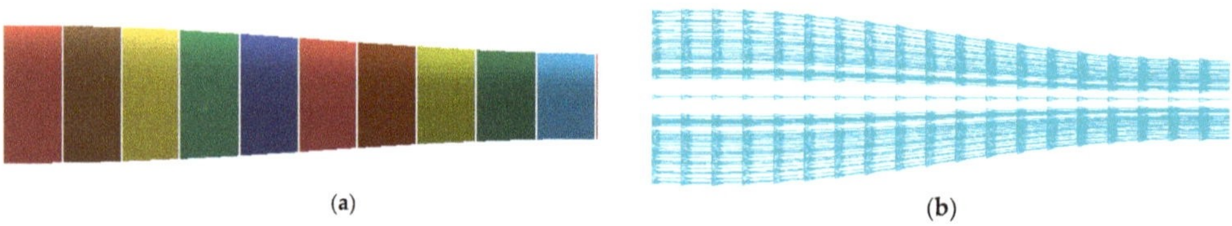

Figure 6. Detail view of yarn model with (**a**) Segmented yarn model, (**b**) Vector field for material orientation definition.

The deviation in the filaments from the linear course due to the tetrahedral geometry was taken into account by corresponding directional vectors in the material model. The required vector field was generated from the geometric description in Equation (4) (see Figure 6b). The tensile properties of each roving segment were correlated with the corresponding vectors.

The tensile tests of the tetrahedral profile rovings were simulated in LS-Dyna. In accordance with the tensile tests, the rovings were fixed at one end and a displacement in the u_z-direction was applied to the other end. Figure 7 shows the preset boundary conditions for the virtual tensile test.

Figure 7. Boundary conditions for simulated tensile test (colors acc. to roving segmentation).

2.5.3. Modeling of the Bond Behavior of Tetrahedral Profiled Rovings

The created FE models of the tetrahedral profiled rovings were transferred to a simulation model of the concrete pullout test. Figure 8a shows an example of the modeled pullout test setup.

To mesh the concrete volume, fully integrated hexahedral elements were chosen. The mesh between the roving and concrete was refined with a 3 mm mesh size (Figure 8c). Since the focus was not on modeling the concrete matrix, an existing, simplified material approach was used according to [40,41] with the material parameters presented in Table 6.

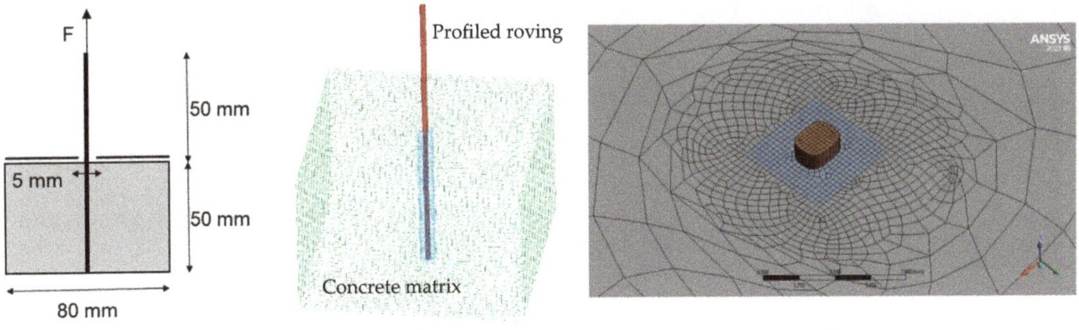

(a) (b) (c)

Figure 8. Simulated pullout test setup. (**a**) Schematic setup, (**b**) Meshed model, (**c**) Detail view of model mesh.

Table 6. Concrete material parameters for bond simulation.

Parameter	Symbol	Value	Unit
RO	$\rho_{concrete}$	2.26	g/cm^3
A0	f_c'	−101	MPa
RSIZE	–	0.03937	In/mm
UCF	–	145	Psi/MPa
MXEPS	ε_{max}	0.08	

Therefore, the parameter $\rho_{concrete}$ describes the density and f_c', the negative of the unconfined compression strength. The parameters RSIZE and UCF are needed for the conversion from imperial units to SI units. The parameter ε_{max} specifies the maximum principal strain and was added in addition to the previously described material model as a variable for element erosion.

The concrete body consists of a rectangle with an edge length of 80 mm in length and width and a height of 50 mm, in the middle of which the geometry of the profiled roving was cut out, and the profiled roving was set into the mold for the bond simulation. The contact area is the surface between the profiled roving and the negative in the concrete mold. The geometry of the profiled roving creates a form-fit-based bond. The mesh of the two surfaces is designed such that the nodes of the finite element mesh of the profiled rovings are congruent with the nodes of the concrete body mesh.

In LS-DYNA, the contact was modeled using the *CONTACT_SURFACE_TO_SURFACE keyword [42]. This model for describing the contact is a two-way contact, i.e., the contact is tested for penetration in both directions, both on the surface of the concrete body and on the surface of the profiled roving.

The model is also based on a penalty contact algorithm. For example, if a node of the profiled roving penetrated the surface of the concrete body, the penetration depth DP was calculated. A force F_N was applied to the node. This penalty force is calculated as follows:

$$F_N = k \cdot DP \tag{8}$$

Hereby, stiffness was also considered. The stiffness constant k is

$$k = \frac{f_s \cdot A_K \cdot K}{V} \tag{9}$$

where f_s is the static penalty factor, A_K is the contact area and K is the compressive modulus of the material with the lowest stiffness. Therefore, K is calculated according to

$$K = \frac{E}{3(1-2v)} \tag{10}$$

when the penalty force was applied, the node was projected back onto the surface of the concrete body. The force balance between the nodes of the profiled roving and the concrete body resulted in a force that acted in the opposite direction of the penalty force on the concrete body, which absorbed the force. In addition, the node was assigned a friction force as a product of F_N and the material dependent friction coefficient μ (without units) according to

$$F_F = \mu \cdot F_N \tag{11}$$

At this point, the friction coefficient was set to 0.5 on the bases of [43,44].

The friction force simulates the friction-based bond. However, this contact description body cannot simulate the adhesion between the profiled roving and the concrete.

Figure 9 shows the boundary conditions for the simulated pullout test.

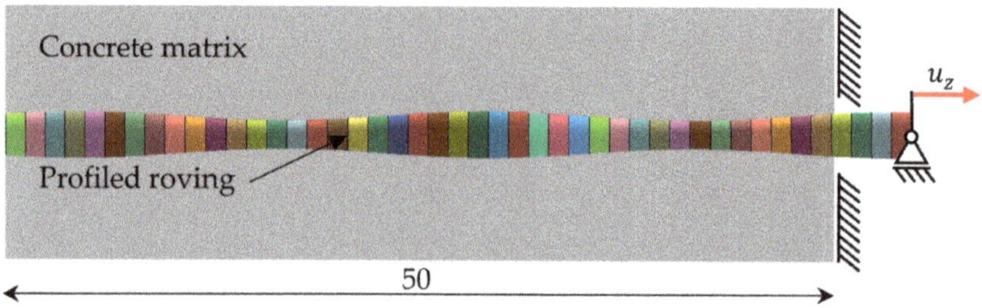

Figure 9. Boundary conditions for simulated pullout test (colors acc. to roving segmentation).

3. Results and Discussions

3.1. Geometric Validation of Tetrahedral Profiled Roving Model

By following the previously described approach, models were derived from scans of the profiled rovings. Based on the optimization, an optimal set of parameters for the Equations (5) and (6) was found. The resulting parameters are provided in Table 7.

Table 7. Fitted geometric parameters (see Equations (5) and (6)).

	a_m	a_p	a_o	a_h	b_m	b_p	b_o	b_h	n
Unit	mm	mm	1	mm	mm	mm	1	mm	1
Value	1.0563	19.9115	−3.4164	0.1995	0.8387	19.7746	−1.9632	0.3611	3.4301

In Figure 10, the fitted curves are compared to the test data by superimposing them. The graph in Figure 10a shows the curves depicting the dimension in the x- and y-directions. The local minima of each measured curve lead to an asymmetric curve trajectory, which therefore could be completely depicted by the trigonometric curves. Since the curves are in overall agreement, the fit was considered sufficient. From Figure 10b, the yarn cross-section perpendicular to the yarn axis is shown.

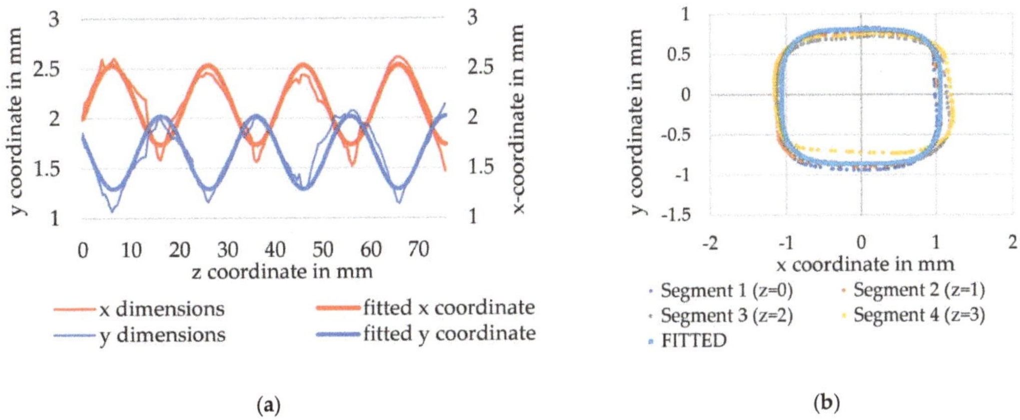

Figure 10. Comparison of derived roving geometry with experimental data. (**a**) Comparison of profile curves, (**b**) Comparison of hull curves.

3.2. Simulation of the Tensile Properties of Tetrahedral Profiled Rovings

The simulation of a yarn tension test provided the force–strain behavior and the geometric changes occurring under load. In addition, the influence of the geometric parameters was analyzed using simulations. Figure 11 shows a comparison of the measured and simulated force–strain behavior of the strong profiled rovings.

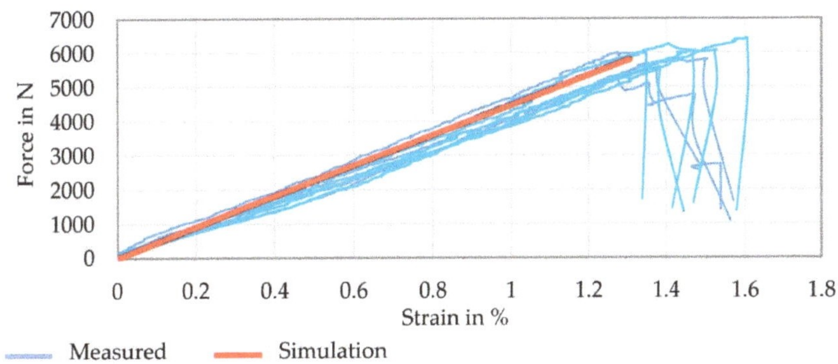

Figure 11. Comparison of the measured and simulated force–strain behavior of profiled rovings with a strong profile.

A good correspondence between the measured and simulated force–strain behavior was achieved. Furthermore, the profiled rovings showed a generally good tensile behavior in the range of the impregnated rovings without a profile (tensile strength approx. 3.100 MPa, Young's modulus approx. 210 GPa). A comparison of rovings with different profile configurations is shown in the diagrams in Figure 12.

Thus, the profiled rovings showed similar tensile properties to the impregnated rovings without a profile, indicating an even load distribution among the single filaments. The configuration of the profile itself had a minor influence, resulting in about 5% less tensile properties of the strong profiled roving in comparison to the light profiled roving. The intensity of the consolidation had an influence on the tensile properties as well. The long consolidated profiled roving showed a 5% higher tensile strength and Young's modulus compared to the profiled roving with a short consolidation.

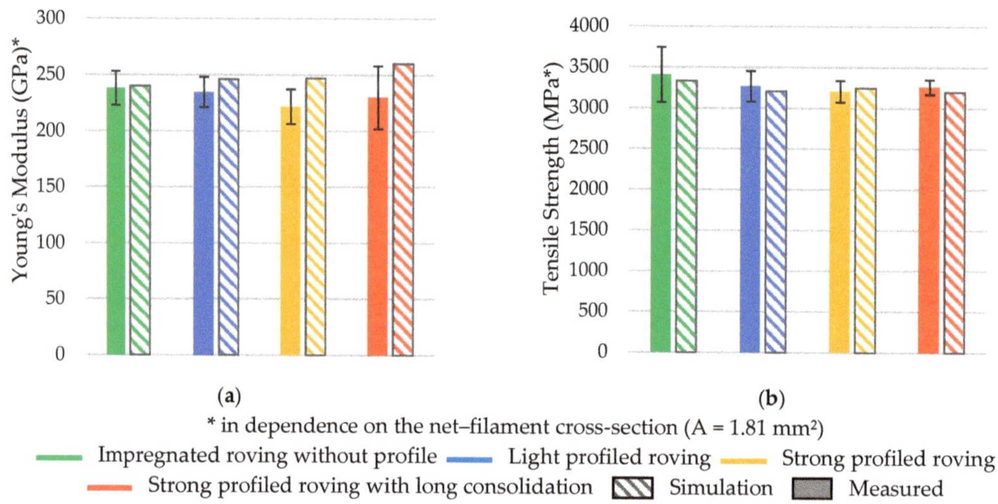

Figure 12. Measured and simulated tensile behavior of different profiled rovings. (**a**) Young's Modulus, (**b**) Tensile strength.

For the increased tensile properties, it was assumed that a more intensive consolidation would result in longer and uniform polymer chains and higher impregnation stiffness, therefore increasing the resistance of the profile against deformation under stress (also called pliability).

The simulation results significantly correlated to the real tensile tests with a 2–5% deviation range. Only Young's modulus of the profiled rovings was about 10% higher than the real tests. A possible reason for this is the linear approach (see Section 2.5.2) of the fiber–volume-based tensile properties and deviation of the fitted profile geometry, resulting in smaller roving areas and, therefore, higher tensile properties.

3.3. Simulation of the Bond Properties of Tetrahedral Profiled Rovings

With the simulated pullout test, the qualitative bond behavior of the different roving variants could be determined. The average measured (five test specimens each) and simulated bond–slip relationships of the profiled rovings with different profile configurations as well as the bond strength at 0.5 mm pullout are shown in Figure 13. The bond strength represents the specific pullout load (in N/mm) and refers to the measured bond force (in N) divided by the bond length (in mm).

It is clearly visible that the profiled rovings transmit much higher pullout loads than the impregnated roving without a profile. The strong profiled roving with a long consolidation achieved a bond strength at about 100 N/mm, an almost 500% higher bond strength than the roving without a profile. The roving with normal consolidation showed a 20% decreased bond behavior at about 80 N/mm. A possible reason for this is the mentioned resistance of the profile against deformation under a load. A higher consolidation increases resistance and allows for the transmission of higher bond forces. The light profiled roving showed a bond strength at about 55 N/mm, with 30% smaller bond properties due to the reduced form-fit effect with the surrounding concrete matrix. Nevertheless, it had 250% higher bond properties than the impregnated roving without a profile.

Furthermore, the dependence of bond properties on profile configuration shows the possibility of a predictable design of bond properties through defined profiling. All rovings showed a similar initial bond stiffness, but in correlation with the maximum bond strength, the linear rise flattened out eventually. For the restriction of the crack openings in the concrete structure, the bond strength at the first 1% pullout, which was 0.5 mm in the case of a 50 mm embedment length, was decisive for the selection and design of the textile reinforcement structure. The profiled rovings thus showed a significant increase compared to the impregnated rovings without profile (see Figure 13b).

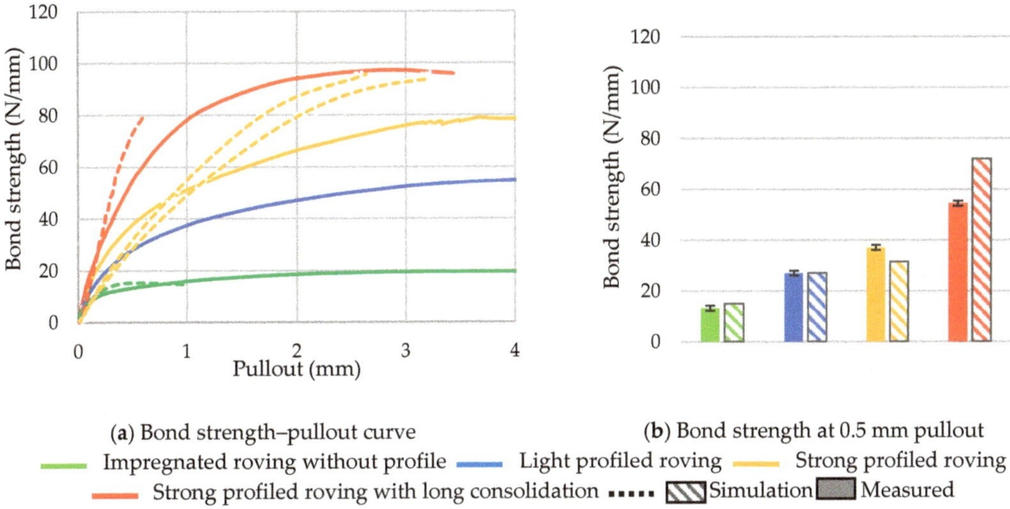

Figure 13. Measured and simulated bond behavior of different profiled rovings. (**a**) Bond-strength–pullout curve, (**b**) Bond strength at 0.5 mm pullout.

In general, the simulated bond test showed a very good correlation with the real bond test. Especially in the initial 0.1–0.8% pullout length, the simulation was very accurate (± 5%). With increasing pullout, the simulated bond test showed a slightly higher bond stiffness. Especially in the case of long consolidated profiled rovings, the bond stiffness was very high, resulting in an early failure. In the case of the profiled rovings with normal consolidation with strong and light profiles, the simulation showed a good correlation until 1.0% pullout, but showed a higher bond strength and stiffness. In the case of the impregnated rovings without profile, a very good correspondence was achieved.

Because only the initial 1.0% pullout is significant, the derived model for the simulated bond test is suitable for well-founded statements and predictions of the bond behavior of profiled rovings. Furthermore, the simulation shows the apparent form-fit effect between the profiled roving and the surrounding concrete matrix, resulting in increased bond properties. This can be clearly illustrated by an exemplary stress distribution between rovings (two profile segments) and the surrounding concrete matrix as is shown in Figure 14.

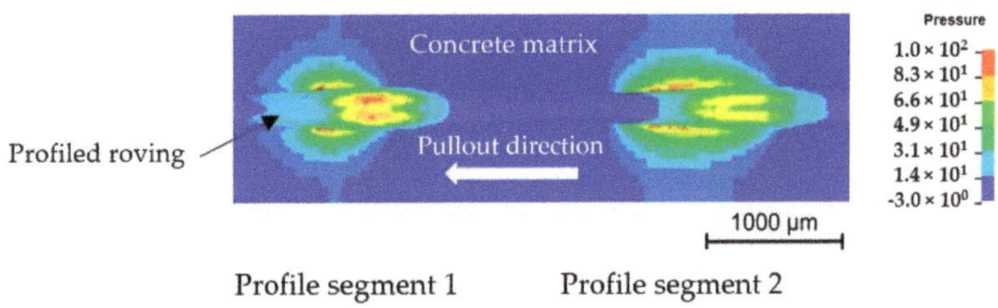

Figure 14. Simulated pressure in a concrete matrix (data in N/mm²).

With the developed FE model, pullout tests with different profile configurations were simulated in order to predict their bond behavior and derive suitable profile geometries. For example, Figure 15 shows the bond behavior of strong tetrahedral profiled rovings with a profile spacing of 10 mm and 7.5 mm between neighboring profile indentations.

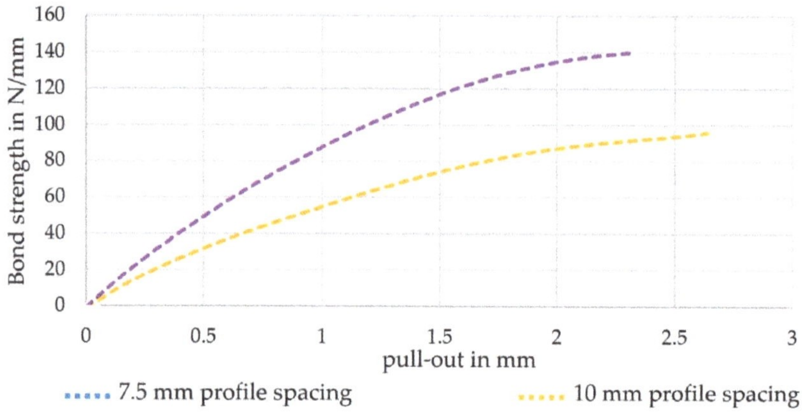

Figure 15. Simulated bond behavior of strong profiled rovings with different profile spacings.

Based on the simulated bond test, profile spacing has an evident influence on bond behavior. On the one hand, a reduction of the profile spacing from 10 mm to 7.5 mm increased the simulated bond strength from 100 N/mm to almost 140 N/mm (+40%). A major reason for this is that the 50 mm long embedment length included six instead of four profile indentations, resulting in an increased mechanical interlock. On the other hand, the tensile strength of the profiled rovings with reduced profile spacing increased by 3.800 MPa due to the mentioned correlation between the fiber–volume content and the tensile properties (see Equation (4) and Section 3.1). A smaller profile spacing results in more profile indentations and therefore higher fiber–volume content, leading to improved tensile properties. This correlation has to be validated in further studies on real profiled rovings with reduced profile spacing.

4. Conclusions

In summary, the presented tetrahedral profiled carbon rovings transmitted up to 500% higher pullout loads compared to impregnated rovings without a profile. In order to predict the tensile and bond behavior of profiled rovings and enable a requirement-based design of textile reinforcement structures, an FE model was derived. For this purpose, a new geometric and material model was developed for the tetrahedral profiled roving. Based on 3D scans and microscopic analyses, a realistic and parametric geometric model with high accuracy was achieved. Due to the profiling and changing roving cross-section with slight deviations of the linear filament orientation, an inhomogeneous material model was implemented. Therefore, correlations between the roving area and fiber–volume content, respectively, and the tensile properties, were derived.

The simulated tensile tests showed a very good correlation with the measured test results. Nevertheless, the FE model showed slightly higher tensile properties due to the overrepresentation of the high fiber–volume content in the profile indentations.

In the simulated bond test, a clear correlation with the real test could be achieved, especially in the initial and most important pullout of about 1.0%. Increased pullouts showed a slight deviation in bond simulation due to higher stiffness of the roving. Thus, the simulation was suitable for predicting the bond behavior of the profiled rovings for concrete applications.

A further parametric study of bond tests with different roving geometries enabled the prediction of the bond behavior of the profiled rovings with reduced profile spacing based on well-founded statements.

In summary, it can be stated that this developed geometric and material model of tetrahedral profiled rovings is suitable for a simulative design and allows for the prediction of their tensile and bond behavior. Furthermore, there is potential in the further development of roving geometry as well as the profiling process to maximize bond strength and bond stiffness by optimizing geometric parameters and the consolidation process. Therefore, extensive further research is planned to find an optimized roving geometry as well as a method for the targeted adjustment of strength and composite properties through the defined and variable profiling of carbon rovings in addition to a numerical description of their bond behavior.

Author Contributions: Conceptualization, T.G., L.H. and C.C.; methodology, P.P., T.G.L., P.B.W. and A.H.; software, P.P., T.G.L. and P.B.W.; validation, P.P., T.G.L. and P.B.W.; formal analysis, P.P., T.G.L. and P.B.W.; investigation, P.P., T.G.L., P.B.W. and A.H.; resources, A.H.; data curation, P.P., T.G.L., P.B.W. and A.H.; writing—original draft preparation, P.P., T.G.L., P.B.W. and A.H.; writing—review and editing, L.H., T.G. and C.C.; visualization, P.P., T.G.L., P.B.W. and A.H.; supervision, L.H., T.G. and C.C.; project administration, C.C.; funding acquisition, P.P., L.H., T.G. and C.C. All authors have read and agreed to the published version of the manuscript.

Funding: This IGF research project 21375 BR of the Forschungsvereinigung Forschungskuratorium Textil e.V. was funded through the AiF within the program for supporting the "Industriellen Gemeinschaftsforschung (IGF)" from funds from the Federal Ministry for Economic Affairs and Climate Action on the basis of a decision by the German Bundestag. "The Article Processing Charges (APCs) were funded by the joint publication fund of TU Dresden, the Medical Faculty Carl Gustav Carus and the SLUB Dresden". All funding is gratefully acknowledged.

Institutional Review Board Statement: Not applicable.

Informed Consent Statement: Not applicable.

Data Availability Statement: The data presented in this study are available upon request from the corresponding author. The data are not publicly available due to privacy concerns.

Acknowledgments: The support from the Institute of Construction Materials of TU Dresden for the specimen production and testing of bond properties is gratefully acknowledged.

Conflicts of Interest: The authors declare no conflict of interest. The funders had no role in the design of the study; in the collection, analyses or interpretation of data; in the writing of the manuscript; or in the decision to publish the results.

References

1. Williams Portal, N.; Lundgren, K.; Wallbaum, H.; Malaga, K. Sustainable Potential of Textile-Reinforced Concrete. *J. Mater. Civ. Eng.* **2015**, *27*, 4014207. [CrossRef]
2. Tietze, M.; Kahnt, A.; Schladitz, F. Marktpotenzial von Carbonbeton—Prognosen zur Substitution von Stahl durch Carbon. *7. Anwendertagung Textilbeton.* In: TUDALIT e.V. Magazin Nr. 13 for the 7. Anwendertagung Textilbeton. Dresden, **2015**, p. 27. Available online: http://tudalit.de/uploads/TUDALIT13%20scrn.pdf (accessed on 1 March 2023).
3. *Sachstandbericht zum Einsatz von Textilien im Massivbau*; Beuth: Berlin, Germany; Wien, Austria; Zürich, Switzerland, 1998; ISBN 9783410656883.
4. Mukhopadhyay, S.; Khatana, S. A review on the use of fibers in reinforced cementitious concrete. *J. Ind. Text.* **2015**, *45*, 239–264. [CrossRef]
5. Peled, A. *Textile Reinforced Concrete*, 1st ed.; Taylor & Francis Group: London, UK, 2017; ISBN 9781482282481.
6. Scholzen, A.; Chudoba, R.; Hegger, J. Thin-walled shell structures made of textile-reinforced concrete. *Struct. Concr.* **2015**, *16*, 106–114. [CrossRef]
7. Hegger, J.; Goralski, C.; Kulas, C. Schlanke Fußgängerbrücke aus Textilbeton. *Beton Stahlbetonbau* **2011**, *106*, 64–71. [CrossRef]
8. Bielak, J.; Spelter, A.; Will, N.; Claßen, M. Verankerungsverhalten textiler Bewehrungen in dünnen Betonbauteilen. *Beton Stahlbetonbau* **2018**, *113*, 515–524. [CrossRef]
9. Preinstorfer, P.; Kromoser, B.; Kollegger, J. Kategorisierung des Verbundverhaltens von Textilbeton/Categorisation of the bond behaviour of textile reinforced concrete. *Bauingenieur* **2019**, *94*, 416–424. [CrossRef]
10. Hartig, J.; Häußler-Combe, U.; Schicktanz, K. Influence of bond properties on the tensile behaviour of Textile Reinforced Concrete. *Cem. Concr. Compos.* **2008**, *30*, 898–906. [CrossRef]
11. Curbach, M. *SFB 528: Textile Bewehrungen zur Bautechnischen Verstärkung und Instandsetzung*; Technische Universität Dresden: Dresden, Germany, 2005.
12. Hahn, L.; Rittner, S.; Nuss, D.; Ashir, M.; Cherif, C. Development of Methods to Improve the Mechanical Performance of Coated Grid-Like Non-Crimp Fabrics for Construction Applications. *Fibres Text. East. Eur.* **2019**, *27*, 51–58. [CrossRef]

13. Lorenz, E.; Ortlepp, R. Bond Behavior of Textile Reinforcements—Development of a Pull-Out Test and Modeling of the Respective Bond versus Slip Relation. In *High Performance Fiber Reinforced Cement Composites 6*; Parra-Montesinos, G.J., Reinhardt, H.W., Naaman, A.E., Eds.; Springer: Dordrecht, The Netherlands, 2012; pp. 479–486. ISBN 978-94-007-2435-8.
14. Donnini, J.; Corinaldesi, V.; Nanni, A. Mechanical properties of FRCM using carbon fabrics with different coating treatments. *Compos. Part B Eng.* **2016**, *88*, 220–228. [CrossRef]
15. Younis, A.; Ebead, U. Bond characteristics of different FRCM systems. *Constr. Build. Mater.* **2018**, *175*, 610–620. [CrossRef]
16. Hahn, L. Entwicklung Einer In-Situ-Beschichtungs- und Trocknungstechnologie für Multiaxiale Gelegestrukturen Mit Hohem Leistungsvermögen. Ph.D. Thesis, Technische Universität Dresden, Dresden, Germany, 2020.
17. Kolkmann, A. Methoden zur Verbesserung des Inneren und Außeren Verbundes Technischer Garne zur Bewehrung Zementgebundener Matrices. Ph.D. Thesis, RWTH Aachen, Aachen, Germany, 2008.
18. Cherif, C.; Curbach, M.; Offermann, P. Patent DE102014102861A1—Bewehrungsgitter für Den Betonbau, Hochleistungsfilamentgarn für den Betonbau und Verfahren zu Deren Herstellung. Available online: https://patents.google.com/patent/DE102014102861A1/it (accessed on 1 March 2023).
19. Waldmann, M.; Rittner, S.; Cherif, C. *Patent DE102017107948A1 - Bewehrungsstab zum Einbringen in eine Betonmatrix sowie dessen Herstellungsverfahren, ein Bewehrungssystem aus Mehreren Bewehrungsstäben Sowie ein Betonbauteil*; Technische Universität Dresden: Dresden, Germany, 2017; Available online: https://patentimages.storage.googleapis.com/8c/e3/60/e2a8a594fd895b/DE102017107948A1.pdf (accessed on 1 March 2023).
20. Penzel, P.; May, M.; Hahn, L.; Cherif, C.; Curbach, M.; Mechtcherine, V. Tetrahedral Profiled Carbon Rovings for Concrete Reinforcements. *SSP* **2022**, *333*, 173–182. [CrossRef]
21. Penzel, P.; May, M.; Hahn, L.; Scheerer, S.; Michler, H.; Butler, M.; Waldmann, M.; Curbach, M.; Cherif, C.; Mechtcherine, V. Bond Modification of Carbon Rovings through Profiling. *Materials* **2022**, *15*, 5581. [CrossRef]
22. Gereke, T.; Döbrich, O.; Hübner, M.; Cherif, C. Experimental and computational composite textile reinforcement forming: A review. *Compos. Part A Appl. Sci. Manuf.* **2013**, *46*, 1–10. [CrossRef]
23. Charmetant, A.; Orliac, J.G.; Vidal-Sallé, E.; Boisse, P. Hyperelastic model for large deformation analyses of 3D interlock composite preforms. *Compos. Sci. Technol.* **2012**, *72*, 1352–1360. [CrossRef]
24. Mathieu, S.; Hamila, N.; Bouillon, F.; Boisse, P. Enhanced modeling of 3D composite preform deformations taking into account local fiber bending stiffness. *Compos. Sci. Technol.* **2015**, *117*, 322–333. [CrossRef]
25. Pazmino, J.; Mathieu, S.; Carvelli, V.; Boisse, P.; Lomov, S.V. Numerical modelling of forming of a non-crimp 3D orthogonal weave E-glass composite reinforcement. *Compos. Part A Appl. Sci. Manuf.* **2015**, *72*, 207–218. [CrossRef]
26. Gereke, T.; Cherif, C. A review of numerical models for 3D woven composite reinforcements. *Compos. Struct.* **2019**, *209*, 60–66. [CrossRef]
27. Stig, F.; Hallström, S. Spatial modelling of 3D-woven textiles. *Compos. Struct.* **2012**, *94*, 1495–1502. [CrossRef]
28. Naouar, N.; Vidal-Sallé, E.; Schneider, J.; Maire, E.; Boisse, P. Meso-scale FE analyses of textile composite reinforcement deformation based on X-ray computed tomography. *Compos. Struct.* **2014**, *116*, 165–176. [CrossRef]
29. Durville, D. Simulation of the mechanical behaviour of woven fabrics at the scale of fibers. *Int. J. Mater. Form.* **2010**, *3*, 1241–1251. [CrossRef]
30. Green, S.D.; Matveev, M.Y.; Long, A.C.; Ivanov, D.; Hallett, S.R. Mechanical modelling of 3D woven composites considering realistic unit cell geometry. *Compos. Struct.* **2014**, *118*, 284–293. [CrossRef]
31. Isart, N.; Mayugo, J.A.; Blanco, N.; Ripoll, L.; Solà, A.; Soler, M. Geometric model for 3D through-thickness orthogonal interlock composites. *Compos. Struct.* **2015**, *119*, 787–798. [CrossRef]
32. *ISO 10618:2004*; Carbon Fibre—Determination of Tensile Properties of Resin-Impregnated Yarn. Beuth Verlag GmbH: Berlin, Germany, 2004.
33. Wendler, J.; Hahn, L.; Farwig, K.; Nocke, A.; Scheerer, S.; Curbach, M.; Cherif, C. Entwicklung eines neuartigen Prüfverfahrens zur Untersuchung der Zugfestigkeit von Fasersträngen für textile Bewehrungsstrukturen. *Bauingenieur* **2020**, *95*, 325–334. [CrossRef]
34. Teijin Limited. Product Data Sheet Tenax™ Filament Yarn No. 1.3. Available online: https://www.teijincarbon.com/products/tenaxr-carbon-fiber/tenaxr-filament-yarn (accessed on 20 December 2022).
35. Hegger, J.; Will, N.; Curbach, M.; Jesse, F. Tragverhalten von textilbewehrtem Beton. *BUST* **2004**, *99*, 452–455. [CrossRef]
36. *DIN EN 196-1*; Methods of Testing Cement—Part 1: Determination of Strength. Beuth Verlag GmbH: Berlin, Germany, 2016.
37. Schneider, K.; Michel, A.; Liebscher, M.; Mechtcherine, V. Verbundverhalten mineralisch gebundener und polymergebundener Bewehrungsstrukturen aus Carbonfasern bei Temperaturen bis 500 °C. *Beton Stahlbetonbau* **2018**, *113*, 886–894. [CrossRef]
38. Barr, A.H. Superquadrics and Angle-Preserving Transformations. *IEEE Comput. Grap. Appl.* **1981**, *1*, 11–23. [CrossRef]
39. Virtanen, P.; Gommers, R.; Oliphant, T.E.; Haberland, M.; Reddy, T.; Cournapeau, D.; Burovski, E.; Peterson, P.; Weckesser, W.; Bright, J.; et al. SciPy 1.0: Fundamental algorithms for scientific computing in Python. *Nat. Methods* **2020**, *17*, 261–272. [CrossRef]
40. Malvar, L.J.; Simons, D. Concrete material modeling in explicit computations. In Proceedings of the Workshop on Recent Advances in Computational Structural Dynamics and High Performance Computing, USAE Waterways Experiment Station, Vicksburg, MS, USA, 24–26 April 1996.
41. Schwer, L.; Malvar, J. Simplified concrete modeling with *MAT_CONCRETE_DAMAGE_REL3. In Proceedings of the JRI LS-DYNA User Week, Nagoya, Japan, 29–30 November 2005.

42. Hallqvist, J. LS-DYNA Keyword User's Manual R13. Available online: https://www.dynamore.de/de/download/manuals/ls-dyna/ls-dyna_manual_volume_ii_r13.pdf (accessed on 20 December 2022).
43. Gao, J.; Xu, P.; Fan, L.; Terrasi, G.P. Study on Bond-Slip Behavior between Seawater Sea-Sand Concrete and Carbon Fiber-Reinforced Polymer (CFRP) Bars with Different Surface Shapes. *Polymers* **2022**, *14*, 2689. [CrossRef]
44. Yuan, J.S.; Hadi, M.N.S. Friction coefficient between FRP pultruded profiles and concrete. *Mater. Struct.* **2018**, *51*, 120. [CrossRef]

materials

MDPI

Article

Alteration of Structure and Characteristics of Concrete with Coconut Shell as a Substitution of a Part of Coarse Aggregate

Sergey A. Stel'makh [1,*], Alexey N. Beskopylny [2,*], Evgenii M. Shcherban' [3], Levon R. Mailyan [1], Besarion Meskhi [4], Alexandr A. Shilov [1], Diana El'shaeva [1], Andrei Chernil'nik [1] and Svetlana Kurilova [5]

1 Department of Unique Buildings and Constructions Engineering, Don State Technical University, 344003 Rostov-on-Don, Russia; lrm@aaanet.ru (L.R.M.); alexandr_shilov@inbox.ru (A.A.S.); diana.elshaeva@yandex.ru (D.E.); chernila_a@mail.ru (A.C.)
2 Department of Transport Systems, Faculty of Roads and Transport Systems, Don State Technical University, 344003 Rostov-on-Don, Russia
3 Department of Engineering Geology, Bases and Foundations, Don State Technical University, 344003 Rostov-on-Don, Russia; au-geen@mail.ru
4 Department of Life Safety and Environmental Protection, Faculty of Life Safety and Environmental Engineering, Don State Technical University, 344003 Rostov-on-Don, Russia; spu-02@donstu.ru
5 Department of Building Materials, Don State Technical University, 344003 Rostov-on-Don, Russia; svet.curilova@yandex.ru
* Correspondence: sergej.stelmax@mail.ru (S.A.S.); besk-an@yandex.ru (A.N.B.); Tel.: +7-86-3273-8454 (A.N.B.)

Abstract: One of the most promising ways to solve the problem of reducing the rate of depletion of natural non-renewable components of concrete is their complete or partial replacement with renewable plant counterparts that are industrial and agricultural waste. The research significance of this article lies in the determination at the micro- and macro-levels of the principles of the relationship between the composition, the process of structure formation and the formation of properties of concrete based on coconut shells (CSs), as well as the substantiation at the micro- and macro-levels of the effectiveness of such a solution from the point of view of fundamental and applied materials science. The aim of this study was to solve the problem of substantiating the feasibility of concrete consisting of a mineral cement–sand matrix and aggregate in the form of crushed CS, as well as finding a rational combination of components and studying the structure and characteristics of concrete. Test samples were manufactured with a partial substitution of natural coarse aggregate with CS in an amount from 0% to 30% in increments of 5% by volume. The following main characteristics have been studied: density, compressive strength, bending strength and prism strength. The study used regulatory testing and scanning electron microscopy. The density of concrete decreased to 9.1% with increasing the CS content to 30%. The highest values for the strength characteristics and coefficient of construction quality (CCQ) were recorded for concretes containing 5% CS: compressive strength—38.0 MPa, prism strength—28.9 MPa, bending strength—6.1 MPa and CCQ—0.01731 MPa × m^3/kg. The increase in compressive strength was 4.1%, prismatic strength—4.0%, bending strength—3.4% and CCQ—6.1% compared with concrete without CS. Increasing the CS content from 10% to 30% inevitably led to a significant drop in the strength characteristics (up to 42%) compared with concrete without CS. Analysis of the microstructure of concrete containing CS instead of part of the natural coarse aggregate revealed that the cement paste penetrates into the pores of the CS, thereby creating good adhesion of this aggregate to the cement–sand matrix.

Keywords: concrete; sustainable concrete; coconut shell; natural coarse aggregate; compressive strength

Citation: Stel'makh, S.A.; Beskopylny, A.N.; Shcherban', E.M.; Mailyan, L.R.; Meskhi, B.; Shilov, A.A.; El'shaeva, D.; Chernil'nik, A.; Kurilova, S. Alteration of Structure and Characteristics of Concrete with Coconut Shell as a Substitution of a Part of Coarse Aggregate. *Materials* **2023**, *16*, 4422. https://doi.org/10.3390/ma16124422

Academic Editors: Hrvoje Smoljanović, Ivan Balić and Nikolina Zivaljic

Received: 28 May 2023
Revised: 10 June 2023
Accepted: 13 June 2023
Published: 15 June 2023

1. Introduction

The relevance of the ongoing research is due to the environmental agenda and the economic feasibility of finding new recipes and technologies for building materials in general, and concrete in particular, that can be used in sustainable construction. The concept of sustainable development implies a full cycle of waste disposal that occurs in various spheres of human activity, enterprise and society [1–3]. Of course, growing fruits and crops are important aspects of the economies of many countries. These countries, in their economic development, reflect the need for the development of agriculture but at the same time all plant production is characterized by an accompanying problem. This problem is the accumulation of waste of plant origin in the absence or shortage of rational methods for their disposal [4,5]. This is why modern scientists direct applied as well as fundamental research to determine the fundamental nature of and, thereby, substantiate the practical expediency of the applicability of plant waste in other areas of the economy. One of the most suitable areas for the disposal of plant waste is in building materials science and the production of building materials [6–12]. At the same time, materials scientists are faced with the question of studying the fundamental nature of the processes and physicochemical processes that take place during the formation of the structure and properties of building materials using plant waste. Of particular difficulty and at the same level of interest is the determination of the most rational quantities and the degree of compatibility and effectiveness of the combination of plant wastes in the body of mineral components of other materials, including concrete [5,13–15]. Recently, the question of the inconsistency of the concept of sustainable development with this approach of using natural mineral resources, for example river sand and crushed stone, for the needs of the construction industry has been increasingly raised [16]. It is becoming promising to develop technologies that can reduce the rate of depletion of natural non-renewable components of concrete [4,17,18]. One of the most promising methods to solve this problem is the complete or partial substitution of the mineral components of concrete with natural renewable plant analogues that are industrial and agricultural waste; this solves a number of problems for these industries [14,15,19,20]. However, new renewable materials should not reduce the strength characteristics of concrete or adversely affect its durability [21].

The promising candidate for partial replacement of the mineral components of concrete is machined coconut shell (CS), which has good potential to become an alternative to coarse aggregate in concrete [22–25]. CS is a by-product of the agricultural processing of coconuts. It is part of the endocarp of the coconut, namely a strong shell that protects the seed from mechanical stress. A schematic composition of the coconut palm fruit is shown in Figure 1.

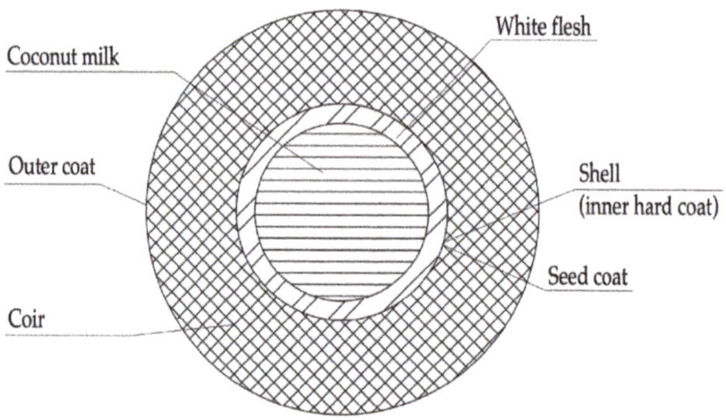

Figure 1. The structure of the fruit of the coconut palm.

After harvesting the fruit of the coconut palm and separating the fibrous part (coir) from the coconut, the CS is mechanically opened and discarded after the pulp has been removed. Both the burial and incineration of the CS waste are often used; both methods, as mentioned earlier, are harmful to the environment [5,26]. The use of CS as a partial or complete replacement for large aggregates in concrete will make it possible to utilize large volumes of this raw material practically without waste or significant damage to the environment [27,28]. The process of making coarse aggregate for concrete from CS is shown in Figure 2.

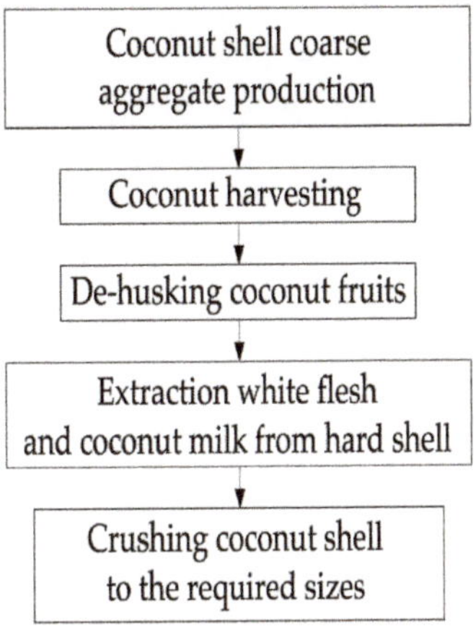

Figure 2. Manufacturing process of coarse aggregate for concrete from CS.

Therefore, CS is a natural raw material with a high degree of renewal, which makes it possible to use it in significant volumes without the risk of depleting it. Additionally, the application of CS in the composition or reinforcement of materials is presented in a number of the following works [13,29–48]. The introduction of CS as a component is studied in detail in [29,30]. It was shown in [29] that using CS as 53% of the fine aggregate was the most optimal amount and made it possible to obtain concrete with a lower thermal conductivity (0.59 W/m K) than concrete with the control composition (0.76 W/m TO). The study in [30] evaluated the permeability of concrete with the introduction of CS as a fine component. It was established that replacement of up to 50% of the fine aggregate with CS is rational and makes it possible to obtain good quality concretes with low penetrating quality (less than 2.7×10^{-11} m/s) and water absorption up to 5%. In the studies [31–35,45], the authors studied the possibility of using CS as a coarse aggregate in the technology of heavy concrete. For example, in [31], the authors found that the use of CS in concrete as a coarse aggregate in an amount of 10–20% "has a negative effect on the strength characteristics of concrete" and contributes to an increase in drying shrinkage. In one study [32], the authors estimated the strength properties of concrete with partial replacement of coarse aggregate with CS. It was found that CS concretes have a moderate permeability to chloride ions. In the works [33–35,45], the authors reported that the inclusion of CS in heavy concrete in rational quantities made it possible to produce concrete without deteriorating performance. In [13,36–40], lightweight concretes with the addition of CS were studied. In study [36], it was found that with an increase in the CS content, the density and strength of lightweight structural concrete decreased. In [37], the authors studied the properties of high-strength lightweight concretes with crushed CS as a coarse component and found that the use of this aggregate in rational amounts made it possible to obtain high-strength lightweight concretes with characteristics in terms of compressive strength, elastic modulus and tensile strength in splitting that were approximately equal to the concretes of the control composition. In [38], lightweight self-compacting concretes with 75% CS content showed good results in properties such as water absorption, sorption capacity and resistance to chloride penetration, which were comparable to the results for concretes of the control composition. The results obtained in studies [13,39] prove the possibility of using CS waste as an aggregate in the technology of lightweight structural concrete without significant deterioration in the properties of these concretes at rationally selected amounts of the CS waste and the other concrete components.

In [41–47], the behavior and properties of various reinforced concrete elements made using CS were studied. For example, in [41], the authors produced and studied two series of reinforced plates for manhole covers that were 600 × 600 × 100 mm in size. One series of covers was made on normal concrete and the second series was made on concrete with CS. Additionally, steel fibers and microsilica were added to the concrete mix. The test results showed that the performance characteristics of concrete caps with coconut filler were approximately comparable with those of conventional concrete caps, meeting the regulatory requirements and being suitable for use. In [42], the authors proved the possibility of using concrete beams containing a 5% replacement of coarse CS and a 10% replacement of part of the cement with coconut ash as an alternative to traditional reinforced concrete beams. In [43], the characteristics of plastic shrinkage and the deflection of concrete slabs with different CS contents were studied. It was found that in slabs with CS there was a decrease in plastic shrinkage cracking; however, at the same time there were deformability increases compared with slabs made of ordinary concrete.

Summarizing the literature and analyzing the references devoted to research on the use of coconut shell in concrete, specific research scientific problems can be identified. One of the main problems is the lack of clearly formulated mechanisms and explanations of the physical and chemical processes that occur during the formation of the structure of an organo-mineral composite, including a cement–sand matrix and a coconut shell aggregate, from the fundamental and applied point of view. Such a generalizing experience will be extremely useful for forming a new scientific and technical basis for future research, as well as for applied developments carried out by engineers and concrete factories, especially in interested countries with a large amount of coconut waste.

The above brief review shows that the use of coconut shell as a replacement for part of the coarse filler is an interesting and relevant scientific problem. However, there is a gap in this area. The processes of the formation of the structure of concrete containing coconut shells has not been studied enough. The strength characteristics of concrete and the optimal compositions of the concrete mix have not been quantitatively evaluated. Thus, the scientific novelty of the study lies in the establishment of new dependencies and relationships of a fundamental and applied nature that occur during the formation of the structure and properties of an artificial organo-mineral conglomerate—a coconut shell in the body of a cement concrete matrix. These new dependencies and relationships are the chains we have obtained:

- "the composition of the conglomerate—the microstructure of the boundaries of the phases of the conglomerate". The macrostructure of the conglomerate—the properties of concrete. The applied methodological and phenomenological approaches made it possible to provide a high degree of verification of the results. These approaches are primarily characterized by the observance of the fundamental principle of materials science "composition—structure—properties", as well as the determination of the most characteristic, relative indicator for non-standard concretes—the coefficient of constructive quality.

In this study, the scientific hypothesis is the possibility of obtaining high-quality building concrete consisting of a mineral cement–sand matrix and aggregate in the form of plant waste, namely crushed CS. The aim of the study is to solve the problem of substantiating the feasibility of such concrete, as well as to find a rational combination of components and study the structure and characteristics of concrete with CS as a replacement for coarse aggregate. Thus, two research problems are solved. Fundamental—study of the mechanisms of structure formation and the formation of properties of concrete using CS. Applied—searching for a rational combination of components and identifying the influence of various recipes and technological factors on the structure and properties of concretes based on CS and, thereby, finding the optimal value of the ratio of concrete strength to its density when using the maximum possible amount of plant waste (in the form of CS) as an aggregate in concrete.

The practical significance of the study thus lies in obtaining an important performance indicator of the proposed concrete, which can be used in the calculations and design of lightweight coco-concrete eco-structures.

2. Materials and Methods

2.1. Materials

Portland cement CEM I 42.5N (CEMROS, Stary Oskol, Russia), which does not contain additives, was used as a binder for the manufacture of prototypes. The main characteristics and mineralogical composition of the Portland cement used in the study are presented in Tables 1 and 2.

Table 1. Physical and mechanical characteristics of Portland cement.

Property	Value
Specific surface area (m^2/kg)	338
Soundness (mm)	0.5
Fineness, passage through a sieve No 008 (%)	98.1
Setting times (min)	
-start	165
-end	230
Compressive strength (MPa):	
-2 days	18.5
-28 days	49.2

Table 2. Mineralogical composition of Portland cement.

Mineral	Content (%)
C_3S (alite)	66
C_2S (belite)	14
C_3A (tricalcium aluminate)	8
C_4AF (tetracalcium aluminoferrite)	12

Table 2 data provided by cement manufacturer. Fine aggregate is represented by sand (LLC "DON-RESURS", Kagalnik, Russia). The characteristics of the fine aggregate are shown in Table 3.

Table 3. Grain composition and physical characteristics of sand.

Residues on Sieves (%)	Sieve Diameter (mm)						Fineness Modulus
	2.5	1.25	0.63	0.315	0.16	<0.16	
Partial	1.5	2.0	10.5	50.5	34.0	1.5	1.82
Total	1.5	3.5	14.0	64.5	98.5		
Bulk density (kg/m^3)					1464		
The content of dust and clay particles (%)					0.3		
Content of clay in lumps (%)					0.1		
Organic and contaminant content					No		

Crushed sandstone (RostMed, Kamensk, Russia) and CS (Auriki Gardens, Yaroslavl, Russia) were used as natural coarse aggregates. The shell was stored in a dry, ventilated room at an air temperature of 25 °C and a humidity of 65% for at least 48 h before use. The characteristics of crushed sandstone are presented in Table 4 and those of CS in Table 5.

Table 4. Characteristics of crushed sandstone.

Indicator Title	Actual Value
Particle size (mm)	5–10
Bulk density (kg/m^3)	1397
Apparent density (kg/m^3)	2548
Resistance to fragmentation (wt.%)	12.6
The content of lamellar and acicular grains (wt.%)	9.1
Voids (%)	45

Table 5. Characteristics of CS filler.

Indicator Title	Actual Value
Particle size (mm)	5–10
Moisture content (%)	4.6
Water absorption (%)	26.1
Bulk density (kg/m^3)	592
Apparent density (kg/m^3)	1082

A general view of the CS used to make the samples is shown in Figure 3.

Figure 3. Appearance of the applied CS.

The superplasticizer Poliplast-SP1 (LLC Poliplast-South, Krasnodar, Russia) was used as a plasticizing additive. This additive is a dark brown aqueous solution with a density of 1.17 g/cm^3 and pH 8 $\pm$ 1.

2.2. Methods

The designs of the concrete mixture of the control composition and the mixtures with different CS contents are compiled in Table 6.

Table 6. Concrete Mix Design.

Composition Type	Concrete Mixture Proportion per 1 m^3							
	Portland Cement (kg/m^3)	Water (L/m^3)	Crushed Stone (kg/m^3)	Coconut Shell (kg/m^3)	Sand (kg/m^3)	Poly-Plast-SP1 (%)	Density (kg/m^3) [49]	Slump (cm) [50]
0 CS	340	195	1002	0	690	0	2227	8.4
5 CS	340	195	952	21	690	0	2198	8.3
10 CS	340	195	901	43	690	0.5	2169	7.9
15 CS	340	195	851	64	690	1.0	2140	7.7
20 CS	340	195	801	85	690	1.0	2111	7.6
25 CS	340	195	751	106	690	1.0	2082	6.8
30 CS	340	195	701	128	690	1.5	2053	6.7

The choice of the proportions of the concrete mixtures (water–cement ratio, percentage of CS and other parameters) was based on the concrete compositions already selected by us in previous works [14,15] and the works of other authors [33,34,46], with their adjustment related to the properties of the components used [51].

With the partial replacement of natural coarse aggregate with coconut shell, the water demand of the concrete mixture increases due to the greater porosity of the structure of the organic component [52]. Accordingly, the slump of the mixture decreases with increasing coconut shell content. Therefore, to maintain the same workability for all mixtures, a superplasticizer was used in the study.

Freshly made sample cubes and prisms of concrete are shown in Figure 4.

(**a**) (**b**)

Figure 4. Freshly made concrete samples: (**a**) cubes; (**b**) prisms.

The manufacture of concrete mixtures and laboratory samples was developed in the following sequence: measurement of the quantities of the raw materials and preliminary mixing of these components in dry mold; the inclusion of mixing water with a plasticizing additive; mixing the mixture until homogeneous; laying the mixture in molds; vibrating molds for 60 s. The finished samples were kept for 1 day and removed from the molds. Then, these samples were kept for 27 days in a normal hardening chamber at a temperature of 20 ± 2 °C and a relative air humidity of 95% [53].

For the manufacture of concrete mixtures and concrete samples, the following technological equipment was used:

- Laboratory concrete mixer BL-10 (ZZBO, Zlatoust, Russia);
- Laboratory scales HT-5000 (NPP Gosmetr, St. Petersburg, Russia);
- Cube shape 2FK-100 and beam shape FB-400 (RNPO RusPribor, St. Petersburg, Russia);
- Normal hardening chamber KNT-1 (RNPO RusPribor, St. Petersburg, Russia).

The program of experimental research is shown in Figure 5.

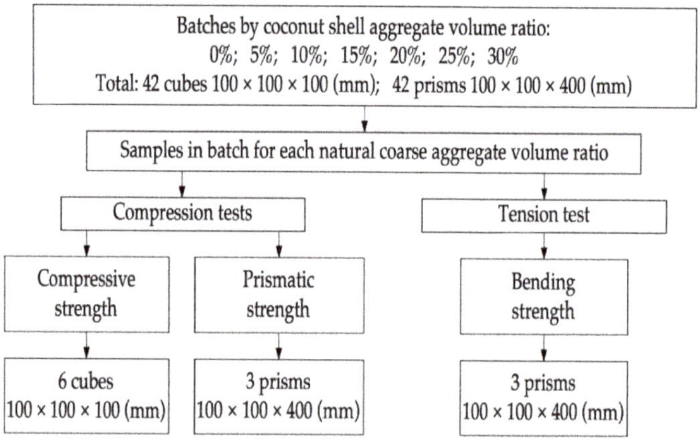

Figure 5. Mechanical test program.

The assessments of the compressive strength and bending strength (four-point) were produced in accordance with document [53], the requirements of which correspond to the main regulations regarding the manufacture and testing of concrete samples given in the following European regional standards [54–58].

Prism strength was determined in accordance with document [59], corresponding to the main regulations given in ASTM C1314-23 "Standard Test Method for Compressive Strength of Masonry Prisms".

The compressive strength was calculated by Formula (1):

$$R = \alpha \frac{F}{A} \tag{1}$$

where F is the breaking load (N), A is the area of the working section of the sample (mm) and α is the scale factor (for cubes with a rib size of 100 mm it is equal to 0.95).

The process of determining the compressive strength, bending strength and prismatic strength of the samples is presented in Figures 6–8, respectively.

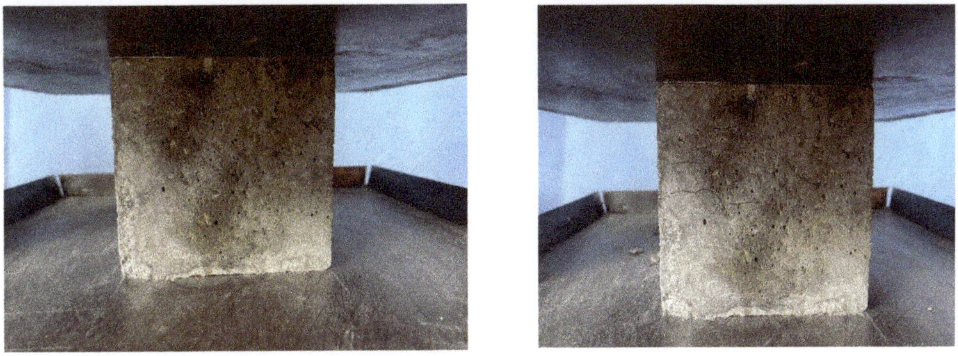

(a) (b)

Figure 6. Compressive strength testing of concrete specimens: (**a**) specimen before failure; (**b**) sample at the time of destruction.

Figure 7. Four-point bend test setup.

(a) (b)

Figure 8. The process of testing concrete specimens for prismatic strength: (**a**) sample before failure; (**b**) sample at the time of destruction.

The bending strength was calculated by Formula (2):

$$R_{tb} = \delta \frac{Fl}{ab^2} \tag{2}$$

where l is the distance between the supports (mm), a and b are the width and height of the cross section of the prism, respectively, and δ is the scale factor (for cubes with a rib size of 100 mm it is equal to 0.92).

Prism strength was calculated by Formula (3):

$$R_{pr} = \frac{F}{A} \tag{3}$$

The density of hardened concrete was determined according to [60]; the main provisions of which are in accordance with EN 12390-7:2019 "Testing hardened concrete—Part 7: Density of hardened concrete".

The density was calculated using Formula (4):

$$\rho_w = 1000 \frac{m}{V} \tag{4}$$

where m is the mass of the sample (g) and V is the sample volume (cm^3).

The coefficient of constructive quality (CCQ_R) determined by compressive strength R was calculated by Formula (5):

$$CCQ_R = \frac{R}{\rho} \tag{5}$$

where R is compressive strength (MPa) and ρ is the density of hardened concrete (g/cm^3) [61,62].

3. Results and Discussion

3.1. Study of the Physical and Mechanical Characteristics of Concrete

The effects of the partial replacement of natural coarse aggregate with organic CS aggregate on concrete density (ρ), compressive strength (R), prismatic strength (R_{pr}) and flexural strength (R_{tb}) are shown in Figures 9–12, respectively.

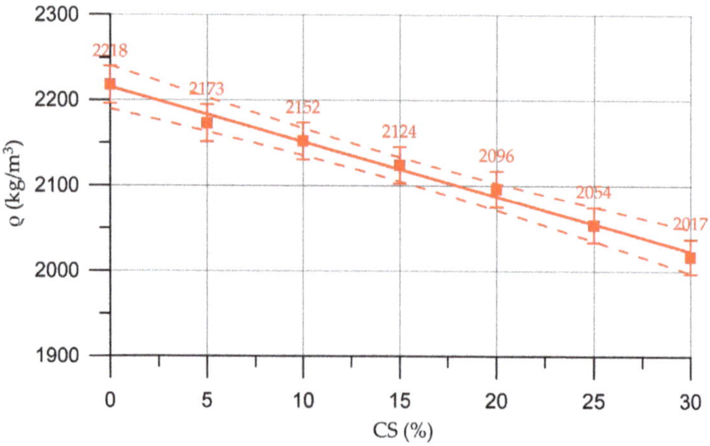

Figure 9. Relationship of concrete density with the proportion of CS.

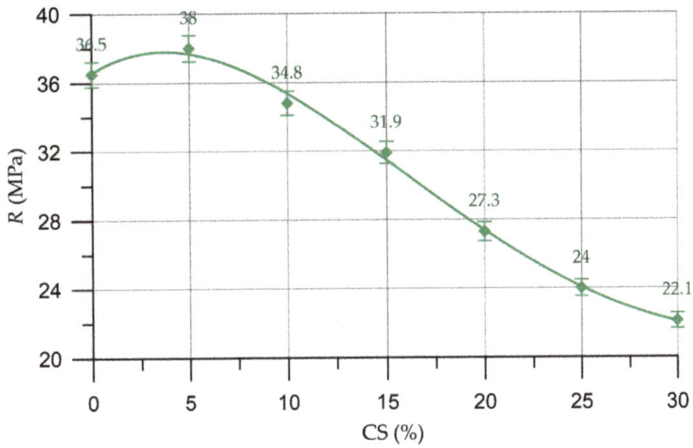

Figure 10. Relationship of concrete compressive strength with the proportion of CS.

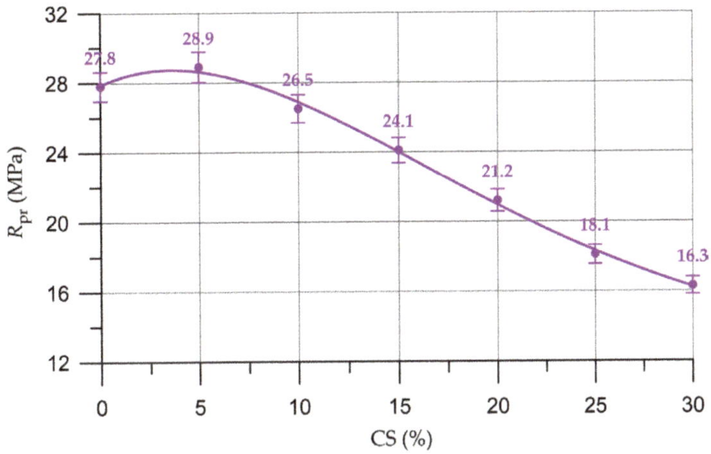

Figure 11. Relationship of the prism strength of concrete with the proportion of CS.

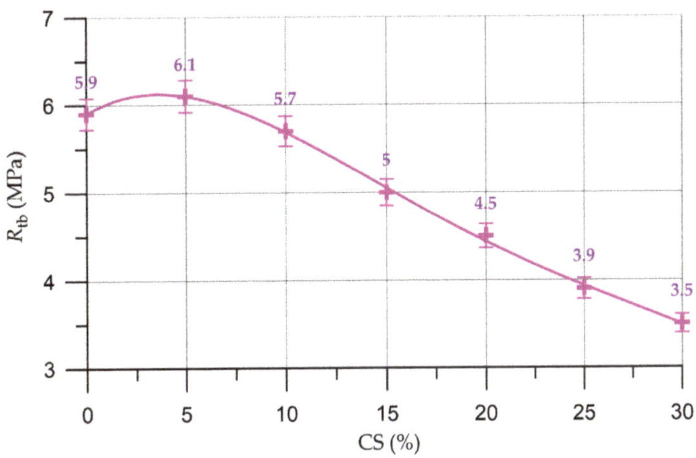

Figure 12. Relationship of concrete flexural strength with the proportion of CS.

Alterations in the characteristics of concrete depending on the proportion of CS introduced to substitute a part of the natural coarse aggregate are shown in Table 7 and are presented as a percentage compared with the control composition.

Table 7. Change in the characteristics of concrete (Δ) in %, depending on the amount of CS replacing natural coarse aggregate in %.

Characteristics of Concrete	**Δ in % with Coarse Aggregate from Coconut Shell; % by Volume of Coarse Aggregate**						
	0	**5**	**10**	**15**	**20**	**25**	**30**
Density (kg/m^3)	0	−2.0	−3.0	−4.2	−5.5	−7.4	−9.1
Compressive strength (MPa)	0	4.1	−4.7	−12.6	−25.2	−34.2	−39.5
Prism strength (MPa)	0	4.0	−4.7	−13.3	−23.7	−34.9	−41.4
Bending strength (MPa)	0	3.4	−3.4	−15.3	−23.7	−33.9	−40.7

The dependencies shown in Figures 9–12 are approximated by polynomial Functions (6)–(9) of the following form:

$$\rho = -6.407\, x + 2215, \quad R_d^2 = 0.991 \tag{6}$$

$$R = 36.570 + 0.7053\, x - 0.112\, x^2 + 0.00328\, x^3 - 2.848 \times 10^{-5}\, x^4, \; R_d^2 = 0.997 \tag{7}$$

$$R_{pr} = 27.87 + 0.5125\, x - 0.0846\, x^2 + 0.00262\, x^3 - 2.667 \times 10^{-5}\, x^4, \; R_d^2 = 0.997 \tag{8}$$

$$R_{tb} = 5.902 + 0.1324\, x - 0.02248\, x^2 + 0.000804\, x^3 - 9.697 \times 10^{-6}\, x^4, \; R_d^2 = 0.998 \tag{9}$$

Equations (6)–(9) show the high level of the determination coefficient R_d^2. The evaluation of the coefficients of the regression equations in the form of Equations (6)–(10) was carried out using the least squares method. The assessment of the significance of the regression equation as a whole was carried out using the Fisher test with the F-criterion at a significance level of $\alpha = 0.05$. In this case, the null hypothesis (H_0) is put forward: the regression coefficient is zero (b = 0); therefore, the factor x does not affect the result y and the regression line is parallel to the x-axis. For all equations, the regression coefficients are significant with a high degree of confidence.

From Figure 9 and Table 7, it can be seen that with an increase in the percentage of CS as a replacement for coarse aggregate, the density of the concrete decreases. The dependence of the change in concrete density on the percentage of CS content is almost linear.

By analyzing the results presented in Figures 10–12 and Table 7, it was established that the highest compressive strength, prismatic strength and bending strength were determined for composition of concrete with a CS content of 5% (which was used instead of a part of the natural coarse aggregate). In comparison with the control composition, the enlargement in the compressive strength was 4.1%, the prismatic strength grew by 4.0% and the bending strength increased by 3.4%. This slight increase in strength properties takes place on the basis of the natural properties of CS. Due to its rough and porous surface, the CS introduced as a replacement for part of the natural aggregate has a fairly high degree of adhesion to the mortar part and acts as an additional sealing component in the cement–sand mortar coarse aggregate. During the production of the concrete mixture, the CS absorbs a small amount of water; then, in the process of the hardening of the cement composite, this water is used as a reserve and the hydration process is more intensive at the phase boundaries [52]. Almost all cement particles enter into a hydration reaction and additional crystallization centers are formed, which, in turn, increase the adhesion strength of CS with the components of the concrete mixture. This is also in good agreement with other studies [33,34,52,63,64]. However, this effect is not so significant, and further increases in the CS content of concrete have no effect and the strength characteristics decrease, which is confirmed by the results of our experimental studies and in [33,63]. With the introduction of 10% CS, the drop in strength characteristics is not so significant. The compressive strength decreased by 4.7%, the prism strength decreased by 4.7% and flexural strength decreased by 3.4%. With the introduction of 15–30% CS significant losses in strength characteristics are observed. Thus, the introduction of more than 10% CS instead of a part of the natural coarse aggregate already leads to significant strength losses from 13 to 42%. The drop in the strength characteristics of concrete with the introduction of more than 5% CS is associated with the characteristics of this material. CS has a significantly lower strength than natural coarse aggregate and is also a lighter material. Therefore, the use of CS in the technology of heavy concrete will inevitably lead to a decrease in the strength characteristics [31–35,46,63,64].

Analyzing the obtained dependences of the strength of concrete on the content of coconut shell in the concrete mixture, the following should be noted. The change in the strength is largely due to the rate of water absorption of the coconut shell and the physical and mechanical characteristics of the shell. Coconut shell as a whole has average characteristics that have been determined experimentally and theoretically; at the same time, there is some difference in the characteristics of specific types of nuts and shells. Thus, a factor may arise that should be learned to manage and regulate it. Thus, the conducted experiment is also useful from the point of view of the analysis of the shell itself as a component of concrete. A very important factor in controlling the strength of concrete is to control the characteristics of the shell from the point of view of preventing increased water absorption of the coconut shell, which can lead to an increase in the water demand of the concrete mixture and, in the case of an excessively low water demand for the coconut shell, there may be a risk of surface underwetting and hence problems for the adhesion of the shell to the composite matrix. Thus, this factor is one of the main ones and deserves special attention.

However, due to its lower density, CS leads to a decrease in the weight of concrete, which contributes to an increase in the CCQ of concrete. The values of the structural quality factors for concrete with different CS contents are shown in Figure 13.

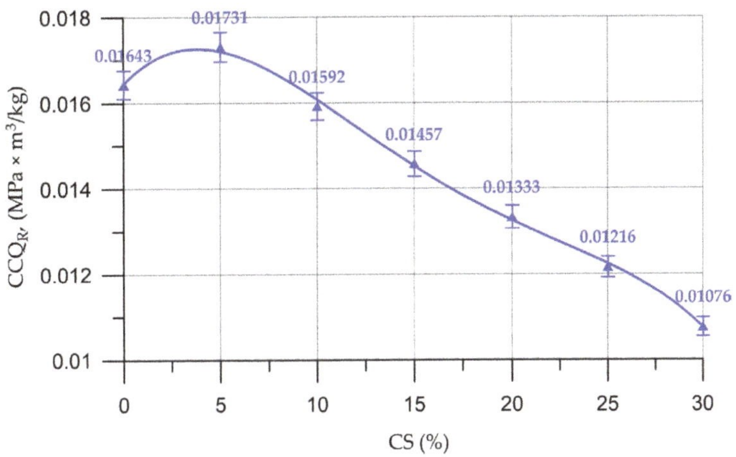

Figure 13. Change in the CCQ of concrete depending on the content of CS.

The dependency of the CCQ shown in Figure 13 is approximated by polynomial Function (10) of the following form:

$$CCQ_R = 0.0164 + 0.000468\,x - 7.97 \times 10^{-5}x^2 + 3.39 \times 10^{-6}\,x^3 - 4.885 \times 10^{-8}\,x^4,\ R_d^2 = 0.998 \quad (10)$$

From Figure 13, it follows that the best value for the coefficient of structural quality is also concrete with a CS content of 5% (up to 6% more than that of the control composition concrete). The CCQ values of the control composition concrete and concrete with 10% CS are approximately equal, which confirms the range of effective percentages of CS obtained in the study of the strength characteristics of concrete.

After analyzing the test results, we can say that the use of up to 10% CS instead of a part of natural coarse aggregate is justified and does not lead to significant loss of strength characteristics. The value of the compressive strength when replacing natural coarse aggregate with 5% CS was 38.0 MPa, it was 34.8 MPa at 10% CS, for prismatic strength—28.9 MPa at 5% CS and 26.5 MPa at 10% CS and for bending strength—6.1 MPa at 5% CS and 5.7 MPa at 10% CS.

Above 10% CS, the compressive, flexural and prismatic strengths begin to decrease significantly. When the content of CS is in a percentage from 15% to 30%, the loss of compressive strength ranged from 13% to 40%, the loss of prismatic strength ranged from 13% to 42% and the loss of bending strength ranged from 15% to 41%.

3.2. *Analysis of the Microstructure of Concrete Samples*

Figures 14 and 15 show images of the microstructure of concrete samples with 5% CS content. Figure 14 shows sections of the phase boundary "natural coarse aggregate—cement-sand matrix", and Figure 15 shows sections of the interface "aggregate of organic origin—cement-sand matrix".

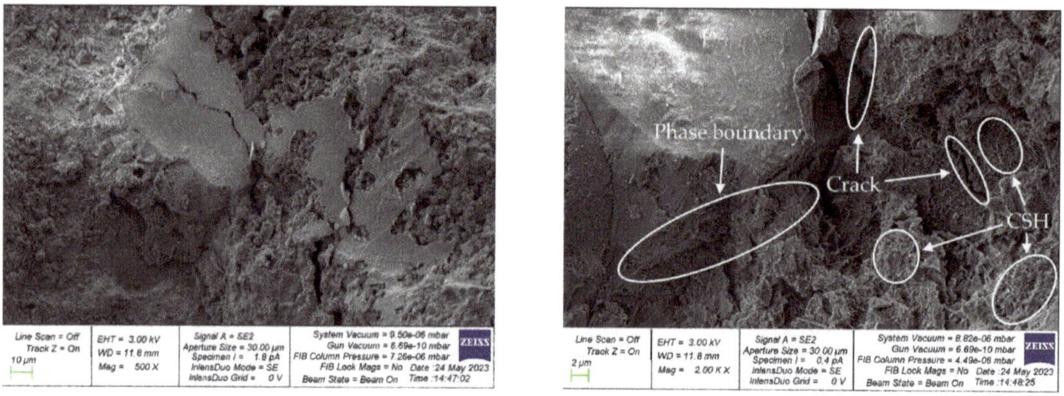

(a) (b)

Figure 14. Photographs of the microstructure of the sample illustrating the phase boundary "natural coarse aggregate—cement-sand matrix": (**a**) 500×; (**b**) 2000×.

At the phase boundary "natural coarse aggregate—cement-sand matrix" shown in Figure 14, there are areas of accumulation of CSH and microcracks. A distinctive feature is that the cement–sand matrix envelops the grain of natural coarse aggregate (Figure 14b). In Figure 15, which illustrates the section of the phase boundary "aggregate of organic origin—cement-sand matrix", one can also see accumulations of calcium hydrosilicates (CSH) and microcracks. Figure 15b clearly illustrates the porous structure of the CS, into which the cement paste penetrates, thereby creating the good adhesion of this aggregate into the cement–sand matrix. Due to the rough and porous surface, CS particles have a high coefficient of adhesion to the cement–sand matrix of concrete, which significantly compacts the material at the interface "aggregate of organic origin—cement-sand matrix". In addition, due to the absorption of a small amount of water by CS particles during the hardening of the composite, the hydration at the phase boundaries proceeds more intensively. These two factors allow, at certain small percentages of such an organic aggregate, the maintenance of the strength characteristics of concrete at approximately the same level as ordinary concrete by compensating for the lower strength characteristics of coconut shell compared with natural coarse aggregate. The drop in the strength characteristics of concrete with the introduction of more than 5% CS as a replacement for part of the natural coarse aggregate is associated with the characteristics of the CS. CS has a significantly lower strength compared with natural coarse aggregate and is also a lighter material, which, at high shell amounts (more than 10%), leads to a significant decrease in the strength of the concrete [31–35,46,63,64]. This makes it possible to involve almost all cement particles in this process and create additional crystallization centers that increase the adhesion strength of CS particles with the cement–sand matrix of concrete, which is in good agreement with the studies reported in [33,34,63,64].

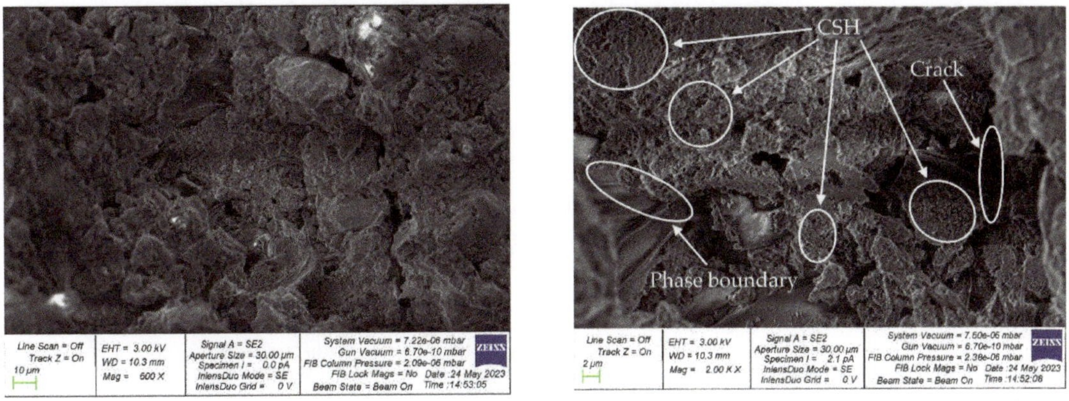

(a) (b)

Figure 15. Photographs of the microstructure of the sample illustrating the phase boundary "aggregate of organic origin—cement-sand matrix": (**a**) 500×; (**b**) 2000×.

Summing up the results of experimental studies, it is necessary to discuss the results obtained. As mentioned above, when establishing rational recipe technological factors and identifying the optimal amounts, it is possible to obtain high-quality concrete using CS as an aggregate. This confirms our working hypothesis that was put forward above; the confirmation of this hypothesis is in good agreement with the works of other authors, namely [33,34,63,64]. In this study, the optimal percentage for replacing a part of the coarse aggregate with CS is considered to be an amount of up to 10%. For example, in [33,63], an amount of CS up to 10% was also optimal. This makes it possible to obtain concrete of the required quality without a significant loss of strength. In [34], the authors managed to obtain concrete without a deterioration in performance with 15% of CS introduced instead of part of the coarse aggregate. Additionally, in [64], based on the results of experimental studies, it was established that it is possible to use up to 20% of CS instead of part of the natural coarse aggregate without a significant deterioration in strength characteristics. In our study, the increase in the obtained characteristics was 4.1% for compressive strength, the density decreased to 9.1% and the CCQ increased by 6.1% compared with the control sample. In [19,20,42], the compressive strength decreased to 25% and the density decreased to 7%; the indicator of CCQ also decreased. In [65], the values of the CCQ varies from 0.00989 to 0.01950 for fine-grained concrete depending on the cement type and the amount of microsilica added, which is comparable with the CCQ values in the current study. In [66], the authors managed to increase the CCQ of lightweight concrete from 0.0252 to 0.0331, which exceeds the CCQ values in the current study. The bending strength values in [48], expressed as a percentage of the compressive strength (16.42% and 17.53%), are comparable with the values obtained in the current study (15.7–16.5%). Moreover, the drops in density and compressive strength in the current study (9.1% and 42%, respectively) were greater than in [33] (7.5% and 22%, respectively), which may be due to differences in the proportions of the mixture and the density of the control composition of concrete, as well as in the characteristics of the coconut shell.

It should be noted that at optimal amounts of CS, an important aspect is not only a slight increase in the strength characteristics but also, to some extent, maintaining them at the same level, provided that the density of concrete is simultaneously reduced. This is achieved through the same mechanism that occurs by replacing a heavy and dense aggregate with a light aggregate of plant origin [67]. Reducing the density of the aggregate automatically leads to a reduction in the weight of the concrete, and this is a great advantage for the concrete created on this basis. Concrete with a carefully selected amount of CS (up to 10%) has strength characteristics that are comparable with ordinary concrete and can be used when high requirements for long-term characteristics are not imposed on the concrete [68,69]. It should be noted that we are the first to propose evaluating the feasibility of such replacements using the parameter "coefficient of constructive quality; other researchers have evaluated the effectiveness of introducing CS into concrete in terms of an absolute indicator such as an increase in strength.

In addition to the obvious advantages of the new concrete, namely its environmental friendliness and economy, we would like to emphasize the features of the structure formation of such concrete. We have carried out studies at the micro and macro levels that confirmed the good joint work of the cement–sand matrix and aggregates of plant origin. At the macro level, the effectiveness of such a recipe and such a composition of concrete is confirmed by good strength characteristics. At the micro level, as can be seen in the photographs of the microstructure, the cement–sand matrix–CS grain interface has good adhesion and there is no or little microcracking. This confirms the good joint work and the high degree of adhesion of the aggregate with the cement–sand matrix, which provide good properties for the resulting concrete. It should be noted that the shell is similar to natural crushed stone and has a rough, angular structure that contributes to the creation of a more developed surface and, thereby, good joint work between the aggregate and the cement–sand concrete matrix. Thus, both practically (at the level of test results) and fundamentally (at the level of structure formation processes), with the help of micro- and macro-approaches, the effectiveness of the developed concrete composition is confirmed.

4. Conclusions

The formulations of concrete mixtures with different CS contents were developed and tested. The main characteristics, such as density, compressive strength, bending strength and prism strength, as well as the coefficient of constructive quality of concretes, were investigated. The study used normative tests and scanning electron microscopy.

It was established that the density of concrete decreases with an increase in the CS content. The best strength characteristics of concrete were recorded at a CS percentage of 5%, which replaced a part of natural coarse aggregate by volume. The effectiveness of the introduction of CS into concrete was also evaluated in terms of the "coefficient of constructive quality", which showed the ratio of the strength of the material to its density. Its highest values were recorded when the CS percentage was 5%.

An analysis of the microstructure of concrete containing CS instead of part of the natural coarse aggregate showed that the cement paste penetrated into the pores of the CS, thereby creating good adhesion of this aggregate to the cement–sand matrix. This factor, at small percentages of CS (up to 10%), compensates for the lower strength of CS compared with natural coarse filler; therefore, the strength of the composite does not significantly decrease.

CS contents in concrete of up to 10% are the most rational; a further increase in the content of CS inevitably leads to a significant drop in strength characteristics in comparison with concrete without CS.

The continuation of this research is planned in the direction of studying the complex partial replacement of the binder and aggregates in concrete with vegetable waste and the use of natural fibers as a reinforcing element, as well as the design and manufacture of more environmentally friendly and economical structures from such concretes.

Author Contributions: Conceptualization, D.E., S.A.S., E.M.S. and S.K.; methodology, A.C., S.A.S., E.M.S. and A.A.S.; software, A.A.S. and D.E.; validation, S.K., S.A.S., E.M.S. and D.E.; formal analysis, A.A.S., S.A.S., E.M.S. and A.C.; investigation, D.E., A.A.S., L.R.M., S.A.S., E.M.S., A.N.B. and B.M.; resources, B.M.; data curation, S.A.S., E.M.S. and S.K.; writing—original draft preparation, S.A.S., E.M.S. and A.N.B.; writing—review and editing, S.A.S., E.M.S. and A.N.B.; visualization, S.A.S., E.M.S. and A.N.B.; supervision, L.R.M. and B.M.; project administration, L.R.M. and B.M.; funding acquisition, A.N.B. and B.M. All authors have read and agreed to the published version of the manuscript.

Funding: This research received no external funding.

Institutional Review Board Statement: Not applicable.

Informed Consent Statement: Not applicable.

Data Availability Statement: The study did not report any data.

Acknowledgments: The authors would like to acknowledge the administration of Don State Technical University for their resources and financial support.

Conflicts of Interest: The authors declare no conflict of interest.

References

1. Khan, S.S.; Ali, M. Short-Term and Long-Term Needs for Sustainable Concrete—An Overview. *Eng. Proc.* **2022**, *22*, 14. [CrossRef]
2. Jang, H.-J.; Ahn, Y.-H.; Tae, S.-H. Proposal of Major Environmental Impact Categories of Construction Materials Based on Life Cycle Impact Assessments. *Materials* **2022**, *15*, 5047. [CrossRef]
3. WMO. *WMO Greenhouse Gas Bulletin—The State of the Greenhouse Gases in the Atmosphere Based on Global Observations through 2018*; WMO: Geneva, Switzerland, 2019; Volume 15. Available online: https://library.wmo.int/doc_num.php?explnum_id=10100 (accessed on 15 May 2023).
4. Katman, H.Y.B.; Khai, W.J.; Bheel, N.; Kırgız, M.S.; Kumar, A.; Khatib, J.; Benjeddou, O. Workability, Strength, Modulus of Elasticity, and Permeability Feature of Wheat Straw Ash-Incorporated Hydraulic Cement Concrete. *Buildings* **2022**, *12*, 1363. [CrossRef]
5. Ramasubramani, R.; Gunasekaran, K. Sustainable Alternate Materials for Concrete Production from Renewable Source and Waste. *Sustainability* **2021**, *13*, 1204. [CrossRef]
6. Shcherban', E.M.; Stel'makh, S.A.; Beskopylny, A.N.; Mailyan, L.R.; Meskhi, B.; Shilov, A.A.; Chernil'nik, A.; Özkılıç, Y.O.; Aksoylu, C. Normal-Weight Concrete with Improved Stress–Strain Characteristics Reinforced with Dispersed Coconut Fibers. *Appl. Sci.* **2022**, *12*, 11734. [CrossRef]
7. Beskopylny, A.N.; Stel'makh, S.A.; Shcherban', E.M.; Mailyan, L.R.; Meskhi, B.; Smolyanichenko, A.S.; Varavka, V.; Beskopylny, N.; Dotsenko, N. Influence of Electromagnetic Activation of Cement Paste and Nano-Modification by Rice Straw Biochar on the Structure and Characteristics of Concrete. *J. Compos. Sci.* **2022**, *6*, 268. [CrossRef]
8. Beskopylny, A.N.; Stel'makh, S.A.; Shcherban, E.M.; Mailyan, L.R.; Meskhi, B.; Shilov, A.A.; Beskopylny, N.; Chernil'nik, A. Enhanced Performance of Concrete Dispersedly Reinforced with Sisal Fibers. *Appl. Sci.* **2022**, *12*, 9102. [CrossRef]
9. Shcherban', E.M.; Stel'makh, S.A.; Beskopylny, A.N.; Mailyan, L.R.; Meskhi, B.; Varavka, V.; Beskopylny, N.; El'shaeva, D. Enhanced Eco-Friendly Concrete Nano-Change with Eggshell Powder. *Appl. Sci.* **2022**, *12*, 6606. [CrossRef]
10. Olanipekun, E.A.; Olusola, K.O.; Ata, O. A comparative study of concrete properties using coconut shell and palm kernel shell as coarse aggregates. *Build. Environ.* **2006**, *41*, 297–301. [CrossRef]
11. Beskopylny, A.N.; Shcherban', E.M.; Stel'makh, S.A.; Mailyan, L.R.; Meskhi, B.; Evtushenko, A.; El'shaeva, D.; Chernil'nik, A. Improving the Physical and Mechanical Characteristics of Modified Aerated Concrete by Reinforcing with Plant Fibers. *Fibers* **2023**, *11*, 33. [CrossRef]
12. Gunasekaran, K.; Annadurai, R.; Kumar, P.S. A study on some durability properties of coconut shell aggregate concrete. *Mater. Struct.* **2015**, *48*, 1253–1264. [CrossRef]
13. Tomar, R.; Kishore, K.; Parihar, H.S.; Gupta, N. A comprehensive study of waste coconut shell aggregate as raw material in concrete. *Mater. Today Proc.* **2021**, *44*, 437–443. [CrossRef]
14. Beskopylny, A.N.; Stel'makh, S.A.; Shcherban', E.M.; Mailyan, L.R.; Meskhi, B.; Shilov, A.A.; Chernil'nik, A.; El'shaeva, D. Effect of Walnut-Shell Additive on the Structure and Characteristics of Concrete. *Materials* **2023**, *16*, 1752. [CrossRef] [PubMed]
15. Beskopylny, A.N.; Shcherban', E.M.; Stel'makh, S.A.; Meskhi, B.; Shilov, A.A.; Varavka, V.; Evtushenko, A.; Özkılıç, Y.O.; Aksoylu, C.; Karalar, M. Composition Component Influence on Concrete Properties with the Additive of Rubber Tree Seed Shells. *Appl. Sci.* **2022**, *12*, 11744. [CrossRef]
16. Coffetti, D.; Crotti, E.; Gazzaniga, G.; Carrara, M.; Pastore, T.; Coppola, L. Pathways towards sustainable concrete. *Cem. Concr. Res.* **2022**, *154*, 106718. [CrossRef]
17. Khan, K.; Salami, B.A.; Jamal, A.; Amin, M.N.; Usman, M.; Al-Faiad, M.A.; Abu-Arab, A.M.; Iqbal, M. Prediction Models for Estimating Compressive Strength of Concrete Made of Manufactured Sand Using Gene Expression Programming Model. *Materials* **2022**, *15*, 5823. [CrossRef]

18. Almadani, M.; Razak, R.A.; Abdullah, M.M.A.B.; Mohamed, R. Geopolymer-Based Artificial Aggregates: A Review on Methods of Producing, Properties, and Improving Techniques. *Materials* **2022**, *15*, 5516. [CrossRef]
19. Gerges, N.N.; Issa, C.A.; Sleiman, E.; Aintrazi, S.; Saadeddine, J.; Abboud, R.; Antoun, M. Eco-Friendly Optimum Structural Concrete Mix Design. *Sustainability* **2022**, *14*, 8660. [CrossRef]
20. Liu, H.; Li, Q.; Quan, H.; Xu, X.; Wang, Q.; Ni, S. Assessment on the Properties of Biomass-Aggregate Geopolymer Concrete. *Appl. Sci.* **2022**, *12*, 3561. [CrossRef]
21. Marey, H.; Kozma, G.; Szabó, G. Effects of Using Green Concrete Materials on the CO_2 Emissions of the Residential Building Sector in Egypt. *Sustainability* **2022**, *14*, 3592. [CrossRef]
22. Liu, H.; Li, Q.; Ni, S. Assessment of the engineering properties of biomass recycled aggregate concrete developed from coconut shells. *Constr. Build. Mater.* **2022**, *342*, 128015. [CrossRef]
23. Nunes, L.A.; Silva, M.L.S.; Gerber, J.Z.; Kalid, R.A. Waste green coconut shells: Diagnosis of the disposal and applications for use in other products. *J. Clean. Prod.* **2020**, *255*, 120169. [CrossRef]
24. Srivani, G.; Vamsi Mohan, U. Study on strength properties of concrete by partial replacement of cement with sugarcane bagasse ash and coarse aggregate with coconut shells. *Mater. Today Proc.* 2023, *in press*. [CrossRef]
25. Bari, H.; Salam, M.A.; Safiuddin, M. Fresh and hardened properties of brick aggregate concrete including coconut shell as a partial replacement of coarse aggregate. *Constr. Build. Mater.* **2021**, *297*, 123745. [CrossRef]
26. Prithika, A.J.; Sekar, S.K. Mechanical and fracture characteristics of Eco-friendly concrete produced using coconut shell, ground granulated blast furnace slag and manufactured sand. *Constr. Build. Mater.* **2016**, *103*, 1–7. [CrossRef]
27. Janani, S.; Kulanthaivel, P.; Sowndarya, G.; Srivishnu, H.; Shanjayvel, P.G. Study of coconut shell as coarse aggregate in light weight concrete- a review. *Mater. Today Proc.* **2022**, *65*, 2003–2006. [CrossRef]
28. Prakash, R.; Divyah, N.; Srividhya, S.; Avudaiappan, S.; Amran, M.; Naidu Raman, S.; Guindos, P.; Vatin, N.I.; Fediuk, R. Effect of Steel Fiber on the Strength and Flexural Characteristics of Coconut Shell Concrete Partially Blended with Fly Ash. *Materials* **2022**, *15*, 4272. [CrossRef]
29. Mhaya, A.M.; Shahidan, S.; Algaifi, H.A.; Zuki, S.S.M.; Benjeddou, O.; Ibrahim, M.H.W.; Huseien, G.F. Thermal Conductivity of Coconut Shell-Incorporated Concrete: A Systematic Assessment via Theory and Experiment. *Sustainability* **2022**, *14*, 16167. [CrossRef]
30. Mhaya, A.M.; Algaifi, H.A.; Shahidan, S.; Zuki, S.S.M.; Azmi, M.A.M.; Ibrahim, M.H.W.; Huseien, G.F. Systematic Evaluation of Permeability of Concrete Incorporating Coconut Shell as Replacement of Fine Aggregate. *Materials* **2022**, *15*, 7944. [CrossRef]
31. Raja, K.C.P.; Thaniarasu, I.; Elkotb, M.A.; Ansari, K.; Saleel, C.A. Shrinkage Study and Strength Aspects of Concrete with Foundry Sand and Coconut Shell as a Partial Replacement for Coarse and Fine Aggregate. *Materials* **2021**, *14*, 7420. [CrossRef]
32. Sekar, A.; Kandasamy, G. Study on Durability Properties of Coconut Shell Concrete with Coconut Fiber. *Buildings* **2019**, *9*, 107. [CrossRef]
33. Kanojia, A.; Jain, S.K. Performance of coconut shell as coarse aggregate in concrete. *Constr. Build. Mater.* **2017**, *140*, 150–156. [CrossRef]
34. Bhoj, S.; Manoj, A.; Bhaskar, S. Usage potential and benefits of processed coconut shells in concrete as coarse aggregates. *Mater. Today Proc.* 2023, *in press*. [CrossRef]
35. Itam, Z.; Johar, A.D.; Syamsir, A.; Zainoodin, M.; Shaikh Ahmad Fadzil, S.M.M.; Beddu, S. Utilization of coconut shell as a supplementary cementitious material in concrete. *Mater. Today Proc.* **2022**, *66*, 2818–2823. [CrossRef]
36. Bari, H.; Safiuddin, M.; Salam, M.A. Microstructure of Structural Lightweight Concrete Incorporating Coconut Shell as a Partial Replacement of Brick Aggregate and Its Influence on Compressive Strength. *Sustainability* **2021**, *13*, 7157. [CrossRef]
37. Sujatha, A.; Deepa Balakrishnan, S. Properties of high strength lightweight concrete incorporating crushed coconut shells as coarse aggregate. *Mater. Today Proc.* 2023, *in press*. [CrossRef]
38. Palanisamy, M.; Kolandasamy, P.; Awoyera, P.; Gobinath, R.; Muthusamy, S.; Krishnasamy, T.R.; Viloria, A. Permeability properties of lightweight self-consolidating concrete made with coconut shell aggregate. *J. Mater. Res. Technol.* **2020**, *9*, 3547–3557. [CrossRef]
39. Aziz, W.; Aslam, M.; Ejaz, M.F.; Ali, M.J.; Ahmad, R.; Wajeeh-ul-Hassan Raza, M.; Khan, A. Mechanical properties, drying shrinkage and structural performance of coconut shell lightweight concrete. *Structures* **2022**, *35*, 26–35. [CrossRef]
40. Krishnaswami, N.; Velusamy, S.; Palanisamy, C.; Prakash, G.; Loganathan, K.K.; Moorthy, J. Experimental studies on light weight concrete using gib & coconut shell in concrete. *Mater. Today Proc.* **2022**, *65*, 1307–1314. [CrossRef]
41. Soumya, S.; Pennarasi, G.; Gunasekaran, K. Study on the reinforced manhole cover slab using coconut shell aggregate concrete. *Mater. Today Proc.* **2019**, *14*, 386–394. [CrossRef]
42. Herring, T.C.; Nyomboi, T.; Thuo, J.N. Ductility and cracking behavior of reinforced coconut shell concrete beams incorporated with coconut shell ash. *Results Eng.* **2022**, *14*, 100401. [CrossRef]
43. Gunasekaran, K.; Annadurai, R.; Kumar, P.S. Plastic shrinkage and deflection characteristics of coconut shell concrete slab. *Constr. Build. Mater.* **2013**, *43*, 203–207. [CrossRef]
44. Sekar, A.; Kandasamy, G. Optimization of Coconut Fiber in Coconut Shell Concrete and Its Mechanical and Bond Properties. *Materials* **2018**, *11*, 1726. [CrossRef] [PubMed]
45. Thangasamy, L.; Kandasamy, G. Behavior of Steel–Coconut Shell Concrete–Steel Composite Beam without and with Shear Studs under Flexural Load. *Materials* **2020**, *13*, 2444. [CrossRef] [PubMed]
46. Gunasekaran, K.; Annadurai, R.; Kumar, P.S. Study on reinforced lightweight coconut shell concrete beam behavior under shear. *Mater. Des.* **2013**, *50*, 293–301. [CrossRef]

47. Gunasekaran, K.; Annadurai, R.; Chandar, S.P.; Anandh, S. Study for the relevance of coconut shell aggregate concrete non-pressure pipe. *Ain Shams Eng. J.* **2017**, *8*, 523–530. [CrossRef]
48. Gunasekaran, K.; Kumar, P.S.; Lakshmipathy, M. Mechanical and bond properties of coconut shell concrete. *Constr. Build. Mater.* **2011**, *25*, 92–98. [CrossRef]
49. *GOST R 57813-2017/EN 12350-6:2009*; Testing Fresh Concrete. Part 6. Density. National Standard of the Russian Federation: Moscow, Russia, 2019. Available online: https://docs.cntd.ru/document/1200157335 (accessed on 15 May 2023).
50. *GOST R 57809-2017/EN 12350-2:2009*; Testing Fresh Concrete. Part 2. Slump Test. National Standard of the Russian Federation: Moscow, Russia, 2019. Available online: https://docs.cntd.ru/document/1200157288 (accessed on 15 May 2023).
51. *GOST R 57345-2016/EN 206-1:2013*; Concrete. General Specifications. National Standard of the Russian Federation: Moscow, Russia, 2017. Available online: https://docs.cntd.ru/document/1200142845 (accessed on 8 June 2023).
52. Thilagashanthi, T.; Gunasekaran, K.; Satyanarayanan, K.S. Microstructural pore analysis using SEM and ImageJ on the absorption of treated coconut shell aggregate. *J. Clean. Prod.* **2021**, *324*, 129217. [CrossRef]
53. *GOST 10180-2012*; Concretes. Methods for Strength Determination Using Reference Specimens. Interstate Standard: Moscow, Russia, 2018. Available online: http://docs.cntd.ru/document/1200100908 (accessed on 15 May 2023).
54. *EN 12390-1:2021*; Testing Hardened Concrete—Part 1: Shape, Dimensions and other Requirements of Specimens and Moulds. CEN-CENELEC Management Centre: Brussels, Belgium, 2021. Available online: https://standards.iteh.ai/catalog/standards/cen/d1c9ccee-2e5a-425e-a964-961da95d2f99/en-12390-1-2021 (accessed on 7 June 2023).
55. *EN 12390-2:2019*; Testing Hardened Concrete—Part 2: Making and Curing Specimens for Strength Tests. CEN-CENELEC Management Centre: Brussels, Belgium, 2021. Available online: https://standards.iteh.ai/catalog/standards/cen/ae7e6a86-1cbc-455e-8b2a-8964be9087f9/en-12390-2-2019 (accessed on 7 June 2023).
56. *EN 12390-3:2019*; Testing Hardened Concrete—Part 3: Compressive Strength of Test Specimens. CEN-CENELEC Management Centre: Brussels, Belgium, 2021. Available online: https://standards.iteh.ai/catalog/standards/cen/7eb738ef-44af-436c-ab8e-e6561571302c/en-12390-3-2019 (accessed on 7 June 2023).
57. *EN 12390-4:2019*; Testing Hardened Concrete—Part 4: Compressive Strength—Specification for Testing Machines. CEN-CENELEC Management Centre: Brussels, Belgium, 2021. Available online: https://standards.iteh.ai/catalog/standards/cen/10b1c613-819b-42d7-8f94-480cd37a666a/en-12390-4-2019 (accessed on 7 June 2023).
58. *EN 12390-5:2019*; Testing Hardened Concrete—Part 5: Flexural Strength of Test Specimens. CEN-CENELEC Management Centre: Brussels, Belgium, 2021. Available online: https://standards.iteh.ai/catalog/standards/cen/5653c2c7-55a9-4bcb-8e13-5b1dfb0e3baf/en-12390-5-2019 (accessed on 7 June 2023).
59. *GOST 24452*; Concretes. Methods of Prismatic, Compressive Strength, Modulus of Elasticity and Poisson's Ratio Determination. Interstate Standard: Moscow, Russia, 2005. Available online: https://docs.cntd.ru/document/9056198 (accessed on 15 May 2022).
60. *GOST 12730.1-2020*; Concretes. Methods of Determination of Density. Interstate Standard: Moscow, Russia, 2021. Available online: https://docs.cntd.ru/document/1200177299 (accessed on 15 May 2023).
61. Stel'makh, S.A.; Shcherban', E.M.; Beskopylny, A.N.; Mailyan, L.R.; Meskhi, B.; Beskopylny, N.; Dotsenko, N.; Kotenko, M. Influence of Recipe Factors on the Structure and Properties of Non-Autoclaved Aerated Concrete of Increased Strength. *Appl. Sci.* **2022**, *12*, 6984. [CrossRef]
62. Shcherban', E.M.; Stel'makh, S.A.; Beskopylny, A.; Mailyan, L.R.; Meskhi, B.; Varavka, V. Nanomodification of Lightweight Fiber Reinforced Concrete with Micro Silica and Its Influence on the Constructive Quality Coefficient. *Materials* **2021**, *14*, 7347. [CrossRef]
63. Prakash, R.; Thenmozhi, R.; Raman, S.N.; Subramanian, C.; Divyah, N. An investigation of key mechanical and durability properties of coconut shell concrete with partial replacement of fly ash. *Struct. Concr.* **2021**, *22*, E985–E996. [CrossRef]
64. Tangadagi, R.B.; Manjunatha, M.; Preethi, S.; Bharath, A.; Reshma, T.V. Strength characteristics of concrete using coconut shell as a coarse aggregate—A sustainable approach. *Mater. Today Proc.* **2021**, *47*, 3845–3851. [CrossRef]
65. Baranova, A.A.; Yazina, O.I.; Bobrova, A.A.; Rudykh, K.N. Effect of quantity of silica fume on the coefficients of constructive quality fine-grained concrete and structural foam concrete. *Mod. Technol. Sci. Technol. Prog.* **2018**, *1*, 95–96.
66. Rylova, T.; Lakhtaryna, S.; Yegorova, O. Lightweight structural concrete with an increased coefficient of structural quality. *Proc. DonNACEA* **2018**, *4–2*, 221–226.
67. Gunasekaran, K.; Annadurai, R.; Kumar, P.S. Study on reinforced lightweight coconut shell concrete beam behavior under flexure. *Mater. Des.* **2013**, *46*, 157–167. [CrossRef]
68. Liu, H.; Li, Q.; Wang, P. Assessment of the engineering properties and economic advantage of recycled aggregate concrete developed from waste clay bricks and coconut shells. *J. Build. Eng.* **2023**, *68*, 106071. [CrossRef]
69. Ni, S.; Liu, H.; Li, Q.; Quan, H.; Gheibi, M.; Fathollahi-Fard, A.M.; Tian, G. Assessment of the engineering properties, carbon dioxide emission and economic of biomass recycled aggregate concrete: A novel approach for building green concretes. *J. Clean. Prod.* **2022**, *365*, 132780. [CrossRef]

materials

MDPI

Article

Optimal Design of Bubble Deck Concrete Slabs: Serviceability Limit State

Tomasz Gajewski [1], Natalia Staszak [2] and Tomasz Garbowski [3,*]

[1] Institute of Structural Analysis, Poznan University of Technology, Piotrowo 5, 60-965 Poznan, Poland; tomasz.gajewski@put.poznan.pl
[2] Doctoral School, Poznan University of Life Sciences, Wojska Polskiego 28, 60-637 Poznan, Poland; natalia.staszak@up.poznan.pl
[3] Department of Biosystems Engineering, Poznan University of Life Sciences, Wojska Polskiego 50, 60-627 Poznan, Poland
* Correspondence: tomasz.garbowski@up.poznan.pl

Abstract: In engineering practice, one can often encounter issues related to optimization, where the goal is to minimize material consumption and minimize stresses or deflections of the structure. In most cases, these issues are addressed with finite element analysis software and simple optimization algorithms. However, in the case of optimization of certain structures, it is not so straightforward. An example of such constructions are bubble deck ceilings, where, in order to reduce the dead weight, air cavities are used, which are regularly arranged over the entire surface of the ceiling. In the case of these slabs, the flexural stiffness is not constant in all its cross-sections, which means that the use of structural finite elements (plate or shell) for static calculations is not possible, and therefore, the optimization process becomes more difficult. This paper presents a minimization procedure of the weight of bubble deck slabs using numerical homogenization and sequential quadratic programming with constraints. Homogenization allows for determining the effective stiffnesses of the floor, which in the next step are sequentially corrected by changing the geometrical parameters of the floor and voids in order to achieve the assumed deflection. The presented procedure allows for minimizing the use of material in a quick and effective way by automatically determining the optimal parameters describing the geometry of the bubble deck floor cross-section. For the optimal solution, the concrete weight of the bubble deck slab was reduced by about 23% in reference to the initial design, and the serviceability limit state was met.

Keywords: lightweight structures; bubble deck concrete slabs; numerical homogenization; weight minimization; sequential quadratic programming

Citation: Gajewski, T.; Staszak, N.; Garbowski, T. Optimal Design of Bubble Deck Concrete Slabs: Serviceability Limit State. *Materials* **2023**, *16*, 4897. https://doi.org/10.3390/ma16144897

Academic Editors: Hrvoje Smoljanović, Ivan Balić and Nikolina Zivaljic

Received: 10 June 2023
Revised: 5 July 2023
Accepted: 7 July 2023
Published: 8 July 2023

1. Introduction

Over the last few decades, concrete structures, in particular prefabricated reinforced concrete structures, have gained popularity and found wide application in many construction sectors around the world [1]. They are used not only in industrial, commercial, or residential facilities, but also in infrastructural construction. Prefabricated reinforced concrete elements, among others, mainly include columns, foundation footings, and retaining walls, as well as prefabricated walls with window and door openings. However, the most commonly used prefabricated concrete elements are girders and floor slabs [2].

This type of construction has numerous advantages, including: (a) saving formwork, (b) high durability and resistance of the structure, (c) high strength, (d) short construction time, (e) quality standards, and (f) reducing the amount of work on construction sites. In addition, prefabricated structures can be shaped in many ways using modern technologies and adapted to local conditions that occur in the designed facilities. Moreover, during the production stage, it is possible to make cuts and openings which allow for carrying out installation, e.g., pipes or cables after mounting the element at the site. It is extremely important to have already planned and anticipated all required openings in the design stage. On the other hand, prefabricated elements also have disadvantages. One of the main disadvantages is their cost, which consists of production, transport, and the need for cranes for their assembly [3]. However, the latter is not a problem, because on large construction sites, in which such structures are used, heavy equipment is available to unload and assemble them.

The main idea of the floor technology has remained unchanged in its general concept for years. Each type of ceiling should meet specific requirements that determine the selection of the appropriate structure and technology [4]. Technical requirements, including thermal and acoustic insulation, adequate strength, stiffness, fire resistance, and durability, have the greatest impact. Another important aspect is economic requirements, which include minimization of costs during construction and the project design stage. In the case of large spans between supports, prefabricated floors are used. The basic types of such floors are solid slabs, filigree slabs, multi-hole slabs, and TT double-rib slabs [5]. In such structures, an important aspect is to increase the span, and at the same time, reduce the weight of the panels [6]. Therefore, lightweight concrete structures are increasingly used, e.g., channel slabs or bubble deck slabs. This approach aims to eliminate the concrete that does not fulfill any structural functions to reduce the weight, and thus, the dead load [7,8].

Over the last few decades, many numerical models have been developed to represent the global behavior of prefabricated concrete structures and to understand their mechanical behavior. These models serve as valuable tools for simulating and analyzing the structural response of prefabricated elements, enabling engineers to evaluate their performance under various loading conditions and optimize their design. For instance, the authors of [9] presented a full 3D model of prefabricated bridge slabs for the purpose of modeling their non-linear behavior using the constitutive model of concrete damage plasticity. In turn, the bending behavior of composite slabs was analyzed by Tzaros et al. [10]. Gholamhossein et al. [11] proposed a three-dimensional solid finite element model to investigate the connection between concrete and steel. Information regarding the puncture resistance of concrete slabs can be found in [12]. Using the finite element method (FEM), it becomes possible to analyze structures with complex shapes and geometries, and to gain insight into the non-linear behavior of concrete and steel [13–15]. However, the detailed modeling of three-dimensional prefabricated slabs requires a lot of work, specialist knowledge, as well as the use of specific software, and it is very time-consuming in terms of calculations. The solution to the problem may be the use of one of the homogenization methods.

Homogenization is a mathematical technique used to analyze and model the behavior of heterogeneous materials or structures. It aims to capture the effective properties of an entire material or structure, taking into account its constituent materials or components. In the context of construction and structural calculations, homogenization refers to the simplification or adoption of uniform material or structure properties in order to facilitate calculations. For materials with an irregular structure or composition, simplifications can be used, where the material is treated as having uniform properties such as strength, stiffness, and density. Homogenization can also refer to simplification in structural analysis, where complex models of structural elements or details are replaced by simpler models that account for similar structural behavior. One can distinguish, among others, the method of periodic homogenization [16], non-linear homogenization [17], or the method of multi-scale homogenization based on genome mechanics [18]. A slightly different approach can be found in the work of Garbowski and Marek [19], where a method based on reverse analysis was used. On the other hand, in [20], a method of homogenization based on strain energy for sandwich panels with a honeycomb core was presented. Furthermore, Biancolini developed a method of strain energy equivalence between the simplified model and the representative volumetric element (RVE) model [21]. This method was then extended in [22]. Homogenization may introduce some simplifications and approximations, so its use should be carefully assessed in the context of a specific project and the fulfillment of relevant load-bearing and safety requirements.

Homogenization methods are also used in structural optimization analyses. This refers to the process of designing and modifying a building or structural elements to achieve the best possible results in terms of strength, cost-effectiveness, and energy efficiency. In addition, structural optimization is a complex process that requires consideration of multiple parameters and various engineering disciplines. It includes the integration of knowledge from structural engineering, materials science, mechanical engineering, and other relevant fields. The goal of the optimization is to find the best compromise between individual design requirements, so that the construction is as efficient, durable, and economical as possible, taking into account factors such as structural loads, material properties, construction techniques, and environmental impacts. This process may include the analysis of various scenarios, e.g., changing the geometry, materials, and configuration of structural elements to find the most suitable solution. The use of advanced tools such as structural analysis software and computer simulations can greatly facilitate the optimization process and help in obtaining optimal design solutions. These tools allow engineers to model and simulate the behavior of the structure under various conditions, accurately predict its performance, and evaluate different design alternatives.

Many studies have been conducted that provide information on the optimization of building construction, e.g., prefabricated elements, steel, or wooden structures. Sotiropoulos et al. presented a conceptual design method based on topology optimization using prefabricated structural elements [23]. The hybrid optimization method was used to optimize cellular beams in [24]. The optimization of thin-walled cross-sections was shown in [25]. Furthermore, the work by Sojobi et al. [26] presented a multi-objective optimization of a prefabricated carbon fiber-reinforced polymer (CFRP) composite sandwich structure. Additionally, Xiao and Bhola [27] presented the design of prefabricated building systems using building information modeling (BIM) technology and structural optimization. In the work by Xie et al. [28], a genetic algorithm was utilized for optimal planning of prefabricated construction projects.

In addition, many studies have been performed on the plates analyzed in this publication. Abishek and Iyappan investigated the bending behavior of a reinforced FRP (fiber-reinforced polymer) bubble deck [29]. Additionally, in [30], the evaluation of the plasticity of reinforced concrete structures with voids was analyzed. Experimental testing of slabs with modified openings was discussed in [31], and the study of concrete using plastic waste applied to hollow plates was presented in [32]. John et al. [33] demonstrated the bubble deck behavior using a model in ANSYS. They analyzed service load deflection, crack pattern, concrete stress distribution, and ultimate load capacity.

In our previous paper [34], a sensitivity study regarding the main bubble deck parameters was performed, i.e., it was checked how the changes of a single engineering parameter (such as bubble dimensions, class of concrete, reinforcement diameter, number of bars, etc.) influence the particular effective stiffnesses of the BD230 and BD340 slabs. In the current paper, the optimal design for a particular slab is sought by employing the minimization technique. According to the prior state-of-the-art review, there are no studies on the optimal design of the bubble deck plate, although this slab technology has many advantages. Determining the optimal design of the bubble deck slab through the numerical homogenization technique is the main novelty of the paper.

This paper presents an algorithm for the optimal design of bubble deck construction in order to minimize the amount of concrete and ensure that the permissible deflection arrow of the structure in the serviceability limit state is not exceeded. To simplify the model and speed up the analysis, the numerical homogenization method based on the equivalence of the strain energy between the simplified shell model and the three-dimensional reference RVE bubble deck model was used. The analyzed concrete slab contains evenly distributed voids over the entire surface and both upper and lower steel reinforcement, which increases the complexity and time of the calculations. Therefore, the original numerical homogenization method, initially developed for shell structures, was modified. In the work, an extension of the method was used to simultaneously include continuum and truss elements, similar to in [34,35].

One of the most important limitations of this study is that currently, the stress analysis cannot be included while using the homogenization technique considered [21,22]. Therefore, the ultimate limit state is beyond the scope of the current study. Another limitation is that the results from optimization with respect to the parameters of continuous domains may not be straightforward to be implemented in the industry. The use of concrete will be less in the optimal solution, but it is likely that bubble deck voids will be more expensive than the traditional bubble deck solution.

2. Materials and Methods

2.1. Lightweight Concrete Slabs

2.1.1. Voided Floor Slabs

The expression "voided floor slabs" refers to a special type of construction where the concrete slab contains a system of regular hollows or empty spaces inside. The hollows may have various shapes, such as spheres, cylinders, clovers, etc. In addition, they can be arranged in a specific pattern or grid. Voided floor slabs are used to achieve greater efficiency in terms of material use and to reduce the weight on the structure, while ensuring the required strength and performance of the floor structure. One of the most common types of voided floor slabs is the bubble-type ceiling.

The bubble deck is a modern ceiling solution compared to traditional slabs, in which the ineffective concrete in the middle of the cross-section is replaced by voids. This floor slab was invented in the 1990s by Danish engineer Jorgen Breuning [36]. It is used in residential, office, industrial, and utility buildings. In addition, the bubble deck ceiling is used in factories, parking structures, schools, and hotels.

Bubble deck has spherical or elliptical voids [37] evenly distributed over the entire surface, without changing the two-way action of the element. This solution allows to reduce the amount of concrete by 33% and the price by 30% compared to traditional solid slabs with the same parameters [38]. One of the main differences between solid slabs and bubble deck is their shear resistance [39]. In addition, by adjusting the appropriate reinforcement mesh and hole geometry, an optimized concrete structure can be achieved, allowing for the simultaneous maximum utilization of both moment and shear zones.

The height of such slabs can achieve up to 600 mm, which allows for a span of up to 20 m. On the other hand, the diameter of the bubble made of HDPE (high-density polyethylene) varies between 180 and 450 mm, depending on the designed thickness of the slab. The HDPE material used comes from recycled plastic waste. Therefore, the above solution contributes to the reduction of pollution and has a positive impact on environmental protection. However, the distance between the individual voids in the bubble deck system must be at least 1/9 of the bubble diameter. This requirement ensures adequate structural integrity and optimal load distribution in the floor slab. By keeping the distance between voids to a minimum, the system can effectively reduce the weight of the structure, while providing sufficient strength and stability. The voids are fixed in special steel baskets between the upper and lower reinforcements of the plate to prevent their displacement during concrete pouring [40,41]. In addition, the slab is directly connected to concrete columns or walls in situ without any beams, providing a wide range of structural costs and benefits. This eliminates the need for additional structural elements, simplifies the construction process, and allows for more flexible design options. Such a direct connection between the slab and columns or walls enhances the structural efficiency and optimizes the construction costs.

There are three methods of manufacturing the bubble deck: (i) the in situ approach—Figure 1a, (ii) semi-prefabricated elements (Filigree element)—Figure 1b, and (iii) prefabricated panels—Figure 1c [42]. In the case of the in situ approach (i), the slabs are made onsite at the place of their installation. Voids are placed in specific places between the bottom and top reinforcements of the floor. In the next step, the finished modules are placed on the prepared formwork. Then, the slab is concreted. It is worth noting that this solution is very effective in buildings where the floor is not flat, e.g., there is a domed or curved ceiling. In the case of approach (ii), the deck is a semi-prefabricated bubble deck. This means that elements are created in the production plant, which then require additional concreting at the construction site. This is a combination of methods (i) and (iii). The lower part of the ceiling, which is also the permanent formwork, is in the form of a concrete slab with a thickness of approximately 6 cm. In addition, it has both lower and upper reinforcements, with spaced voids. Then, the whole element is transported to the construction site and concreted after moving to the appropriate place. In contrast, approach (iii) deals with components that are fully manufactured. The finished prefabricated elements are transported to the construction site. This solution partially limits the ability of the plates to function as bi-directional. The solution to this problem may be the proper design of the connection between the prefabricated slabs, which will enable the use of prefabricated elements as two-way floors, similar to approaches (i) and (ii).

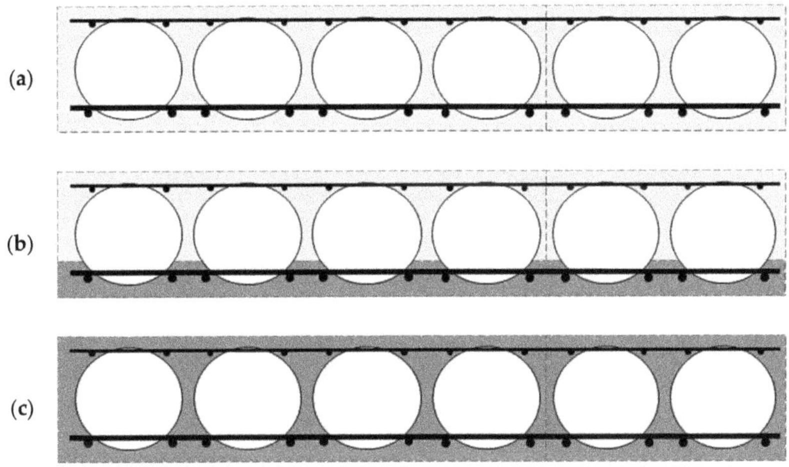

Figure 1. Types of bubble deck: (**a**) in situ element, (**b**) semi-prefabricated element, and (**c**) fully prefabricated element.

The bubble deck is generally designed using conventional solid-ceiling design methods, in accordance with applicable international and local design standards. The above floor system is a true bidirectional monolithic slab and behaves as a solid slab in both elastic and plastic modes. This means it can effectively carry and distribute loads in a manner similar to traditional solid slabs, ensuring structural integrity and performance. Besides, it mainly uses two analysis methods in its design, such as the linear elastic and yield line methods. Thanks to the optimized geometry and spherical bubbles, every part of the concrete in the slab is actively involved and relevant in the calculation of different types of forces.

2.1.2. Serviceability Limit State of Bubble Deck Concrete Slabs

During the design and analysis of structures, it is important to check two main limit states to ensure the safety and strength of the structure: the ultimate limit state (ULS) and the serviceability limit state (SLS). The first one concerns the assessment of the load capacity of the structure and checking whether it is able to withstand loads, in accordance with the adopted standards, building regulations, and design requirements. For different types of structures, such as floors, columns, foundations, bridges, etc., the ULS refers to the evaluation of their strength in response to various forces, such as compression, bending, tension, bending moments, shear forces, etc. The strength of the bubble deck is typically determined by structural analysis and calculation, taking into account factors such as the flexural strength, shear capacity, and the ability to efficiently distribute loads.

The serviceability limit state (SLS) refers to the conditions in which the structure can be operated without unacceptable deformations or damage that may affect its functionality and safety. In the case of bubble boards, the key aspect is to ensure that the bubbles inside the board are stable and do not undergo deformation or damage that could affect the load capacity and stiffness of the board. The SLS covers various aspects, including deflections and cracks. In the case of slabs, there are certain permissible deflection limits that should not be exceeded to ensure the stability and functionality of the structure. Excessive deflections can lead to improper functioning of finishing elements, problems with water drainage, or deterioration of interior aesthetics. Usually, when assessing the SLS, it is recommended to examine the cracks to determine their nature, size, and their impact on the safety and functionality of the structure.

In this work, analyses of the optimization of the bubble deck ceiling in terms of the SLS were carried out. It was assumed that the maximum plate deflection cannot exceed the permissible value equal to 1/250 of the span between supports, according to [43], for the quasi-permanent load case. In addition, it was assumed that the floor is located in an office building and is subjected to the following loads, evenly distributed over the entire surface of the slab. One of them is the useful load with a characteristic value of: $q_k = 3\ \mathrm{kN/m^2}$ (according to [44] for office rooms). In addition, it was assumed that an equivalent load from partition walls equal to: $q_k = 0.8\ \mathrm{kN/m^2}$, and a permanent load outside the weight of the slab structure with a value of: $g_k = 1.5\ \mathrm{kN/m^2}$, are applied to the ceiling. The ceiling weight is a variable value and depends on the geometrical parameters of the slab and bubble cross-section, so it was directly taken into account in the cost function of the optimization algorithm. The deflection arrow was determined for the case of quasi-permanent loads in the serviceability limit state; therefore, the factors reducing the load values according to [44] were applied.

2.2. *Numerical Homogenization of the Slab*

In the following work, the method of numerical homogenization based on the equivalence of strain energy between the three-dimensional reference model and the simplified shell model was used [19,34,35,45–48]. The above method has already been adapted to prefabricated concrete slabs reinforced with spatial trusses [35] and reinforced slabs with voids, such as the bubble deck [34]. This approach uses the classical formulation of the displacement-based finite element method, extracting individual values for internal nodes—subscript "i", and external nodes—subscript "e":

$$\mathbf{Ku} = \mathbf{F} \rightarrow \begin{bmatrix} \mathbf{K}_{ee} & \mathbf{K}_{ei} \\ \mathbf{K}_{ie} & \mathbf{K}_{ii} \end{bmatrix} \begin{bmatrix} \mathbf{u}_e \\ \mathbf{u}_i \end{bmatrix} = \begin{bmatrix} \mathbf{F}_e \\ \mathbf{0} \end{bmatrix}. \quad (1)$$

where: $\mathbf{K}$ is the stiffness matrix, $\mathbf{u}$ is a displacement vector of nodes, and $\mathbf{F}$ is the external nodal load vector.

For this purpose, it is necessary to separate a representative volume element (RVE) from the model and perform static condensation. Condensation is the elimination of secondary degrees of freedom; in this case, internal nodes. Then, it is necessary to redefine the stiffness matrix with a reduced number of degrees of freedom only at the external nodes (see Figure 2a).

Additionally, the presented method of homogenization uses the relationship between the total energy of elastic deformation stored in the system after static condensation and the work of external forces on appropriate displacements:

$$E = \frac{1}{2}\mathbf{u}_e^T \mathbf{F}_e \quad (2)$$

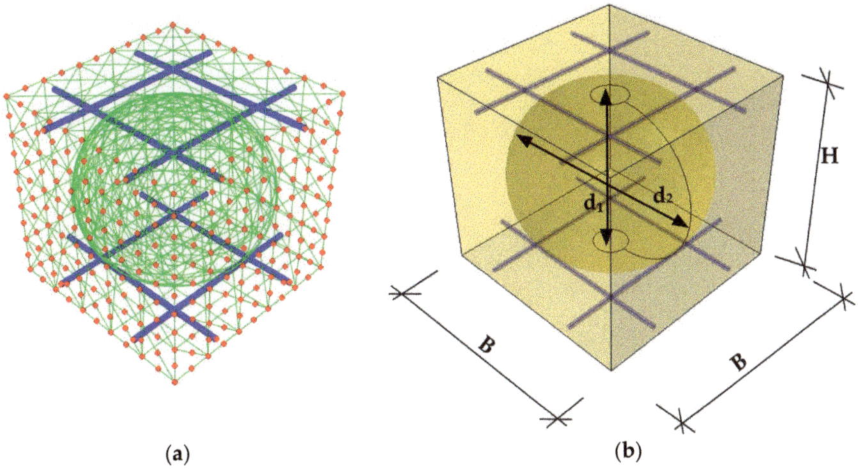

Figure 2. RVE: (**a**) external (in red color) and internal nodes and (**b**) parametrized for optimization purposes.

The homogenization method, as in [34,35], has been modified here to include only translational degrees of freedom (for two types of finite elements used in the models—truss and continuum). Therefore, the relationship between the generalized strain constants and the location of the external RVE nodes is expressed by the following transformation:

$$\begin{bmatrix} u_x \\ u_y \\ u_z \end{bmatrix}_i = \begin{bmatrix} x & 0 & y/2 & z/2 & 0 & xz & 0 & yz/2 \\ 0 & y & x/2 & 0 & z/2 & 0 & yz & xz/2 \\ 0 & 0 & 0 & x/2 & y/2 & -x^2/2 & -y^2/2 & -xy/2 \end{bmatrix}_i \begin{bmatrix} \varepsilon_x \\ \varepsilon_y \\ \gamma_{xy} \\ \gamma_{xz} \\ \gamma_{yz} \\ \kappa_x \\ \kappa_y \\ \kappa_{xy} \end{bmatrix}_i, \tag{3}$$

Using the definition of the elastic strain energy for a discrete model:

$$E = \frac{1}{2}\mathbf{u}_e^T\mathbf{K}\mathbf{u}_e = \frac{1}{2}\boldsymbol{\epsilon}_e^T\mathbf{A}_e^T\mathbf{K}\mathbf{A}_e\boldsymbol{\epsilon}_e. \tag{4}$$

Taking into account the finite element model including bending, tension, and transverse shear, the elastic internal energy for the plate or shell can be expressed as:

$$E = \frac{1}{2}\boldsymbol{\epsilon}_e^T\mathbf{A}_k\boldsymbol{\epsilon}_e\{area\} \tag{5}$$

Thanks to this, the stiffness matrix for the homogenization method can be extracted from the discrete matrix:

$$\mathbf{A}_k = \frac{\mathbf{A}_e^T\mathbf{K}\mathbf{A}_e}{area}. \tag{6}$$

Then, after appropriate transformations, we can obtain the stiffness matrix $\mathbf{A}_k$, which is the ABDR matrix, consisting of all the required compression, bending, and shear stiffnesses:

$$\mathbf{A}_k = \begin{bmatrix} \mathbf{A}_{3\times3} & \mathbf{B}_{3\times3} & 0 \\ \mathbf{B}_{3\times3} & \mathbf{D}_{3\times3} & 0 \\ 0 & 0 & \mathbf{R}_{2\times2} \end{bmatrix} \tag{7}$$

More information on numerical homogenization based on strain energy equivalence can be found in [19,34,35,45–48].

The design parameters, $\overline{x}$, that were used to determine the representative RVE for a specific design of the bubble bridge slab in the analyzed optimization problem are shown in Figure 2b. The assumed designed parameters read:

$$\overline{x} = \{B, H, d_1, d_2\} \tag{8}$$

in which B is the width and length of the RVE concrete unit, H is the height of the RVE concrete unit, and d_1, d_2 are the dimensions of the ellipsoidal void, height, and horizontal diameter, respectively.

2.3. Study Framework and Optimization Problem Definition

In everyday challenges, structural engineers tackle various problems, and one of the most common is the optimal design of the structure. In this paper, the optimal design of the bubble deck slab in regard to not exceeding the serviceability limit state (SLS) and minimal use of the concrete is analyzed. The problem is not trivial since the bubble deck slabs have variable cross-sections. In regions with full concrete cross-sections, the plate has a higher stiffness, while for bubble void regions, the plate is less stiff. The properties of the plate periodically vary across the span, which makes it difficult to calculate the displacement field of such slab. Additionally, the minimal use of the concrete is opposed to limiting plate deflection. Therefore, the typical approach for computations must be extended in order to meet the requirements of the SLS and limit the use of material.

In this paper, the numerical homogenization technique was used to determine the effective bending stiffness for computing the plate displacement via the analytical formula. The homogenization technique used here was that presented by Garbowski and Gajewski [22]. The bending stiffnesses of D_{11}, D_{22}, D_{12}, and D_{33} were determined by the homogenization technique in [22], which was used in multiple papers [34,45–48].

The square and symmetric bubble deck concrete slab was considered here, and the structure was reinforced with upper and lower steel mesh with $\phi 10$ steel bars. The cross-section design of the concrete bubble deck was described by a parametric RVE model with the design parameters gathered in $\overline{x}$ (see Section 2.2). The span dimensions of the slab were assumed to be 12×12 m^2, with evenly distributed load q_0. See Section 2.1.2 for more details of the assumed load.

The governing equation for the Kirchhoff–Love plate takes the following form [49,50]:

$$D_{11}\frac{\partial^4 w}{\partial x^4} + 2(D_{12} + 2D_{33})\frac{\partial^4 w}{\partial x^2 \partial y^2} + D_{22}\frac{\partial^4 w}{\partial y^4} = q. \tag{9}$$

in which w is the transverse deflection, x and y are the in-plane coordinates of the plate, and q is the transverse load.

The assumed plate was simply supported for all edges; therefore:

$$\begin{aligned} a&: \ x_1 = 0, w = 0, M_1 = 0. \\ b&: \ x_2 = 0, w = 0, M_2 = 0. \end{aligned} \tag{10}$$

in which a, b are the dimensions of the floor slab (here, $a = 12$ m and $b = 12$ m were assumed), and x_1, x_2 are the orthogonal coordinates along the perpendicular edges, respectively.

It was assumed that the orthotropic plate was subjected to a transverse, uniformly distributed load, labeled as q_0:

$$q(x_1, x_2) = q_0 = q_0(\overline{x}). \tag{11}$$

For more details regarding the determination of the uniformly distributed load, q_0, please refer to Section 2.1.2.

The final form of the plate deflection read:

$$w(\overline{x}) = \frac{16q_0}{\pi^6} \sum_{m=1}^{\infty} \sum_{n=1}^{\infty} \frac{\sin\frac{m\pi x_1}{a} \sin\frac{n\pi x_2}{b}}{mn\left[D_{11}(\overline{x})\left(\frac{m}{a}\right)^4 + 2(D_{12}(\overline{x}) + 2D_{33}(\overline{x}))\left(\frac{mn}{ab}\right)^2 + D_{22}(\overline{x})\left(\frac{n}{b}\right)^4\right]}. \quad (12)$$

in which m, n are the odd numbers.

The total cost function, F, in the optimization problem to be solved takes the two following components:

$$F(\overline{x}) = \omega F_{vol}(\overline{x}) + F_{defl}(\overline{x}). \quad (13)$$

in which F_{vol} is responsible for decreasing the concrete use and F_{defl} regards not exceeding the serviceability limit state due to the Eurocode standard [43]. The dimensionless factor ω is the scaling factor of the previous two and was selected by trial and error, with the aim of balancing the influence of both components on the objective function. Therefore, in this study, ω was set to 0.2×10^{-9}. The mathematical details of the optimization algorithm used for minimizing the cost function, $F(\overline{x})$, were included in Section 2.4.

The total volume of the bubble deck floor slabs, F_{vol}, was calculated by using a single representative volume element of the bubble deck unit:

$$V_{RVE}(\overline{x}) = B \cdot L \cdot H - \frac{4}{3}\pi\frac{d_1}{2}\left(\frac{d_2}{2}\right)^2. \quad (14)$$

Therefore, the first component of the cost function, F_{vol}, read:

$$F_{vol}(\overline{x}) = \frac{ab}{BL} V_{RVE}(\overline{x}). \quad (15)$$

The second component of the cost function, F_{defl}, was computed based on the maximum slab deflection (computed by Equation (12)), combined with the serviceability limit state:

$$F_{defl}(\overline{x}) = \left| w(\overline{x}) - \frac{\min(a,b)}{250} \right|. \quad (16)$$

In the optimization problem, the boundary limits of each design parameter of the concrete bubble deck were assumed and are presented in Table 1, where b_{min} is the lower, while b_{max} is the upper boundary of the physical dimensions. In Section 2.2, full details regarding the meaning of the symbols are presented, including exemplary graphics.

Table 1. The lower and upper boundary values of the parameters selected for optimization of the concrete part of the bubble deck slab.

Boundary	B (mm)	H (mm)	d_1 (mm)	d_2 (mm)
b_{min}	100	100	50	50
b_{max}	500	500	500	500

Since the dimensions of the bubble changed in the optimization process and the bubble was immersed in concrete with a variable height and width of the RVE module, physical inequality restrictions should be introduced. Therefore, the following inequality constraints were adopted:

$$\begin{aligned} d_1 - H + 40 \leq 0 \\ d_2 - B + 40 \leq 0 \end{aligned} \quad (17)$$

In Equation (17), 40 represents the concrete bubble cover in mm that is, the distance between the surface of the bubble void and the outer surface of the concrete at cardinal points. The assumed nominal value of the concrete cover regarding the steel mesh was 35 mm.

Local search algorithms, such as the one used in this paper, are vulnerable to find the locally optimal solutions. Therefore, in order to minimize the probability that a globally and not locally optimal solution will be found, the optimization algorithm was run many times from different starting points (initial guesses) (see Table 2). This is a typical approach for better exploration of the multi-dimensional space of the design parameters.

Table 2. The initial guesses of the design parameters selected for optimization of the concrete part of the bubble deck slab.

No.	*B* (mm)	*H* (mm)	d_1 (mm)	d_2 (mm)
$\overline{x}^0{}_1$	150	300	100	70
$\overline{x}^0{}_2$	250	200	150	80
$\overline{x}^0{}_3$	220	200	160	100
$\overline{x}^0{}_4$	200	250	110	180
$\overline{x}^0{}_5$	170	150	90	90

All computations in the research, apart from computing the stiffness matrices of RVEs (see Section 2.2), were performed using MATLAB software (ver. 2023a) [51].

2.4. Mathematical Optimization Procedure

Among the many available methods of optimization, the sequential quadratic programming (SQP) method is one of the most reliable and trustworthy. This mainly regards its efficiency, namely, the smallest number of cost function evaluations is obtained in the benchmark examples and sufficient accuracy is maintained [52–56]. Therefore, in this study, the SQP method was used, as in [57].

The classical optimization problem read [56]:

$$\min F(\overline{x}), \tag{18}$$

in which $F(\overline{x})$ is the cost function of the sought parameters, $\overline{x}$. The sough parameters may be constrained with equalities:

$$\begin{gathered} C_{eq}(\overline{x}) = 0, \\ A_{eq} \cdot \overline{x} = b_{eq}, \end{gathered} \tag{19}$$

and/or more complex constraints may be more adequate. For instance, the nonequality constraints:

$$\begin{gathered} C(\overline{x}) \leq 0, \\ A \cdot \overline{x} \leq b, \\ b_{min} \leq \overline{x} \leq b_{max}, \end{gathered} \tag{20}$$

in which b, b_{eq} are one-column matrices, A, A_{eq} are matrices, C, C_{eq} are functions, and b_{min} and b_{max} represent the lower and upper boundaries of the sought parameters, $\overline{x}$.

Constraints in the function $F(\overline{x})$ were computed by utilizing the Lagrange's function approach, L. Thus, the mathematically equivalent subproblem is defined by the following:

$$L(\overline{x}, \lambda) = F(\overline{x}) + \sum_{i=1}^{m} \lambda_i \cdot g(\overline{x}), \tag{21}$$

in which λ_i are the so-called Lagrange multipliers, while $g_i(\overline{x})$ are the constraints of nonequality.

In the SQP method, the following form of quadratic programming was solved:

$$\min_{d \in R^n} \frac{1}{2} d^T H_k + \nabla F(\overline{x}_k)^T d, \tag{22}$$

in which H_k is the positive definite approximation of the Hessian matrix, which approximates Equation (21). The approximation of the Hessian matrix was modified at each primary iteration by the Broyden–Fletcher–Goldfarb–Shanno (BFGS) method:

$$H_{k+1} = H_k + \frac{q_k q_k^T}{q_k^T s_k} - \frac{H_k s_k s_k^T H_k^T}{s_k^T H_k s_k}, \tag{23}$$

in which:

$$s_k = \overline{x}_{k+1} - \overline{x}_k, \tag{24}$$

$$q_k = \left(\nabla F(\overline{x}_{k+1}) + \sum_{i=1}^{m} \lambda_i \nabla g_i(\overline{x}_{k+1})\right) - \left(\nabla F(\overline{x}_k) + \sum_{i=1}^{m} \lambda_i \nabla g_i(\overline{x}_k)\right). \tag{25}$$

A new step was computed based on the solution of the quadratic programming problem:

$$\overline{x}_{k+1} = \overline{x}_k + \alpha_k d_k, \tag{26}$$

in which α_k is the step length obtained by minimization of the objective function [52–56].

3. Results

In local search optimization algorithms, it is recommended to solve multiple optimization problems to determine the solution, which is not locally but globally optimal. Therefore, the optimization procedure was conducted for several initial guesses of the design parameters to find the best solution, and the solutions of the initial assumed guesses were presented in Section 2.3. The results obtained from solving the optimization problem stated in Section 2.3 by the optimization method shown in Section 2.4 are summarized in Table 3. The second to fifth columns present the optimal parameters of the concrete bubble deck designs. In column six, the slab deflection obtained for the optimal designs due to the uniformly distributed load can be found. Moreover, the seventh to ninth columns show the components of the cost function and the total value of the cost function.

More details of the convergence of the solutions are presented for selected examples from Table 3 in Figures 3–5. In each figure, the minimization of the cost function, F, is demonstrated with its components for iterations of the optimization algorithm, i.e., ωF_{vol}—component of minimizing the volume of the concrete, and F_{defl}—component of minimizing the maximum plate deflection (see Figures 3a, 4a and 5a). Additionally, in Figures 3b, 4b and 5b, the maximum plate deflection was confronted with the SLS condition from the Eurocode standard [43]. For the analyzed case of the slab, namely, 12×12 m^2, the limit computed from the 1/250 condition was equal to 48 mm. In Figures 3b, 4b and 5b, the limit is marked with a dashed line. In addition, in Figures 3c, 4c and 5c, the changes of the sought parameters of B, H, d_1, and d_2 within the optimization are shown, showing the convergence to the final sought parameters of the bubble deck slab.

Table 3. Optimal designs of the concrete bubble deck slab with corresponding cost function values obtained by the optimization algorithm.

No.	B (mm)	H (mm)	d_1 (mm)	d_2 (mm)	w (mm)	ωF_{vol} (–)	F_{defl} (–)	F (–)
$\overline{x}^0{}_1$	100.0	158.6	115.8	50.0	47.94	4.1314	0.0616	4.1930
$\overline{x}^0{}_2$	122.1	158.5	109.1	55.9	48.02	4.2193	0.0164	4.2357
$\overline{x}^0{}_3$	158.2	157.5	99.7	89.1	48.01	4.0593	0.0121	4.0714
$\overline{x}^0{}_4$	110.7	156.0	52.8	70.5	48.12	4.2505	0.1165	4.3669
$\overline{x}^0{}_5$	144.8	155.6	53.8	104.4	48.04	4.0600	0.0390	4.0990

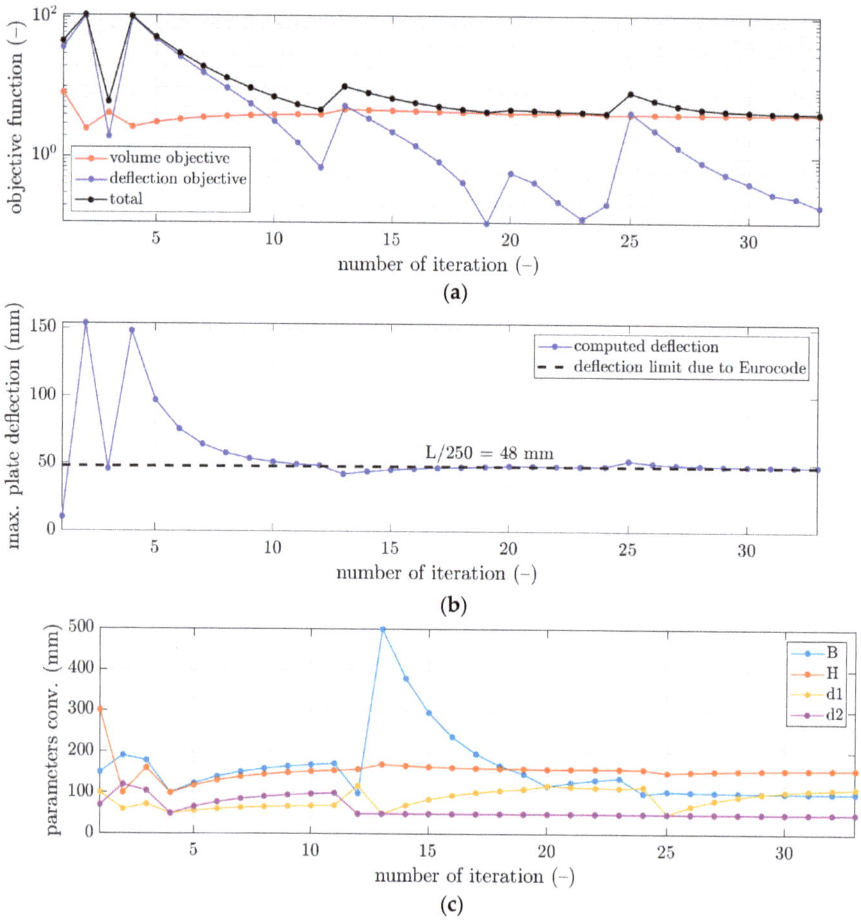

Figure 3. The convergence of the optimal selection of geometrical parameters of the bubble deck slab for initial guess $\bar{x}^0{}_1$: (**a**) cost function, (**b**) verification of the serviceability limit state, and (**c**) derived bubble deck parameters.

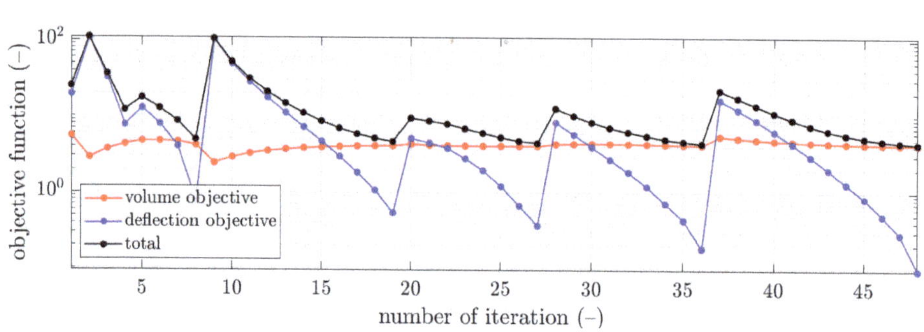

Figure 4. *Cont.*

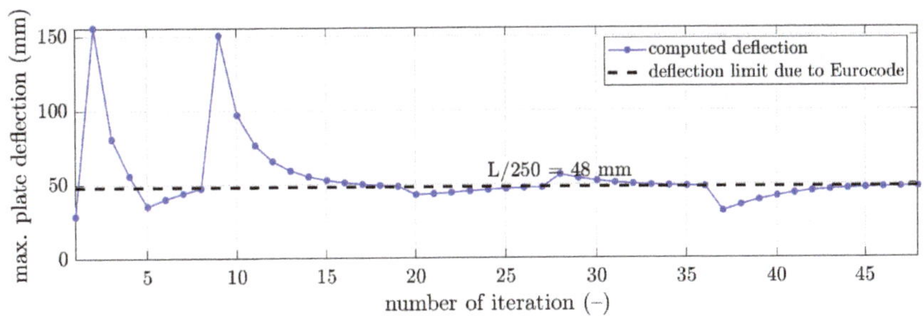

(b)

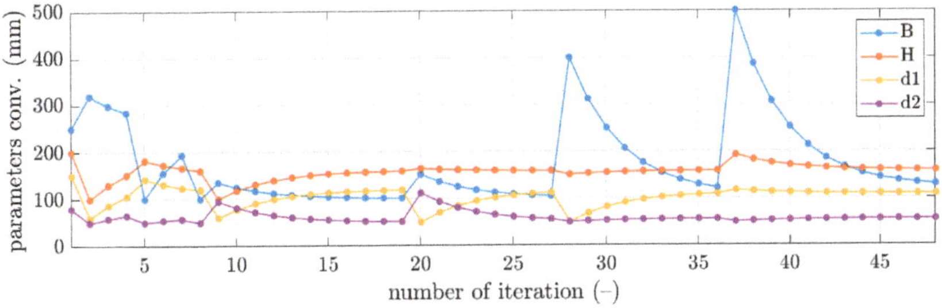

(c)

Figure 4. The convergence of the optimal selection of geometrical parameters of the bubble deck slab for initial guess $\overline{x}^0{}_2$: (**a**) cost function, (**b**) verification of the serviceability limit state, and (**c**) derived bubble deck parameters.

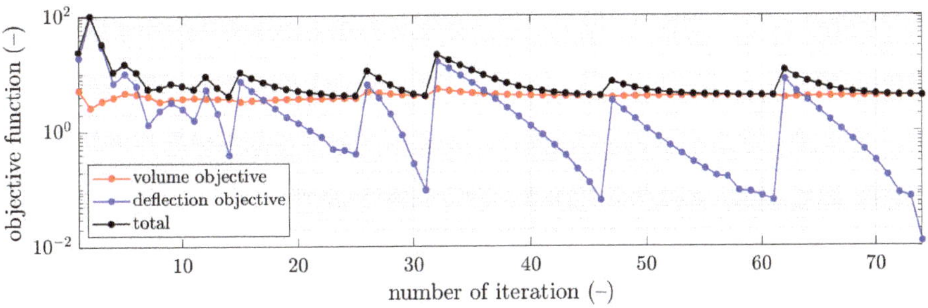

(a)

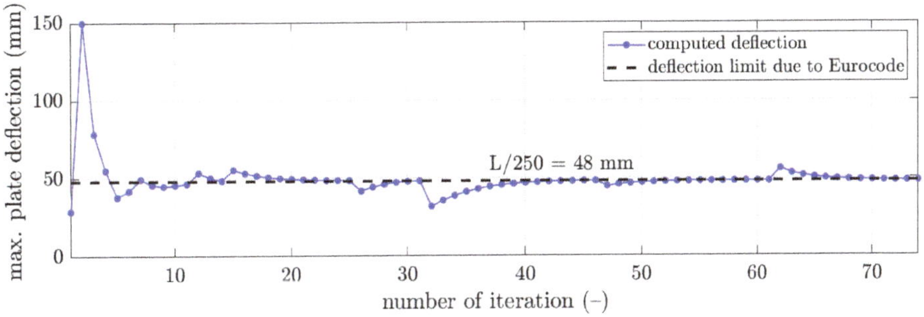

(b)

Figure 5. *Cont.*

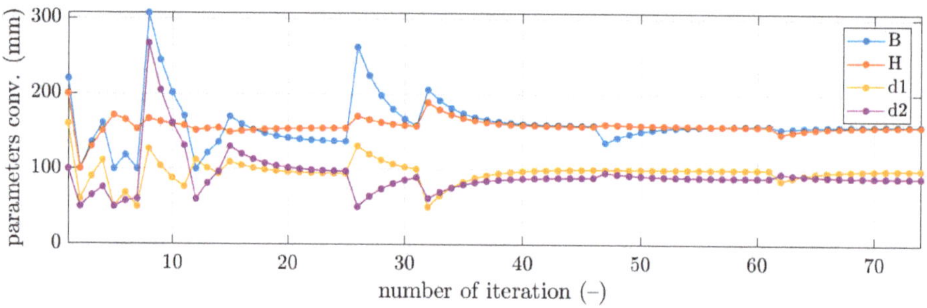

(c)

Figure 5. The convergence of the optimal selection of geometrical parameters of the bubble deck slab for initial guess $\overline{x}^0{}_3$: (**a**) cost function, (**b**) verification of the serviceability limit state, and (**c**) derived bubble deck parameters.

4. Discussion

The optimal design of the bubble deck slab floor regarding concrete use and the SLS is not a trivial task. The main difficulty is determining the mechanical properties of the periodically changing cross-section of the plate. Fully detailed finite element modeling of such structures is time-consuming in modeling and computations. Therefore, for this reason, for engineering purposes, the method presented in the paper is highly attractive. It does not require full formal finite element analysis of the floor slab, but only building the global stiffness matrix of the single periodic RVE unit and straightforward post-computations to obtain effective stiffnesses.

Therefore, in this paper, the complex structure of the bubble deck slab was considered to determine the optimal solution without using a typical, less accurate method. In this paper, the minimization of F_{vol} and F_{defl} components was in contrast; therefore, it was typical that the F_{vol} component decreased, while the F_{defl} component increased (for instance, see iterations 2 and 9 in Figure 4a). The rapid increase of the deflection plots was related to the principle of operation of the optimization algorithm, which changed the slab parameters so that the concrete volume was reduced. This increased the deflection, which, when L/250 was exceeded, activated a kind of penalty function (F_{defl}) that increased the total value of the objective function.

However, as presented in the optimization summary in Section 3, it was possible to obtain the deflection very close to the design standard limit, i.e., 48 mm. As shown in Table 3, all differences in the deflections achieved in relation to the design standard limit were approximately not bigger than 0.15 mm. Therefore, those components in the cost function outcomes were relatively small, not bigger than 0.12; however, the smallest was computed for the initial guess of $\overline{x}^0{}_3$, i.e., 0.0121. In Figures 3b, 4b and 5b, it can be observed that occasionally, the optimization algorithm broke the deflection limitation (for instance, see iterations 2 and 4 in Figure 3b), but it returned to respecting the limit after one or a few iterations. Similar features are visible in Figure 4b, iterations 2 and 9, and in Figure 5b, iterations 2, 12, and 15.

On the other hand, it was observed that the component related to the minimization of the concrete use yielded much greater values, that is, between 4.06 and 4.25. Since this component is the scaled volume of the concrete of the slab, it cannot be minimized to 0. Still, significant decreases of this component were observed compared to the initial guess values (see red plots in Figures 3a, 4a and 5a). Here, the lowest value was obtained for the initial guess of $\overline{x}^0{}_3$, i.e., 4.0593.

The best solution, that is, the globally optimal solution, was obtained for the initial guess of $\overline{x}^0{}_3$. Compared to the initial design of the bubble deck, the weight loss of the concrete was 23% (4.06, compared to 5.26 of F_{vol}). Therefore, the optimal parameters of the simply supported bubble deck slab of 12×12 m^2 for the uniformly distributed load were:

$B = 158.2$ mm, $H = 157.5$ mm, $d_1 = 99.7$ mm, and $d_2 = 89.1$ mm. Optimal parameter H was similar for all locally optimal solutions, as can be observed in Table 3, where it changed from 155.6 mm to 158.6 mm.

The main advantage of the methodology shown in the paper is the computational time of the analysis. The single evaluation of the cost function lasted less than 15 s. Therefore, in less than approximately 20 min, the single optimization procedure was finished. Going further, after about 2 h, the reasonable exploration of the design space can be achieved, and finally, the globally optimal solution can be expected.

5. Conclusions

The main aim of this paper was to find the optimal designs of the bubble deck slab subjected to a uniformly distributed load with regard to minimal concrete use and not exceeding the serviceability limit state of the Eurocode standard. In the research study, the numerical homogenization technique was used to determine the effective properties of the bubble deck slab within the cost function. Moreover, the local search algorithm of sequential quadratic programming was used in the minimization problem with linear constraints to derive the bubble deck slab module; that is, its length/width and height, but also the geometry of the ellipsoidal bubble void, i.e., its height and horizontal diameter.

As confirmed in the research, the optimization allowed to determine designs of the bubble deck slabs that ensured the minimum mass of concrete and met the serviceability limit state. Compared to the initial design of the bubble deck, the weight loss of the concrete was 23%. It was shown that the homogenization method used in this paper is highly attractive because it does not require solving complex structural problems through the computationally expensive finite element method. Achieving an optimal bubble deck design from a computational point of view would be carried out in just a few hours, instead of the heavy calculations of a single design case that would take the same amount of time. Due to the method used, the complex structure of the bubble deck slab was considered without using the typical method of substituting the cross-section of the concrete, which is less accurate due to shape simplification.

Author Contributions: Conceptualization, T.G. (Tomasz Garbowski); methodology, T.G. (Tomasz Gajewski) and T.G. (Tomasz Garbowski); software, T.G. (Tomasz Gajewski) and N.S.; validation, T.G. (Tomasz Garbowski), T.G. (Tomasz Gajewski), and N.S.; formal analysis, T.G. (Tomasz Gajewski) and N.S.; investigation, N.S. and T.G. (Tomasz Gajewski); writing—original draft preparation, T.G. (Tomasz Gajewski) and N.S.; writing—review and editing, T.G. (Tomasz Gajewski), T.G. (Tomasz Garbowski), and N.S.; visualization, T.G. (Tomasz Gajewski) and N.S.; supervision, T.G. (Tomasz Garbowski) and T.G. (Tomasz Gajewski); funding acquisition, T.G. (Tomasz Garbowski). All authors have read and agreed to the published version of the manuscript.

Funding: This research received no external funding.

Institutional Review Board Statement: Not applicable.

Informed Consent Statement: Not applicable.

Data Availability Statement: The data presented in this study are available upon request from the corresponding author.

Acknowledgments: The authors also acknowledge the grant of the Ministry of Education and Science, Poland, from Poznan University of Technology for Young Researchers; grant number 0411/SBAD/0009.

Conflicts of Interest: The authors declare no conflict of interest. The funders had no role in the design of the study; in the collection, analyses, or interpretation of data; in the writing of the manuscript, or in the decision to publish the results.

References

1. Feng, D.; Wu, G.; Lu, Y. Finite element modelling approach for precast reinforced concrete beam-to-column connections under cyclic loading. *Eng. Struct.* **2018**, *174*, 49–66. [CrossRef]
2. Baran, E. Effects of cast-in-place concrete topping on flexural response of precast concrete hollow-core slabs. *Eng. Struct.* **2015**, *98*, 109–117. [CrossRef]
3. Ramírez-Torres, A.; Di Stefano, S.; Grillo, A.; Rodríguez-Ramos, R.; Merodio, J.; Penta, R. An asymptotic homogenization approach to the microstructural evolution of heterogeneous media. *Int. J. Non-Linear Mech.* **2018**, *106*, 245–257. [CrossRef]
4. Szymczak-Graczyk, A.; Ksit, B.; Laks, I. Operational Problems in Structural Nodes of Reinforced Concrete Constructions. *IOP Conf. Ser. Mater. Sci. Eng.* **2019**, *603*, 032096. [CrossRef]
5. Lichołai, L. *Budownictwo Ogólne. Tom 3. Elementy Budynków Podstawy Projektowania*; Arkady: Warszawa, Poland, 2008. (In Polish)
6. Ksit, B.; Szymczak-Graczyk, A. Rare weather phenomena and the work of large-format roof coverings. *Civ. Environ. Eng. Rep.* **2019**, *29*, 123–133. [CrossRef]
7. Mirajkar, S.; Balapure, M.; Kshirsagar, T. Study of Bubble Deck Slab. *Int. J. Res. Sci. Eng.* **2017**, *7*, 1–5.
8. Quraisyah, A.D.S.; Kartini, K.; Hamidah, M.S.; Daiana, K. Bubble Deck Slab as an Innovative Biaxial Hollow Slab—A Review. *J. Phys. Conf. Ser.* **2020**, *1711*, 012003. [CrossRef]
9. Ren, W.; Sneed, L.H.; Yang, Y.; He, R. Numerical Simulation of Prestressed Precast Concrete Bridge Deck Panels Using Damage Plasticity Model. *Int. J. Concr. Struct. Mater.* **2015**, *9*, 45–54. [CrossRef]
10. Tzaros, K.A.; Mistakidis, E.S.; Perdikaris, P.C. A numerical model based on nonconvex–nonsmooth optimization for the simulation of bending tests on composite slabs with profiled steel sheeting. *Eng. Struct.* **2010**, *32*, 843–853. [CrossRef]
11. Gholamhoseini, A.; Gilbert, R.; Bradford, M.; Chang, Z. Longitudinal shear stress and bond–slip relationships in composite concrete slabs. *Eng. Struct.* **2014**, *69*, 37–48. [CrossRef]
12. Clement, T.; Ramos, A.P.; Ruiz, M.F.; Muttoni, A. Design for punching of prestressed concrete slabs. *Structural Concrete* **2013**, *14*, 157–167. [CrossRef]
13. Van Greunen, J.; Scordelis, A.C. Nonlinear analysis of prestressed concrete slabs. *J. Struct. Eng.* **1983**, *109*, 1742–1760. [CrossRef]
14. da Silva, A.R.; de Souza Rosa, J.P. Nonlinear numerical analysis of prestressed concrete beams and slabs. *Eng. Struct.* **2020**, *223*, 111187. [CrossRef]
15. Sathurappan, G.; Rajogopalan, N.; Krishnamoorthy, C.S. Nonlinear finite element analysis of reinforced and prestressed concrete slabs with reinforcement (inclusive of prestressing steel) modelled as discrete integral components. *Comput. Struct.* **1992**, *44*, 575–584. [CrossRef]
16. Buannic, N.; Cartraud, P.; Quesnel, T. Homogenization of corrugated core sandwich panels. *Compos. Struct.* **2003**, *59*, 299–312. [CrossRef]
17. Terada, K.; Kikuchi, N. Nonlinear homogenization method for practical applications. *Am. Soc. Mech. Eng. Appl. Mech. Div. AMD* **1995**, *212*, 103324.
18. Xin, L.; Khizar, R.; Peng, B.; Wenbin, Y. Two-Step Homogenization of Textile Composites Using Mechanics of Structure Genome. *Compos. Struct.* **2017**, *171*, 252–262. [CrossRef]
19. Garbowski, T.; Marek, A. Homogenization of corrugated boards through inverse analysis. In Proceedings of the 1st International Conference on Engineering and Applied Sciences Optimization, Kos Island, Greece, 4–6 June 2014; pp. 1751–1766.
20. Hohe, J. A direct homogenization approach for determination of the stiffness matrix for microheterogeneous plates with application to sandwich panels. *Compos. Part. B* **2003**, *34*, 615–626. [CrossRef]
21. Biancolini, M.E. Evaluation of equivalent stiffness properties of corrugated board. *Comp. Struct.* **2005**, *69*, 322–328. [CrossRef]
22. Garbowski, T.; Gajewski, T. Determination of transverse shear stiffness of sandwich panels with a corrugated core by numerical homogenization. *Materials* **2021**, *14*, 1976. [CrossRef]
23. Sotiropoulos, S.; Kazakis, G.; Lagaros, N.D. Conceptual design of structural systems based on topology optimization and prefabricated components. *Comput. Struct.* **2020**, *226*, 106136. [CrossRef]
24. Lian, Y.; Uzzaman, A.; Lim, J.B.; Abdelal, G.; Nash, D.; Young, B. Effect of web holes on web crippling strength of cold-formed steel channel sections under end-one-flange loading condition–Part I: Tests and finite element analysis. *Thin-Walled Struct.* **2016**, *107*, 443–452. [CrossRef]
25. Parastesh, H.; Hajirasouliha, I.; Taji, H.; Sabbagh, A.B. Shape optimization of cold-formed steel beam-columns with practical and manufacturing constraints. *J. Constr. Steel Res.* **2019**, *155*, 249–259. [CrossRef]
26. Sojobi, A.O.; Liew, K.M. Multi-objective optimization of high performance bio-inspired prefabricated composites for sustainable and resilient construction. *Compos. Struct.* **2022**, *279*, 114732. [CrossRef]
27. Xiao, Y.; Bhola, J. Design and optimization of prefabricated building system based on BIM technology. *Int. J. Syst. Assur. Eng. Manag.* **2022**, *13*, 111–120. [CrossRef]
28. Xie, L.; Chen, Y.; Chang, R. Scheduling optimization of prefabricated construction projects by genetic algorithm. *Appl. Sci.* **2021**, *11*, 5531. [CrossRef]
29. Abishek, V.; Iyappan, G.R. Study on flexural behavior of bubble deck slab strengthened with FRP. *J. Phys. Conf. Ser.* **2021**, *2040*, 012018. [CrossRef]
30. Hashemi, S.S.; Sadeghi, K.; Vagheri, M.; Siadat, S.A. Evaluation of ductility of RC structures constructed with bubble deck system. *Int. J. Civ. Eng.* **2018**, *16*, 513–526. [CrossRef]

31. Hai, L.V.; Hung, V.D.; Thi, T.M.; Nguyen-Thoi, T.; Phuoc, N.T. The experimental analysis of bubbledeck slab using modified elliptical balls. In Proceedings of the Thirteenth East Asia-Pacific Conference on Structural Engineering and Construction (EASEC-13), Sapporo, Japan, 11–13 September 2013; p. G-6-1.
32. Dinesh, A.; Kumar, R.P.; Abijith, S.R. Experimental Investigation on Bubble Deck Concrete Using Plastic Waste. In *Recent Developments in Waste Management: Select Proceedings of Recycle*; Springer: Singapore, 2018. [CrossRef]
33. John, R.; Varghese, J. A study on behavior of bubble deck slab using ansys. *Int. J. Innov. Sci. Eng. Technol.* **2015**, *2*, 2136–2139.
34. Staszak, N.; Garbowski, T.; Ksit, B. Optimal Design of Bubble Deck Concrete Slabs: Sensitivity Analysis and Numerical Homogenization. *Materials* **2023**, *16*, 2320. [CrossRef]
35. Staszak, N.; Garbowski, T.; Szymczak-Graczyk, A. Solid Truss to Shell Numerical Homogenization of Prefabricated Composite Slabs. *Materials* **2021**, *14*, 4120. [CrossRef] [PubMed]
36. Reich, E.V.; Lima, M.; Strelets, K.I. Efficiency of Applying Sustainable Technology of Bubbledeck Technology in Concrete in Russia. *Constr. Unique Build. Struct.* **2017**, *11*, 83–92.
37. Shetkar, A.; Hanche, N. An Experimental Study on Bubble Deck Slab System with Elliptical Balls. *Indian J. Sci. Res.* **2015**, *12*, 21–27.
38. Hokrane, N.S.; Saha, S. Comparative Studies of Conventional Slab and Bubble Deck Slab Based on Stiffness and Economy. *Int. J. Sci. Res. Dev.* **2017**, *5*, 1396–1398.
39. Fuchs, A.C. *BubbleDeck Floor System—An Innovative Sustainable Floor System*; BubbleDeck Netherlands B.V.: Leiden, The Netherlands, 2009.
40. Jamal, J.; Jolly, J. A study on structural behaviour of bubble deck slab using spherical and elliptical balls. *Int. Res. J. Eng. Technol.* **2017**, *4*, 2090–2095.
41. Konuri, S.; Varalakshmi, T.V.S. Review on Bubble Deck Slabs Technology and their Applications. *Int. J. Sci. Technol. Res.* **2019**, *8*, 427–432.
42. Kyng Consulting PTY Ltd. BubbleDeck Design Guide for Compliance with BCA Using AS3600 and EC2; DG-V.1.2; BubbleDeck Australia and New Zealand. 2008. Available online: https://www.bubbledeck.com.au/images/BubbleDeck-Design-Guide-v1.2.pdf (accessed on 1 May 2023).
43. *PN-EN 1992-1-1:2008*; Eurocode 2: Design of Concrete Structures-Part 1-1: General Rules, and Rules for Buildings. European Union: Brussels, Belgium, 2008.
44. *PN-EN 1990:2004*; Eurocode 0: Basics of Structural Design. European Union: Brussels, Belgium, 2004.
45. Garbowski, T.; Knitter-Piątkowska, A.; Mrówczyński, D. Numerical Homogenization of Multi-Layered Corrugated Cardboard with Creasing or Perforation. *Materials* **2021**, *14*, 3786. [CrossRef]
46. Mrówczyński, D.; Knitter-Piątkowska, A.; Garbowski, T. Non-Local Sensitivity Analysis and Numerical Homogenization in Optimal Design of Single-Wall Corrugated Board Packaging. *Materials* **2022**, *15*, 720. [CrossRef]
47. Mrówczyński, D.; Knitter-Piątkowska, A.; Garbowski, T. Numerical Homogenization of Single-Walled Corrugated Board with Imperfections. *Appl. Sci.* **2022**, *12*, 9632. [CrossRef]
48. Staszak, N.; Szymczak-Graczyk, A.; Garbowski, T. Elastic Analysis of Three-Layer Concrete Slab Based on Numerical Homogenization with an Analytical Shear Correction Factor. *Appl. Sci.* **2022**, *12*, 9918. [CrossRef]
49. Reddy, J.N. *Theory and Analysis of Elastic Plates and Shells*; CRC Press: Boca Raton, FL, USA; Taylor and Francis: Abingdon, UK, 2007.
50. Timoshenko, S.; Woinowsky-Krieger, S. *Theory of Plates and Shells*; McGraw-Hill: New York, NY, USA, 1959.
51. *Software*; Version 2023a; Mathworks: Natick, MA, USA, 2023.
52. Biggs, M.C. Constrained minimization using recursive quadratic programming. In *Towards Global Optimization*; North-Holland: Amsterdam, The Netherlands, 1975.
53. Han, S.P. A globally convergent method for nonlinear programming. *J. Optim. Theory Appl.* **1977**, *22*, 297–309. [CrossRef]
54. Powell, M.J.D. The convergence of variable metric methods for nonlinearly constrained optimization calculations. In Proceedings of the Special Interest Group on Mathematical Programming Symposium Conducted by the Computer Sciences Department at the University of Wisconsin–Madison, Madison, WI, USA, 11–13 July 1977, Academic Press: Cambridge, MA, USA. [CrossRef]
55. Powell, M.J.D. A fast algorithm for nonlinearly constrained optimization calculations. *Numer. Anal.* **1978**, *630*, 144–157.
56. Nocedal, J.; Wright, S.J. *Numerical Optimization*, 2nd ed.; Springer Series in Operations Research; Springer: Berlin/Heidelberg, Germany, 2006.
57. Gajewski, T.; Staszak, N.; Garbowski, T. Parametric optimization of thin-walled 3D beams with perforation based on homogenization and soft computing. *Materials* **2022**, *15*, 2520. [CrossRef] [PubMed]

materials

Article

Formulation and Performance of Model Concrete in Reduced-Scale Physical Model Tests

Gang Zheng [1,2,3], Boyang Xia [1,2], Haizuo Zhou [1,2,3,*], Yu Diao [1,2,3], Jianyou Huang [1,2], Junbo Zhang [1,2] and Xiaoxuan Yu [1,2]

1 School of Civil Engineering, Tianjin University, Tianjin 300072, China; zhenggang1967@163.com (G.Z.); boyang_1027@tju.edu.cn (B.X.); yudiao@tju.edu.cn (Y.D.); jianyou_huang@163.com (J.H.); zhangjunbo@tju.edu.cn (J.Z.); yuxiaoxuan@tju.edu.cn (X.Y.)

2 Key Laboratory of Coast Civil Structure Safety, Tianjin University, Ministry of Education, Tianjin 300072, China

3 State Key Laboratory of Hydraulic Engineering Simulation and Safety, Tianjin University, Tianjin 300072, China

* Correspondence: hzzhou@tju.edu.cn

Abstract: The utility of geotechnical centrifuge tests depends on how correctly they predict the physical and mechanical behaviour of concrete. In this study, a model concrete material that consisted of α-gypsum plaster, fine silica sand, and water was developed. An orthogonal test design was used to evaluate the effect of the mix proportion on the model concrete performance. The physical (i.e., flowability and bleeding rate) and mechanical (i.e., compressive and flexural strength) characteristics were considered as indices. Various mix ratios resulted in remarkable relative contributions to model concrete performance, and each raw material dosage exhibited positive or negative synergy. The water–plaster ratio (W/P) and aggregate–plaster ratio (A/P) strongly influenced the mechanical and physical characteristics, respectively. Multiple linear regression analysis (MLRA) was carried out to determine a forecast model for various small-scale test demands. Finally, the applicability and outlines of the presented forecasting method in proportioning design were evaluated by typical use of model concrete in small-scale model tests.

Keywords: model concrete; small-scale physical test; α-gypsum plaster; hydration; mechanical strength; prediction model

Citation: Zheng, G.; Xia, B.; Zhou, H.; Diao, Y.; Huang, J.; Zhang, J.; Yu, X. Formulation and Performance of Model Concrete in Reduced-Scale Physical Model Tests. *Materials* **2023**, *16*, 5784. https://doi.org/10.3390/ma16175784

Academic Editors: Hrvoje Smoljanović, Ivan Balić and Nikolina Zivaljic

Received: 22 July 2023
Revised: 17 August 2023
Accepted: 22 August 2023
Published: 24 August 2023

1. Introduction

Simple concrete, which has relatively high strength and stiffness, is an efficient and effective solution for geotechnical infrastructure in soft soil [1–5]. In addition to full-scale field testing, small-scale modelling is pertinent to investigating the mechanical and failure behaviour of simple concrete structures and soil–structure interactions [6–9]. Appropriate modelling of the prototype stress level is crucial in small-scale physical tests for capturing the stress-dependent characteristics in geotechnical engineering. Centrifuge modelling can produce a prototype stress by using centrifugal acceleration (g). On this basis, researchers can test small-scale physical models at stress levels identical to those experienced at field-prototype scale [10]. To reliably simulate the behaviour of concrete structures in small-scale model tests, the material properties of the element should be correctly scaled. However, the particle size of coarse aggregates in concrete is difficult to scale accurately (e.g., for the 1:40 scale, a 5 mm grain diameter of the aggregate is the same as the size of a 0.2 m prototype), which overestimates the strength of the model material. Furthermore, conventional simple concrete usually requires 28 d to reach the expected strength, which is inefficient for model tests.

A gypsum-based mixture mortar is useful in small-scale modelling and can reflect the quasi-brittle nature of concrete; especially in centrifuge model tests, which require large-scale factors [11,12]. Knappett et al. [11] evaluated various types of gypsum-based mixture mortar and uniformly graded silica sand, which they used to geometrically scale the coarse aggregate. Such model concrete was able to simulate the quasi-brittle failure behaviour of concrete and exhibited typical mechanical strengths. This type of mixture mortar was recently used in modelling concrete in centrifuge model tests [13–15]. However, the formulations of model concrete in literature studies are only applicable to specific test conditions and cannot be extended in generalized applications. Generally, excellent performance of mixture mortar requires favourable fluidity, stability, and sufficient mechanical strength [16–18]. Therefore, an approach that can describe the relationship between the various mix proportions and performance is necessary; a quantitative analysis should be applied to evaluate the effects of each mix factor and determine the optimal formulations for use in different small-scale tests.

In this study, a total of 20 groups of gypsum-based mortar samples with various water–plaster ratios and silica sand dosages were tested by orthogonal test design. The flowability, bleeding rate, and compressive and flexural strength were considered as evaluation indices of model concrete materials. Prediction models of the relationship between the evaluation indices and various mix proportions are proposed by using multiple linear regression analysis (MLRA). Finally, by typical application in a centrifuge model test, guidance for general application of gypsum-based materials for modelling concrete elements in small-scale physical tests are provided.

2. Raw Materials and Methodologies

2.1. Raw Materials

To satisfy the spatial and physical scaling laws between small-scale test and those of a corresponding prototype, the raw materials of the emulation model concrete column consisted of α-gypsum plaster (model concrete; Jing-men, Hubei, China), water, and fine silica sand (coarse aggregate; Zhengzhou, Henan, China). The chemical composition of the α-gypsum plaster was analysed with an X-ray fluorescence spectrometer (model Panalytical Axios, Malvern Panalytical, Malvern, UK) [19], as shown in Tables 1 and 2. Figure 1 shows the grading curve of the fine silica sand. The particle size distribution was consistent with a typical coarse aggregate grading specified in BS 882 [20], scaled between 1:30 and 1:50.

Table 1. Chemical compositions of α-gypsum plaster (wt%).

Material	MgO	Al_2O_3	SiO_2	SO_3	CaO	SrO
α-gypsum	1.28	0.084	0.143	38.291	59.605	0.597

Table 2. Physical properties of α-gypsum plaster.

Material	Consistency (%)	Initial Setting Time (min)	Final Setting Time (min)	Bulk Density (kg/m^3)	7 Days Compressive Strength (MPa)
α-gypsum	58.0	8	17	1200	15

2.2. Orthogonal Test Approach

To investigate the effect of various mixing ratios on the performance of model concrete, an orthogonal test design was used. The advantage of orthogonal testing was to select representative test points from a comprehensive test plan for minimizing the experimental quantity and obtaining comprehensive test results. Thus, the water/plaster (W/P) ratio and coarse aggregate content were considered as the pertinent variables in the orthogonal tests. Remarkably, the mass fraction of plaster (A/P) was used as the content of fine silica sand, which served as the coarse aggregate for model concrete. In accordance with relevant research and previous experiments, the standards of each mixing ratio were determined by choosing typical values [11]. Table 3 shows the experiment composition of 20 groups of test samples. The influence of the W/P and A/P ratios on the flowability, bleeding, and compressive and flexural strength performance is discussed in subsequent paragraphs.

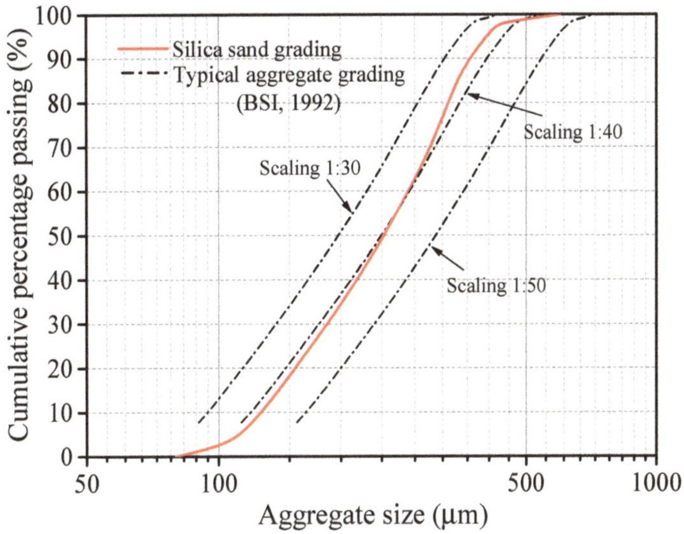

Figure 1. Particle size distributions of coarse aggregate (fine silica sand) [20].

Table 3. Composition of the orthogonal tests.

Parameters		Water Quantity (g)	Plaster Quantity (g)	Aggregate Quantity (g)	W/P	A/P
A	A1	300	500	500	0.6	1
	A2	300		600		1.2
	A3	300		700		1.4
	A4	300		800		1.6
	A5	300		900		1.8
B	B1	350	500	500	0.7	1
	B2	350		600		1.2
	B3	350		700		1.4
	B4	350		800		1.6
	B5	350		900		1.8
C	C1	400	500	500	0.8	1
	C2	400		600		1.2
	C3	400		700		1.4
	C4	400		800		1.6
	C5	400		900		1.8
D	D1	450	500	500	0.9	1
	D2	450		600		1.2
	D3	450		700		1.4
	D4	450		800		1.6
	D5	450		900		1.8

2.3. *Preparation of Model Concrete Specimens*

All 20 groups of model concrete mortar samples (Table 3) consisted of α-gypsum plaster, fine silica sand, and water. The preparation process was as follows.

The weighted powder, including the alpha-gypsum plaster and fine silica sand, was added into the mixer and dry-mixed for about 60 s.

The sand plaster mixture was poured into the water and quickly stirred in the mixer for 120 s.

Fresh paste and casted samples were prepared for further testing.

2.4. Test Methods

The flowability characteristic of the model concrete mortar was determined by measuring the spread diameter of fresh mixture slurry by a mini-slump test [21,22]. A 40 × 40 cm glossy glass plate was used in the test and a typical steel mould was placed in the centre of the glossy glass. The mould filled with the slurry was rapidly lifted from the glass plate and then the slurry flowed freely onto the glass plate for measuring the maximum expanded diameter length of the slurry. Figure 2 shows typical results of mini-slump tests.

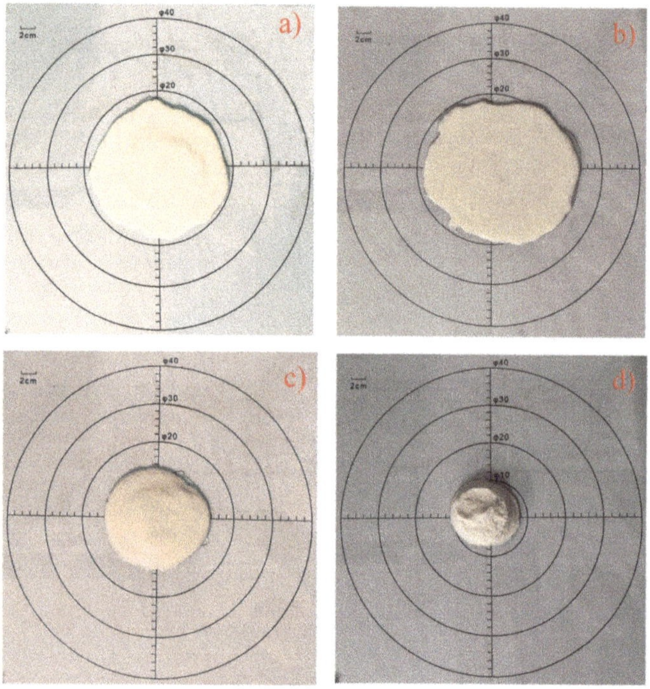

Figure 2. Flowability characteristic of mini-slump tests: ((**a**) A2; (**b**) B2; (**c**) C5; (**d**) A5).

The standing observation method indicates the stability of the mixture paste by measurement of the percentage of the upper bleeding water volume of fresh slurry compared with the total volume (Figure 3). A larger water bleeding ratio corresponds to lesser stability of the fresh mixture slurry [17,18,23].

Cubic model concrete specimens of a standard size (50 × 50 × 50 mm) were prepared for unconfined compression tests. The specimens were made by pouring the sand plaster mixture into the mould and were subsequently demoulded. Afterwards, the standard cubes were cured for 3, 7, 14, and 28 days under standard curing conditions at a temperature of 25 ± 2 °C and a relative humidity of 95% [24]. The compressive strength of the specimens was tested with an MTS servo-hydraulic testing machine under a loading of 2.4 kN/s (Figure 4) [23,24].

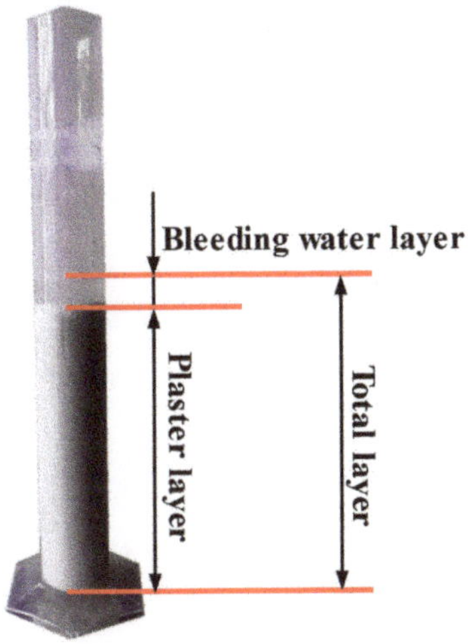

Figure 3. Schematic of bleeding rate test.

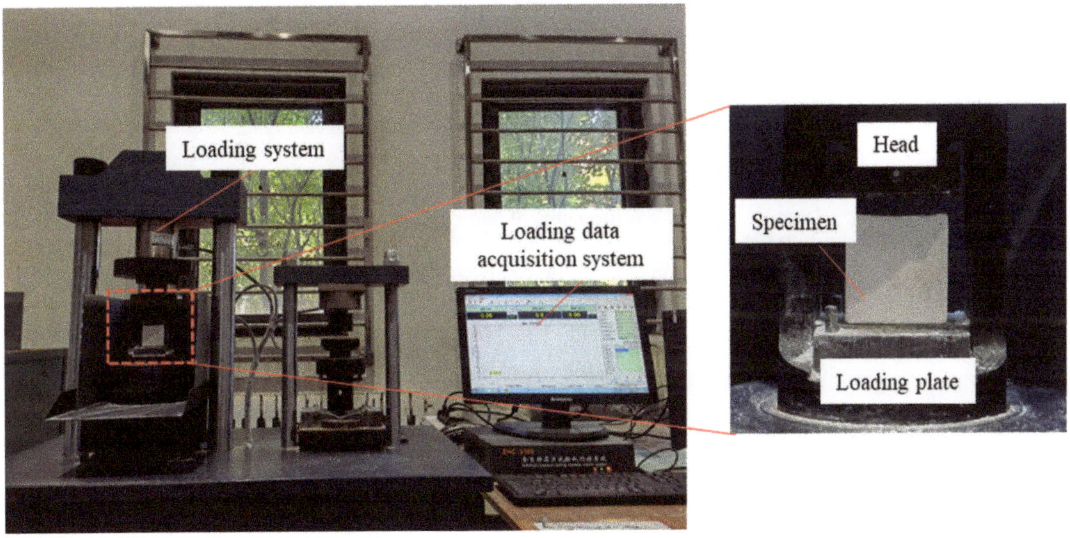

Figure 4. Schematic of unconfined compressive strength test system.

Compared with the direct flexural strength using standard specimens, the modulus of rupture (f_r) was considered to be more suitable for representing the flexural behaviour of concrete under bending-induced flexural stress [11,25,26]. The f_r has a substantial size effect. For each bending test, prismoid-shaped specimens (size: 20 mm × 20 mm × 450 mm) were used to obtain the average f_r for each mixing composition. These dimensions can provide corresponding values of f_r for subsequent centrifuge model tests of a model concrete pile at 1:40 scale [26]. Figure 5 shows installation details of the test. Bending tests were performed with a microcomputer-controlled electronic universal testing machine (model WDW-100E); vertical concentrated load was applied in a displacement-controlled mode (0.5 mm/min) until specimen rupture [27].

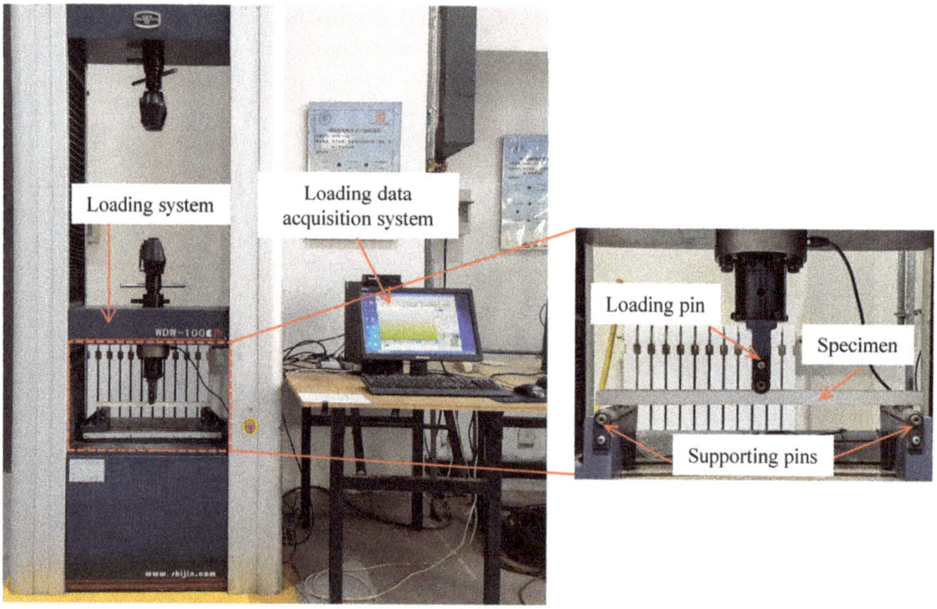

Figure 5. Schematic of three-point bending test.

3. Orthogonal Experiment Results and Correlation Analysis

3.1. Flowability

Flowability is a significant index for fresh mixture mortar and must be sufficiently large to meet the requirements of homogeneity [28,29]. Figure 6 shows the influence of the A/P ratio on the flowability of mixture slurries with various W/P ratios. The flowability decreased gradually with increasing A/P ratio, and more substantially so at lower W/P ratios (e.g., W/P = 0.6, 0.7). At a given A/P ratio, the flowability depended on the W/P ratio; the aggregate content required to achieve a decreasing flowability was lower with decreasing W/P ratio. In accordance with the range analysis, the significance of the influence of the aggregate content on the flowability decreased with the increasing W/P ratio. Higher W/P ratios led to increasing relative content of free water in the slurry, and the free water caused an increase in the overall flowability by reducing the friction contact among the particles [18]. As a result, the mixing proportion of the maximum flowability of the slurry was W/P = 0.9 and A/P = 1.

3.2. Bleeding

Bleeding refers to separation of water from the pores between filled particles. A higher bleeding rate results in instability of the mixture mortar, which leads to large volume change of the specimen during hardening and deviation from the design volume [25]. Figure 7 shows the variation of the bleeding rate for all the specimens with various ratios. The bleeding rate decreased with increasing A/P ratio. This result is possibly because free water in the paste was absorbed by sand particles with a large specific surface area. Therefore, the stability of the slurry can be improved by reducing the content of water in the formulation. When the A/P ratio exceeded 1.6, the effect of the aggregate content on reducing the bleeding rate lessened. Moreover, the range analysis indicates that the significance of the impact of the A/P ratio on the bleeding rate was almost invariable under various W/P ratios.

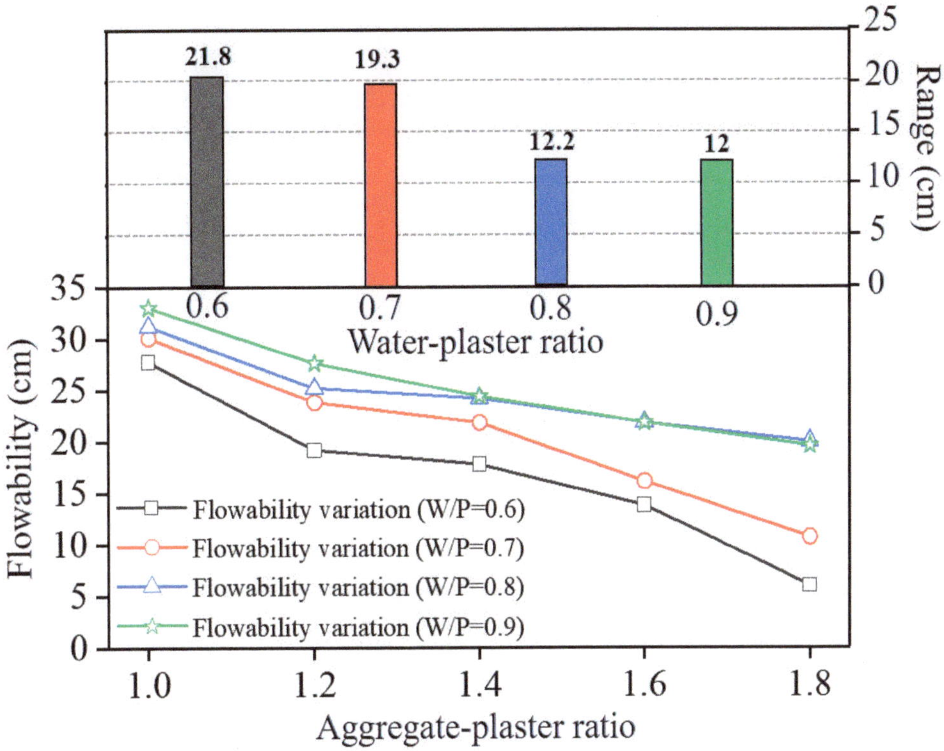

Figure 6. Influence of each composition on the flowability of the mixture slurry.

3.3. Compressive Strength

Figure 8 shows the variation in the compressive strength of all of the specimens from curing 3–28 days. A remarkable compression strength diversity was found between 3 and 7 days for all sets of test specimens. For example, regarding group No. 6, the 3 d compressive strength value only accounted for 51.2% of the 7 d compressive strength. In addition, the compressive strength exhibited a significant correspondence to the W/P ratio; the specimens with W/P = 0.6 (no. 1–5) exhibited the highest compressive strength after 28 days of curing.

Figure 9 shows the influence of pertinent variables on the compressive strength. The compressive strength increased significantly with the decreasing W/P ratio, whereas the A/P ratio exhibited a negligible effect on the compressive strength. This phenomenon is because the corresponding content of plaster particles in the paste increased with smaller water/plaster ratios, and the active ingredients in the plaster formed more high-strength hydration products after adequate hydration, which led to an increase in the compressive strength [19,30]. The optimum for the 28 d compressive strength of the sample was group No. 1. The range analysis results indicate that the significance of each composition on compressive strength was $(R_{W/P}) > (R_{A/P})$.

3.4. Flexural Strength

To describe the behaviour of the concrete under bending-induced flexural stress (i.e., flexural strength of the model concrete), the modulus of rupture was used [31]. Figure 10 shows the variation of the modulus of rupture from curing 3–28 days. Compared with the compressive strength, the flexural strength of the model concrete was relatively low, which is in accordance with a quasi-brittle material that exhibits low flexural fracture energy. After curing for 14 days, the modulus of rupture value tended to be stable, indicating that the hardening of the model concrete was nearly complete.

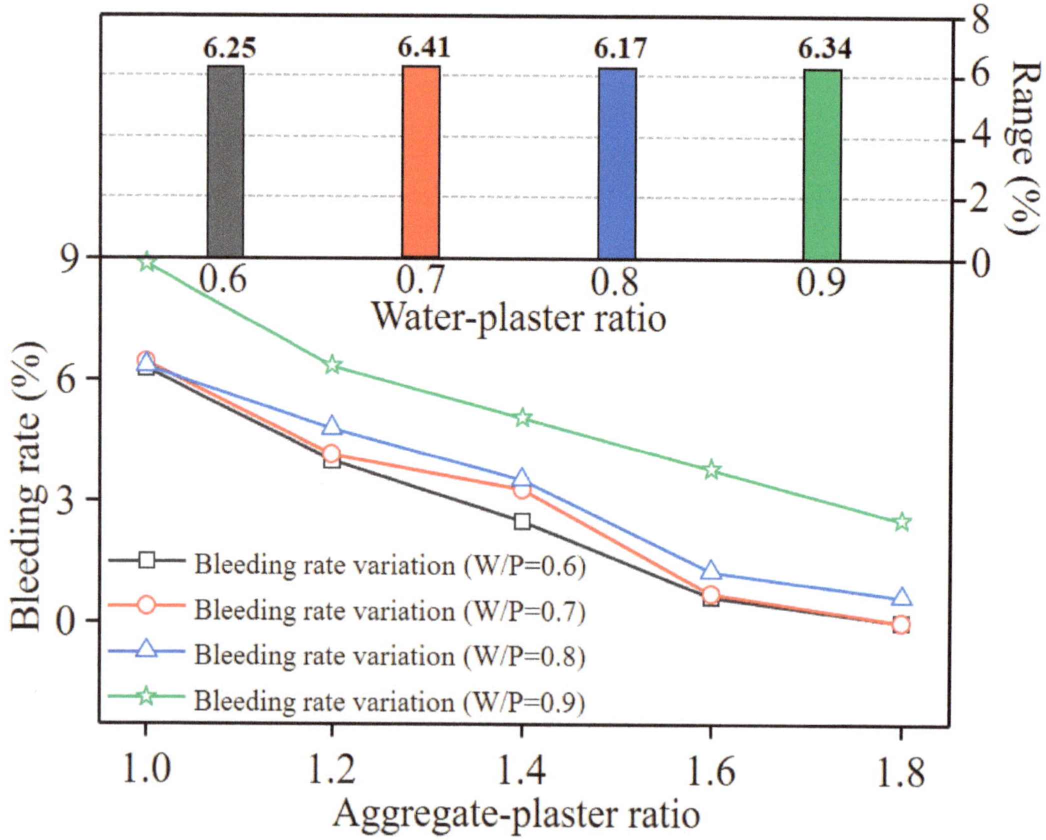

Figure 7. Influence of each composition on the bleeding rate of the mixture slurry.

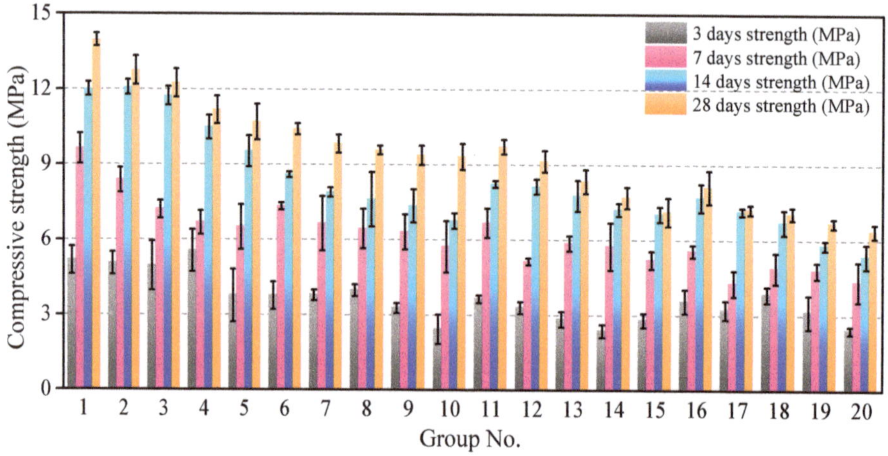

Figure 8. Variation in compressive strength of specimens from curing 3–28 days.

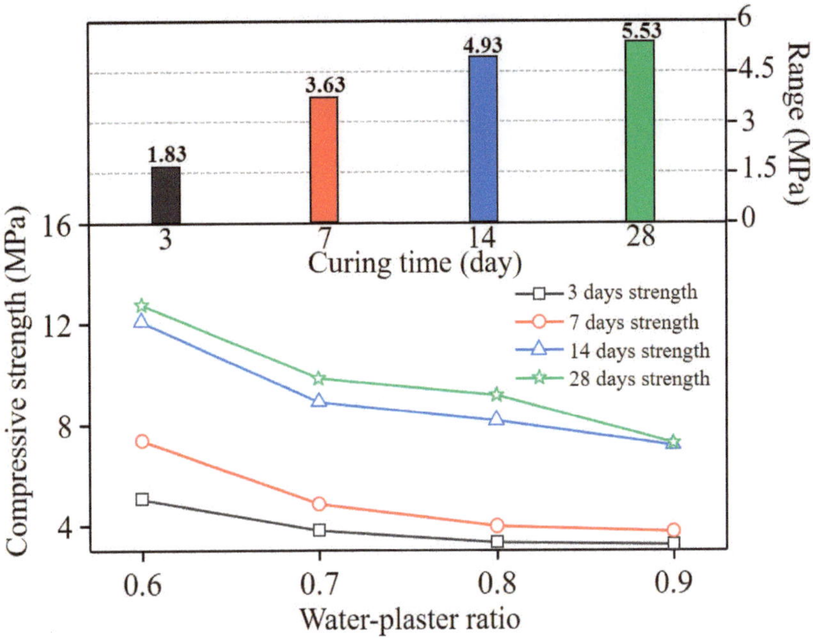

(a)

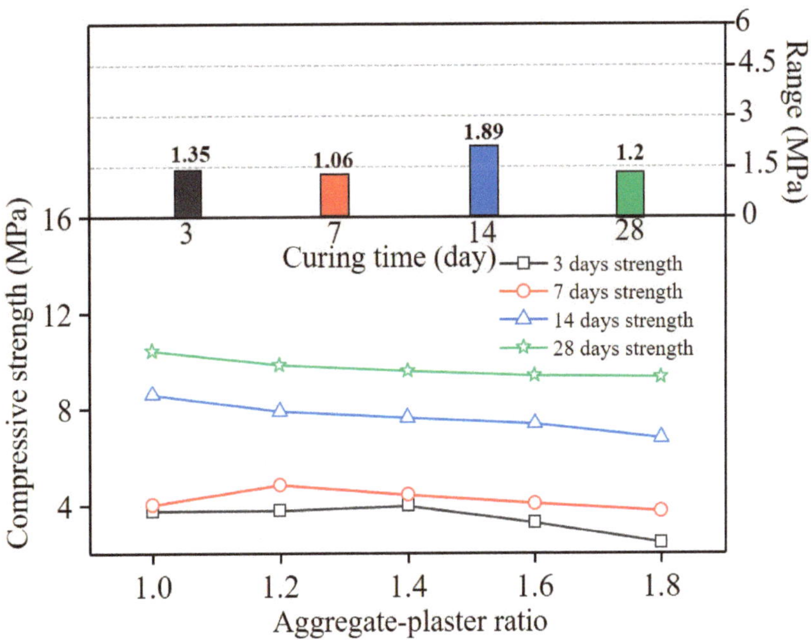

(b)

Figure 9. Influence of each mix proportion on the compressive strength: (**a**) water/plaster ratio; (**b**) aggregate/plaster ratio.

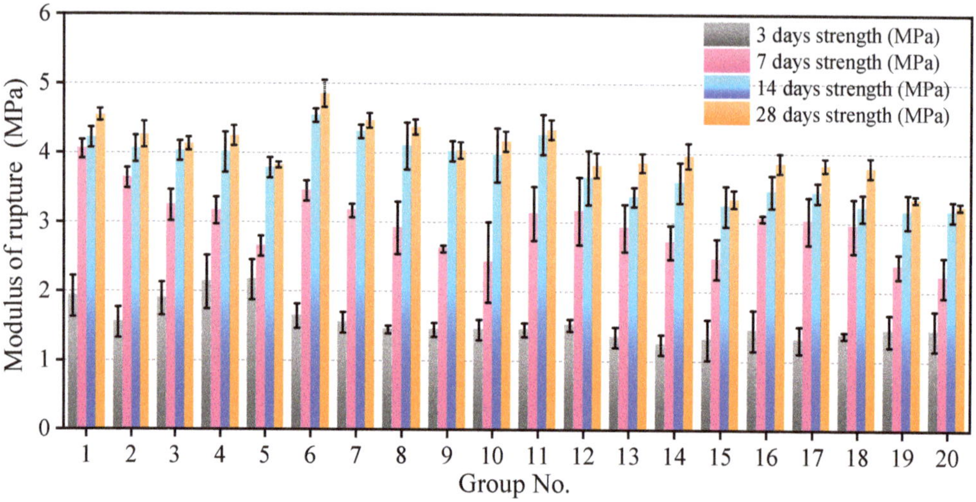

Figure 10. Variation in flexural strength of specimens from curing 3–28 days.

Figure 11 shows the effects of pertinent variables on the flexural strength. After curing for 3 days, the influence of the mix proportion on the flexural strength was slight; the maximum range differences of the W/P ratio and A/P ratio were 0.24 and 0.14 MPa, respectively. With increasing curing time, the influence of each composition on the flexural strength became obvious. After curing for 28 days, the significance of each composition on the flexural strength lessened, illustrating that the hydration was nearly complete in the specimens and stable strength was achieved.

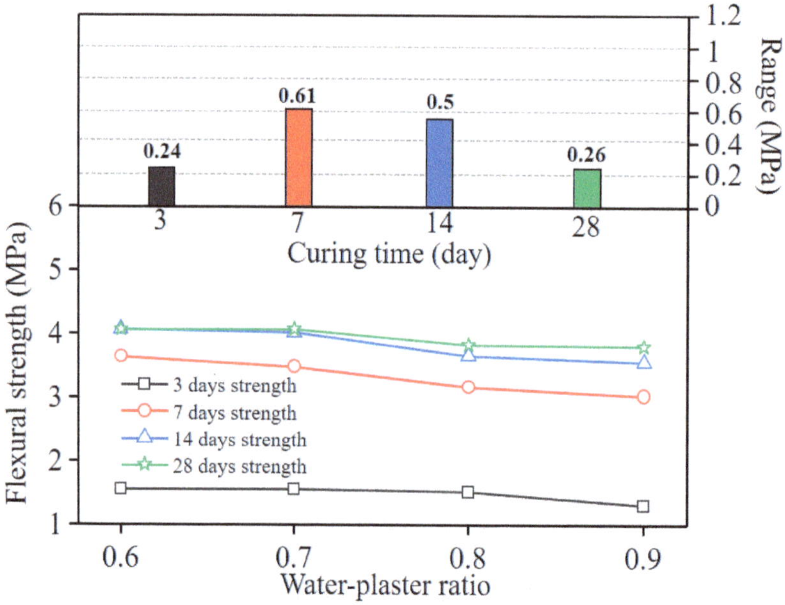

(**a**)

Figure 11. *Cont.*

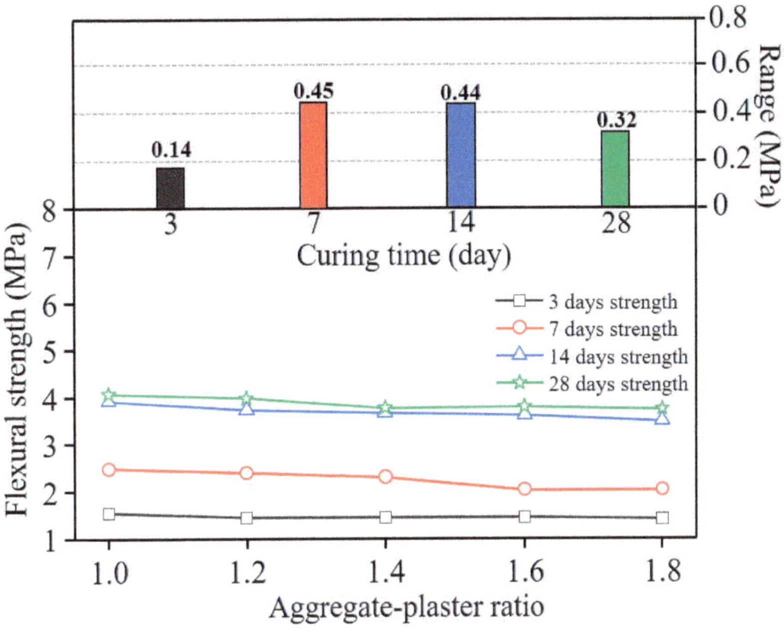

(**b**)

Figure 11. Influence of each variable on the flexural strength: (**a**) water/plaster ratio; (**b**) aggregate/plaster ratio.

3.5. Correlation Analysis

The correlation between various indices in this study was further investigated by Pearson correlation analysis [32]. Pearson's correlation coefficient is defined as

$$\rho(X, Y) = \frac{\mathrm{cov}(X, Y)}{\sigma_X \sigma_Y} \quad (1)$$

where ρ is the population correlation coefficient; *X*, *Y* are the two datasets of indices; and σ_x and σ_y are the standard deviation of *X*, *Y*, respectively. A large coefficient indicates a significant relationship between the two indices.

Figure 12 shows a heatmap of the interrelation between each index. A Pearson's coefficient value of $|\rho| > 0.7$ indicates a significant correlation, $0.4 < |\rho| < 0.7$ indicates a medium correlation, and $|\rho| < 0.4$, indicates a weak correlation [33]. The aforementioned correlation analysis demonstrates that the correlation between the flowability and bleeding rate was significant (Pearson's coefficient: 0.92). This phenomenon is attributable to the fact that the flowability and bleeding rate are determined by the free water content in the mixture paste. For the mechanical properties, there is a positive correlation between the compressive and flexural strength at different curing times. The highest correlation between these two strength indicators is observed after a curing period of 14 days, with a corresponding Pearson correlation coefficient of 0.79. Furthermore, there is a relatively weak correlation between the physical and mechanical characteristics of the model concrete, indicating that there is no inherent connection between them and they are independent of each other.

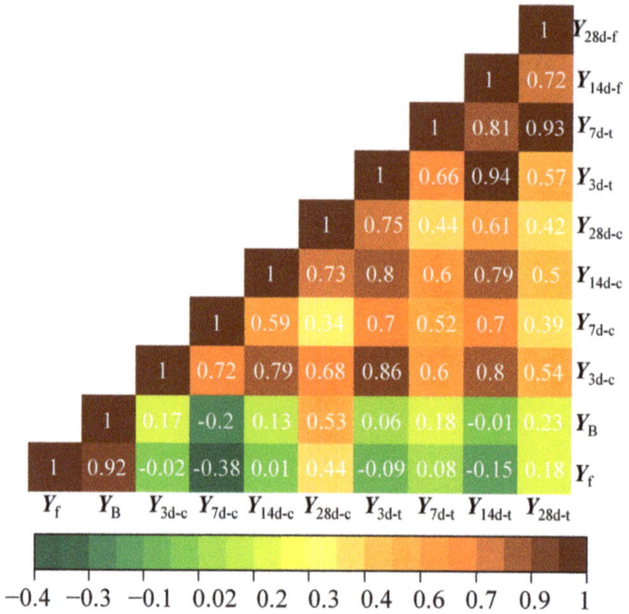

Figure 12. Heatmap of Pearson's correlation coefficients.

4. MLRA

The significance of each composition on the performance indices could be determined by an analysis of the orthogonal test results, and then the most appropriate ratio can be selected to simulate the concrete structure behaviour in small-scale models. However, the specific mixing ratio of model concrete is frequently modified to satisfy the requirements for various small-scale model test conditions. To conveniently determine the optimum mixing ratio, MLRA was performed to forecast the ratio parameters of the model concrete in this study. SPSS statistics (version 26; IBM, Armonk, NY, USA) were used to develop the predicting regression models by MLRA by 20 set results of orthogonal tests.

4.1. Validation of MLRA Models

To verify the feasibility of the MLRA models developed in this study, all of the test indices (flowability; bleeding rate; 3-, 7-, 14-, and 28 days compressive strength; and 3-, 7-, 14-, and 28 days flexural strength) should be statistically verified. Table 4 shows the correlation coefficients between the dependent variables (orthogonal test results) and the independent variable (W/P and A/P ratios). The fitted linear regression coefficients (R^2) reveal excellent linear relationships between the independent and dependent variables. Moreover, the significance of variance analysis in Table 5 was all <0.05, confirming the highly statistical significance.

4.2. Multiple Regression Coefficients

Table 5 shows the results of the regression coefficients; the importance of various mix proportions on the indices can be revealed by standardized regression coefficients. For example, in the model $F_{\text{flowability}}$, the standardized coefficients of the A/P and W/P ratios were −0.82 and 0.48, respectively. Thus, the aggregate contents negatively influenced the flowability of the slurry, whereas the water/plaster ratio exhibited positive synergy. In addition, the significance of the two mix proportions on the flowability—which considers the absolute value of standardized coefficients—was A/P > W/P, and it is consistent with the orthogonal test result analysis. On the basis of the regression coefficients in Table 5, linear relationship models between the mix proportion and indices of model concrete were developed. For example, the relationships between flowability, bleeding rate, and 7 days compressive and flexural strength with various compositions of model concrete were proposed as

$$F_{\text{flowability}}(\text{cm}) = 27.3 + 26.67 \times \text{W/P} - 18.09 \times \text{A/P} \quad (2)$$

$$F_{\text{bleeding}}(\%) = 8.2 + 8.33 \times \text{W/P} - 7.78 \times \text{A/P} \quad (3)$$

$$F_{7-\text{d compressive}}\,(\text{MPa}) = 18.038 - 9.499 \times \text{W/P} - 1.962 \times \text{A/P} \quad (4)$$

$$F_{7-\text{d flexural}}\,(\text{MPa}) = 6.22 - 1.942 \times \text{W/P} - 1.268 \times \text{A/P} \quad (5)$$

where $F_{\text{flowability}}$, F_{bleeding}, $F_{\text{7-d compressive}}$, and $F_{\text{7-d flexural}}$ are the prediction equations of the flowability, bleeding rate, and 7 days compressive and flexural strength of model concrete, respectively.

Table 4. Description of fitting models.

Model	R	R^2	Adjusted R^2	Std. Error of the Estimate
$F_{\text{flowability}}$	0.956	0.914	0.904	1.983
F_{bleeding}	0.933	0.871	0.855	0.992
$F_{\text{3-d compressive}}$	0.972	0.945	0.939	0.509
$F_{\text{3-d flexural}}$	0.85	0.722	0.69	0.2539
$F_{\text{7-d compressive}}$	0.926	0.858	0.841	0.916
$F_{\text{7-d flexural}}$	0.958	0.91	0.907	0.136
$F_{\text{14-dcompressive}}$	0.877	0.769	0.723	1.172
$F_{\text{14-d flexural}}$	0.846	0.715	0.679	0.367
$F_{\text{28-d compressive}}$	0.884	0.781	0.724	0.61
$F_{\text{28-d flexural}}$	0.843	0.71	0.677	0.188

Table 5. Results of regression coefficients.

Model		Unstandardized Coefficients		Standardize Coefficients (Beta)	Statistic (t)	Sig.
		Regression Coefficients	Std. Error			
$F_{\text{flowability}}$	Constant	27.30	3.805	/	7.715	<0.001
	A/P	−18.09	1.602	−0.82	−11.29	<0.001
	W/P	26.67	4.054	0.48	6.58	<0.001
F_{bleeding}	Constant	8.20	1.219	/	6.73	<0.001
	A/P	−7.78	0.513	−0.89	−15.16	<0.001
	W/P	8.33	1.298	0.38	6.41	<0.001
$F_{\text{3-d compressive}}$	Constant	9.54	1.147	/	8.32	<0.001
	A/P	−1.30	0.483	−0.41	−2.69	0.015
	W/P	−5.44	1.222	−0.67	−4.45	<0.001
$F_{\text{3-d flexural}}$	Constant	2.82	0.355	/	6.07	<0.001
	A/P	0.02	0.387	−0.33	−1.85	0.031
	W/P	−1.73	0.149	−0.61	−3.46	0.003
$F_{\text{7-d compressive}}$	Constant	18.038	1.72	/	12.89	<0.001
	A/P	−1.962	0.724	−0.33	−3.47	0.003
	W/P	−9.499	1.832	−0.61	−7.56	<0.001
$F_{\text{7-d flexural}}$	Constant	6.22	0.256	/	8.09	<0.001
	A/P	−1.942	0.108	−0.819	−7.11	<0.001
	W/P	−1.268	0.273	−0.496	−11.74	<0.001

Table 5. *Cont.*

Model		Unstandardized Coefficients		Standardize Coefficients (Beta)	Statistic (t)	Sig.
		Regression Coefficients	Std. Error			
$F_{14\text{-d compressive}}$	Constant	22.16	0.355	/	7.96	<0.001
	A/P	−2.51	0.149	−0.37	0.12	0.003
	W/P	−13.85	0.378	−0.81	−4.57	<0.001
$F_{14\text{-d flexural}}$	Constant	7.73	0.688	/	8.09	<0.001
	A/P	−0.96	0.29	−0.69	−4.27	0.001
	W/P	−3.44	0.733	−0.86	−1.77	0.024
$F_{28\text{-d compressive}}$	Constant	25.04	0.471	/	16.38	<0.001
	A/P	−2.4	0.198	−0.34	−5.236	<0.001
	W/P	−16.3	0.502	−0.91	−6.491	<0.001
$F_{28\text{-d flexural}}$	Constant	7.15	0.921	/	7.048	<0.001
	A/P	−0.88	0.388	−0.638	−1.57	0.035
	W/P	−2.54	0.981	−0.562	−3.245	0.005

5. Guidelines for the Optimal Parameter Selection

To verify the feasibility of the prediction model and further explain the guidelines for parameter selection of the model concrete for small-scale modelling, a typical application of model concrete piles in a geotechnical centrifuge test is presented. The aim of the presented centrifuge test is to investigate the failure behaviour of a concrete column-supported embankment. In general, concrete columns undergo bending failure in practice, which causes global failure of the embankment [5,34]. Therefore, the flexural strength of model concrete is the most important performance index and must satisfy the requirements put forward by the design criterion in prototypes. The target flexural strength of the model concrete in this centrifuge test case was about 2.81 MPa and the bleeding rate was $\leq 5\%$, which satisfies engineering criteria [5,34–37]. Combined with predicted model and orthogonal test results, two formulations (W/P = 0.7; A/P = 1.6 and W/P = 0.8; A/P = 1.4) that satisfied the requirements were selected.

Figure 13 shows the relationship between the model concrete performance and the minimum–maximum index values of the orthogonal tests. Considering the multi-performance criteria of model concrete, such as high flowability, good structural stability (low bleeding rate), and sufficient compressive strength, the recommended water/plaster and aggregate/plaster ratios by the prediction model are 0.7 and 1.6, respectively. The optimal formulation achieves sufficient flexural strength after 7 days of curing, whereas other indices could also meet the material performance requirements.

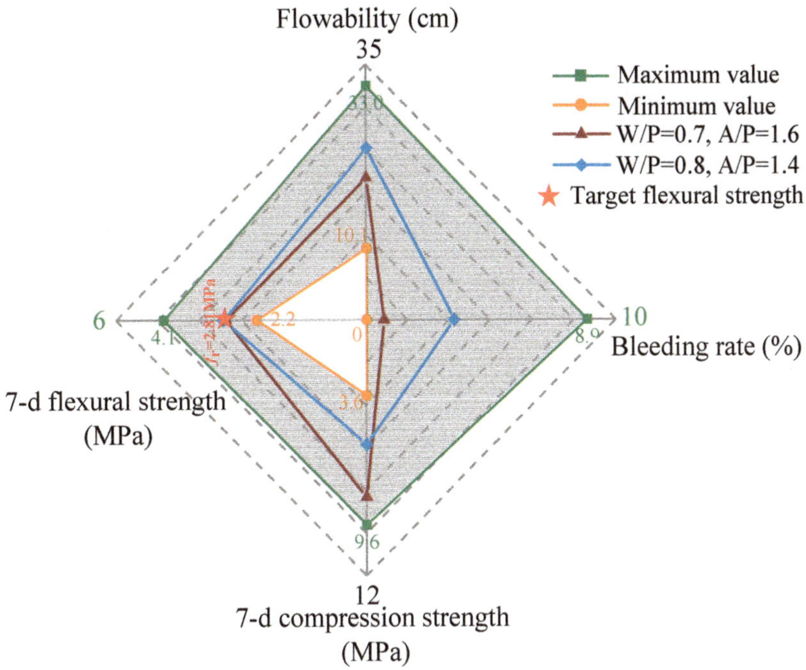

Figure 13. Range of optional aggregate dosages and W/P ratios for target flexural strength.

6. Conclusions

In this study, orthogonal test design was applied to quantitatively evaluate the effects of various compositions on model concrete performance. On the basis of the test results, a forecast model for determining the optimal formulations of model concrete was proposed to satisfy various small-scale model test requirements. The following conclusions were drawn.

(i) The effects of aggregate content and water/plaster ratio on the performance of model concrete were interdependent. The aggregate content negatively influenced the flowability and the bleeding rate of the slurry, whereas the water/plaster ratio exhibited positive synergy. Increasing the aggregate content and water/plaster ratio negatively corresponded to the mechanical characteristics, which was pertinent to compressive and flexural strength.

(ii) The correlation between the flowability and bleeding rate was significant (Pearson correlation coefficient: 0.92) because of the high content of free water in the mixture slurry. Moreover, the correlation between the compressive and flexural strength was positive; a significant correlation coefficient was found after curing for 14 days.

(iii) On the basis of the MLRA, a prediction model for the performance of model concrete was proposed. The significance level of each mix was proposed in the following sequence: for optimum flowability and bleeding rate, A/P > W/P. However, regarding the compressive and flexural strength, A/P < W/P.

(iv) Outlines for parameter selection of the model concrete were summarized based on a typical application case in geotechnical centrifuge testing. Combined with the specific test target, a reasonable range of water/gypsum ratios and aggregate content can be precisely determined by the prediction model.

Author Contributions: G.Z.: methodology, resources, supervision. B.X.: methodology, investigation, writing—original draft, writing—review and editing. H.Z.: writing—review and editing, visualization, supervision. Y.D.: visualization, supervision. J.H.: formal analysis. J.Z.: formal analysis, orthogonal tests and centrifuge model tests. X.Y.: formal analysis, orthogonal tests. All authors have read and agreed to the published version of the manuscript.

Funding: This research was funded by the National Natural Science Foundation of China (No. 52078337 and No. 52208363), the Project of Tianjin Science and Technology Plan (No. 21JCZXJC00070), the Natural Science Foundation of Hebei, China (No. E2021202215) and the China National Postdoctoral Program for Innovative Talents (Grant No. BX20220225).

Institutional Review Board Statement: Not applicable.

Informed Consent Statement: Not applicable.

Data Availability Statement: Some or all data and code that support the findings of this study are available from the corresponding author upon reasonable request.

Conflicts of Interest: The authors declare that they have no known competing financial interests or personal relationships that could have appeared to influence the work reported in this paper.

References

1. *BS8006-1:2010*; Code of Practice for Strengthened/Reinforced Soils and Other Fills. British Standards Institution: London, UK, 2010.
2. *JGJ 79-2002*; Technical Code for Ground Treatment of Buildings. Standardization Administration of the People's Republic of China: Beijing, China, 2002.
3. *ACI 318-14*; Building Code Requirements for Structural Concrete and Commentary. American Concrete Institute: Farmington Hills, MI, USA, 2014.
4. Han, J. *Principles and Practice of Ground Improvement*; John Wiley & Sons: Hoboken, NJ, USA, 2015.
5. Zheng, G.; Yang, X.; Zhou, H.; Chai, J. Numerical modeling of progressive failure of rigid piles under embankment load. *Can. Geotech. J.* **2019**, *56*, 23–34. [CrossRef]
6. Ng CW, W.; Shi, C.; Gunawan, A.; Laloui, L.; Liu, H.L. Centrifuge modelling of heating effects on energy pile performance in saturated sand. *Can. Geotech. J.* **2015**, *52*, 1045–1057.
7. Knappett, J.A.; Brown, M.J.; Shields, L.; Al-Defae, A.H.; Loli, M. Variability of small scale model reinforced concrete and implications for geotechnical centrifuge testing. In Proceedings of the 9th International Conference on Physical Modelling in Geotechnics, London, UK, 17–20 July 2018; pp. 241–246.
8. Shen, P.; Xu, C.; Han, J. Centrifuge tests to investigate global performance of geosynthetic-reinforced pile-supported embankments with side slopes. *Geotext. Geomembr.* **2020**, *48*, 120–127. [CrossRef]
9. Yu, J.L.; Zhou, J.J.; Gong, X.N.; Xu, R.Q.; Li, J.Y.; Xu, S.D. Centrifuge study on behavior of rigid pile composite foundation under embankment in soft soil. *Acta Geotech.* **2021**, *16*, 1909–1921. [CrossRef]
10. Schofield, A.N. Cambridge geotechnical centrifuge operations. *Geotechnique* **1980**, *30*, 227–268. [CrossRef]
11. Knappett, J.A.; Reid, C.; Kinmond, S.; O'Reilly, K. Small-scale modeling of reinforced concrete structural elements for use in a geotechnical centrifuge. *J. Struct. Eng.* **2011**, *137*, 1263–1271. [CrossRef]
12. Lirola, J.M.; Castañeda, E.; Lauret, B.; Khayet, M. A review on experimental research using scale models for buildings: Application and methodologies. *Energy Build.* **2017**, *142*, 72–110. [CrossRef]
13. Al-Defae, A.H.; Knappett, J.A. Centrifuge modeling of the seismic performance of pile-reinforced slopes. *J. Geotech. Geoenvironmental Eng.* **2014**, *140*, 04014014. [CrossRef]
14. Loli, M.; Knappett, J.A.; Brown, M.J.; Anastasopoulos, I.; Gazetas, G. Centrifuge modeling of rocking-isolated inelastic RC bridge piers. *Earthq. Eng. Struct. Dyn.* **2014**, *43*, 2341–2359. [CrossRef] [PubMed]
15. Vitali, D.; Leung, A.K.; Feng, S.; Knappett, J.A.; Ma, L. Centrifuge modelling of the use of discretely spaced energy pile row to reinforce unsaturated silt. *Géotechnique* **2022**, *72*, 618–631. [CrossRef]
16. Zhang, C.; Yang, J.; Ou, X.; Fu, J.; Xie, Y.; Liang, X. Clay dosage and water/cement ratio of clay-cement grout for optimal engineering performance. *Appl. Clay Sci.* **2018**, *163*, 312–318. [CrossRef]
17. Zheng, G.; Huang, J.; Diao, Y.; Ma, A.; Su, Y.; Chen, H. Formulation and performance of slow-setting cement-based grouting paste (SCGP) for capsule grouting technology using orthogonal test. *Constr. Build. Mater.* **2021**, *302*, 124204. [CrossRef]
18. Gustin, E.J.G.; Karim, U.F.A.; Brouwers, H.J.H. Bleeding characteristics for viscous cement and cement–bentonite grouts. *Géotechnique* **2007**, *57*, 391–395. [CrossRef]
19. Lei, D.Y.; Guo, L.P.; Sun, W.; Liu, J.P.; Miao, C.W. Study on properties of untreated FGD gypsum-based high-strength building materials. *Constr. Build. Mater.* **2017**, *153*, 765–773. [CrossRef]
20. *BS 882:1992*; Specification for Aggregates from Natural Sources for Concrete. British Standards Institution (BSI): London, UK, 1992.

21. *GB/T 28627-2012*; Gypsum Plaster. Standardization Administration of the People's Republic of China: Beijing, China, 2012; (In Chinese).
22. *JTG/E30-2005*; Test Methods of Cement and Concrete for Highway Engineering. Ministry of Communications of the People's Republic of China: Beijing, China, 2005.
23. Jiang, J.; Lu, Z.; Li, J.; Fan, Y.; Niu, Y. Preparation and hardened properties of lightweight gypsum plaster based on pre-swelled bentonite. *Constr. Build. Mater.* **2019**, *215*, 360–370. [CrossRef]
24. *GB 50081-2019*; Standard for Test Methods of Concrete Physical and Mechanical Properties. Standardization Administration of the People's Republic of China: Beijing, China, 2019.
25. Cherki, A.B.; Remy, B.; Khabbazi, A.; Jannot, Y.; Baillis, D. Experimental thermal properties characterization of insulating cork–gypsum composite. *Constr. Build. Mater.* **2014**, *54*, 202–209. [CrossRef]
26. Zheng, G.; Xia, B.; Zhou, H.; Yu, X.; Diao, Y. Influence of deep-cement-mixing column rows on the performance of geosynthetics-reinforced column-supported railway embankment. *Transp. Geotech.* **2023**, *41*, 101012. [CrossRef]
27. *GB/T 9776-2008*; Calcined Gypsum. Standardization Administration of the People's Republic of China: Beijing, China, 2008; (In Chinese).
28. Choi, M.S.; Lee, J.S.; Ryu, K.S.; Koh, K.T.; Kwon, S.H. Estimation of rheological properties of UHPC using mini slump test. *Constr. Build. Mater.* **2016**, *106*, 632–639. [CrossRef]
29. Jiménez-Quero, V.G.; León-Martínez, F.M.; Montes-Garcia, P.; Gaona-Tiburcio, C.; Chacón-Nava, J.G. Influence of sugar-cane bagasse ash and fly ash on the rheological behavior of cement pastes and mortars. *Constr. Build. Mater.* **2013**, *40*, 691–701. [CrossRef]
30. Zhang, Y.; Yang, J.; Cao, X. Effects of several retarders on setting time and strength of building gypsum. *Constr. Build. Mater.* **2020**, *240*, 117927. [CrossRef]
31. Bal, B.C. Flexural properties, bonding performance and splitting strength of LVL reinforced with woven glass fiber. *Constr. Build. Mater.* **2014**, *51*, 9–14. [CrossRef]
32. Benesty, J.; Chen, J.; Huang, Y.; Cohen, I. Pearson correlation coefficient. In *Noise Reduction in Speech Processing*; Springer: Berlin/Heidelberg, Germany, 2009; pp. 1–4.
33. Cleophas, T.J.; Zwinderman, A.H.; Cleophas, T.J.; Zwinderman, A.H. Bayesian Pearson correlation analysis. In *Modern Bayesian Statistics in Clinical Research*; Springer: Cham, Switzerland, 2018; pp. 111–118.
34. Zhou, H.; Shi, Y.; Yu, X.; Xu, H.; Zheng, G.; Yang, S.; He, Y. Failure mechanism and bearing capacity of rigid footings placed on top of cohesive soil slopes in spatially random soil. *Int. J. Geomech.* **2023**, *23*, 04023110. [CrossRef]
35. Zhou, H.; Liu, X.; Tan, J.; Zhao, J.; Zheng, G. Seismic fragility evaluation of embankments on liquefiable soils and remedial countermeasures. *Soil Dyn. Earthq. Eng.* **2023**, *164*, 107631. [CrossRef]
36. *GB/T 50783-2012*; Technical Code for Composite Foundation. Standardization Administration of the People's Republic of China: Beijing, China, 2012.
37. Ma, H.; Luo, Q.; Wang, T.; Jiang, H.; Lu, Q. Numerical stability analysis of piled embankments reinforced with ground beams. *Transp. Geotech.* **2021**, *26*, 100427. [CrossRef]

MDPI
St. Alban-Anlage 66
4052 Basel
Switzerland
www.mdpi.com

Materials Editorial Office
E-mail: materials@mdpi.com
www.mdpi.com/journal/materials

www.ingramcontent.com/pod-product-compliance
Lightning Source LLC
Chambersburg PA
CBHW061839210926
51850CB00006B/928
* 9 7 8 3 0 3 6 5 8 3 9 6 9 *